A Comprehensive Guide to Safety and Aging

This book is a comprehensive survey on safety for older adults. It contains contributions by experts from over a dozen disciplines, including physicians, audiologists, optometrists, blindness and low vision specialists, mental health professionals, lawyers, occupational therapists, and policy makers. This multi-disciplinary approach provides a new and expansive conceptual framework for health care professionals, students, policymakers, and others who care for older adults, and promotes an understanding of the many challenges that adults face as they age.

This book describes the complex range of issues that need to be considered when safeguarding older adults. We hope that this book will be of benefit to anyone currently working or training to work with older adults, helping them to fully appreciate the many safety issues that can arise. The book will be also be useful for both older adults and their caregivers, helping them to identify and address areas of concern. Our goal is to mitigate injury or other harm through an increased understanding of the risks encountered by older adults. This text will also appeal to professionals and graduate students in the fields of human factors and ergonomics, occupational health, and safety.

A Comprehensive Guide to Safety and Aging

Minimizing Risk, Maximizing Security

Edited by
Robert Wolf, Barry S. Eckert, and Amy R. Ehrlich

CRC Press
Taylor & Francis Group
Boca Raton London New York

CRC Press is an imprint of the
Taylor & Francis Group, an **informa** business

Cover image: Rawpixel.com/Shutterstock

First edition published 2023
by CRC Press
6000 Broken Sound Parkway NW, Suite 300, Boca Raton, FL 33487-2742

and by CRC Press
4 Park Square, Milton Park, Abingdon, Oxon, OX14 4RN

CRC Press is an imprint of Taylor & Francis Group, LLC

ISBN: 978-1-032-05502-2 (hbk)
ISBN: 978-1-032-05505-3 (pbk)
ISBN: 978-1-003-19784-3 (ebk)

DOI: 10.1201/9781003197843

Typeset in Times
by Newgen Publishing UK

Contents

Editors' Note

When we consider safety, we are connecting to one of our most basic and primitive instincts – the desire to protect those we care for and ensure that they are as safe as possible. It is what we worry about when we check on babies in the middle of the night, it's what we wish our friends when they are setting off to travel. For professionals who care for older adults, the commitment to safety spans all disciplines and all fields.

Given the basic role safety plays in our lives, we were surprised to discover that there were no books devoted to the topic of keeping older adults safe. This is particularly surprising when one considers that there are dozens of books and educational initiatives devoted to the safety of children. However, it is also true that older adults are very different from children; they are presumed to have capacity, autonomy, and freedom in their activities and choices. In fact, one of our most important considerations in putting this book together was to not characterize older adults as helpless, vulnerable, or a societal burden. Accordingly, we asked all our contributors to consider the recommendations adopted by the American Geriatrics Society (JAGS), Leaders of Aging Organizations, the FrameWorks Institute, and the American Medical Association Manual of Style and use non- "ageist" language to describe older people, with the goal of building better public perceptions of aging. (See *Journal of the American Geriatrics Society*, July 2017–Vol. 65, No. 7)

Safety covers a lot of territory. Threats to safety may result from a dangerous environment, poor personal decisions, emergencies, or from vulnerabilities stemming from physical, cognitive, or sensory changes in later life, or even because of the malevolent behavior of others. We also note that some confusion still exists when it comes to safety-related terminology. We now understand injuries as potentially preventable events, and the term accident is primarily used to describe chance events. Injuries in older adults can be physical, financial, or emotional. Maximizing safety can necessitate changing environments and/or personal behaviors. But to do so we must first improve both professional and public education – the primary goal of the book you are now reading.

When we considered the topics we would include in this book, some decisions were obvious. You cannot consider a book on older adult safety without addressing the issues of falls, elder abuse and neglect, driving, and polypharmacy. We included psychosocial topics such as depression, anxiety, and the growing number of older adult suicides. We also wanted to address non-medical issues such as financial, legal, and emergency preparedness. Safety considerations can also change over time, which is reflected in topics such as internet scams or social isolation, a safety issue that, as a consequence of COVID-19, resulted in the seclusion of millions of older adults. And perhaps, in the not-so-distant future, autonomous cars might eliminate the need to discuss driving as a safety issue.

We want to thank all our authors for their excellent contributions. The authors come from many different professions with a wide range of educational backgrounds. They also bring their unique personal experiences and perspectives in working with different cohorts of older adults in a broad range of both institutional and community settings. Despite these differences, they all share a passion about improving safety for older adults. We have already begun discussing what topics we would like to consider in future additions of this book and are aware of several key areas that are not included in this book. We hope in reading these chapters, particularly those from disciplines outside of your own daily experience, that you will share with us your thoughts about topics that should be included in any future editions.

Sincerely,

Bob, Amy, and Barry

About the Editors

Robert Wolf, JD, MUP
Bob Wolf has had a distinguished career in aging, health, law and philanthropy. He has served as a senior advisor to the SC Group, one of the largest funders of aging programs in the United States, for more than 25 years. He also consults with other leading national organizations in the field, including the New York Academy of Medicine, Rockefeller Philanthropy Associates, and the National Council on Aging. Prior to that Bob had leadership roles at AARP, UJA-Federation of NY, and Healthcare Chaplaincy. Bob was a pioneer in the field of elder law, at the Brookdale Center on Aging at Hunter College and at Strauss and Wolf, one of the nation's first law firms devoted to eldercare. He has authored publications on aging policy, law, and the rights of caregivers. Bob has a JD degree from Brooklyn Law School, an MUP from Hunter College, and a postgraduate certificate in not-for-profit management from the Columbia School of Business.

Barry S. Eckert, PhD
Dr. Eckert has served as Provost at Salus University in Elkins Park, Pennsylvania, Dean of the School of Health Professions at Long Island University, Brooklyn Campus, Dean of the College of Health Professions at Armstrong Atlantic State University in Savannah, Georgia, and Dean of the School of Health Related Professions at the University at Buffalo (SUNY).

Dr. Eckert earned his PhD in Anatomy from the University of Miami School of Medicine and did postdoctoral research in the Department of Cellular, Molecular and Developmental Biology at the University of Colorado. As a faculty member at the University of Buffalo, School of Medicine and Biomedical Sciences, he taught human anatomy to medical, dental, and allied health students and focused his research in cell biology and biochemistry. He currently concentrates on health care delivery, particularly in relation to workforce, new program development, and interprofessional education.

Dr. Eckert served on the Task Force for Health Education for the University System of Georgia, which focused upon the health workforce needs of that state. He is the recipient of a President's Grant from the Josiah Macy Foundation for his work in Interprofessional Education. He is a fellow of the Association of Schools of Allied Health Professions and was on the Board of Directors of that organization. Dr. Eckert has served on the Board of Directors of the Association of Specialized and Professional Accreditors (ASPA) and is active in the Health Professions Accreditors Collaborative (HPAC). He also has served as a Board Member and Treasurer of the Commission on Accreditation of Allied Health Education Programs (CAAHEP) and is currently a CAAHEP Commissioner.

Amy R. Ehrlich, MD

Dr. Amy Ehrlich is the Associate Chief of the Division of Geriatrics at Montefiore Medical Center and Professor of Medicine at Albert Einstein College of Medicine, Bronx, New York. She is board certified in Internal Medicine, Geriatrics, and Hospice and Palliative Care. Dr. Ehrlich received her MD from Harvard Medical School, and trained in Internal Medicine at the Beth Israel Hospital in Boston, Massachusetts. She is a Fellow of the American Geriatrics Society.

Dr. Ehrlich has been involved in program design and development in geriatrics education for medical students, residents, and fellows at Albert Einstein College of Medicine. Dr. Ehrlich's research interests include screening for mild cognitive impairment and dementia in ethnically and racially diverse primary care settings and the prevention of burns and fires in older adults. In her role as Medical Director of Montefiore Home Care, she works with the interdisciplinary home care team to develop innovative disease-specific programs which help address challenges in transitions of care in the Montefiore network.

Contributors

Maysoon Agarib
Albert Einstein College of Medicine
Bronx, New York, United States of
America

Sarah D. Appel
Salus University
Elkins Park, PA, United States of
America

Radhika Aravamudhan
Salus University
Elkins Park, PA, United States of
America

Michael Bogaisky
Albert Einstein College of Medicine
Bronx, New York, United States of
America

Nicole Braccio
National Patient Advocate Foundation
Washington DC, United States of
America

Leonard Brennan
Harvard Dental Geriatric Fellowship
Program
Cambridge, MA, United States of
America

Brianna Brim
Salus University
Elkins Park, PA, United States of
America

Zahava Nilly Brodt-Ciner
University of Maryland Baltimore
Washington Medical Center
Glen Burnie, MD, United States of
America

Kathleen A. Cameron
National Council on Aging
Washington DC, United States of
America

Rachel Chalmer
Albert Einstein College of Medicine
Bronx, New York, United States of
America

Elise B. Ciner
Salus University
Elkins Park, PA, United States of
America

Ann Marie Cook
Lifespan of Greater Rochester
Rochester, New York, United States of
America

Jennifer A. Crittenden
University of Maine School of
Social Work
Orono, ME, United States of
America

Sara J. Czaja
Weill Medical College of Cornell
University
New York, United States of
America

Beth E. Davidoff
Salus University
Elkins Park, PA, United States of
America

Caitlyn Foy
Salus University
Elkins Park, PA, United States of
America

David M. Godfrey
ABA Commission on Law and Aging
Washington DC, United States of
  America

Marcy Graboyes
Salus University
Elkins Park, PA, United States of
  America

Anna Y. Grasso
Salus University
Elkins Park, PA, United States of
  America

Alice Guo
Albert Einstein College of Medicine
Bronx, New York, United States of
  America

Wanda Horn
Albert Einstein College of Medicine
Bronx, New York, United States of
  America

Janet Kasoff
Albert Einstein College of Medicine
Bronx, New York, United States of
  America

Erin Kenny
Salus University
Elkins Park, PA, United States of
  America

Rebecca Kirch
National Patient Advocate
  Foundation
Washington DC, United States of
  America

Michelle Knapp
Wholistic Psychiatry PLLC,
Venice, Florida, United States of
  America

James Konopack
Salus University
Elkins Park, PA, United States of
  America

Kerry Lueders
Salus University
Elkins Park, PA, United States of
  America

Daniel Lyon
Lifespan of Greater Rochester
Rochester, New York, United States of
  America

Jamie Maffit
Salus University
Elkins Park, PA, United States of
  America

Rubina Malik
Albert Einstein College of Medicine
Bronx, New York, United States of
  America

Donna E. McCabe
New York University Meyers College of
  Nursing
New York, United States of America

George Mois
University of Illinois
  Urbana-Champaign
Champaign, IL, United States of
  America

Juliana M. Mosley-Williams
Salus University
Elkins Park, PA, United States of
  America

Madeline A. Naegle
New York University Meyers College of
  Nursing
New York, United States of America

Tracy Offerdahl-McGowan
Salus University
Elkins Park, PA, United States of
America

Jean Marie Pagani
Salus University
Elkins Park, PA, United States of
America

Marvin Pathrose
Salus University
Elkins Park, PA, United States of
America

Fabiana Perla
Salus University
Elkins Park, PA, United States of
America

Alice Pomidor
Florida State University College of
Medicine
Tallahassee, FL, United States of
America

Daniel A. Reingold
RiverSpring Living
Bronx, New York, United States of
America

Wendy A. Rogers
University of Illinois
Urbana-Champaign
Champaign, IL, United States of
America

Randi Rothbaum
Albert Einstein College of Medicine
Bronx, New York, United States of
America

Casey Rust
Florida State University College of
Medicine

Tallahassee, FL, United States of
America

Charles P. Sabatino
ABA Commission on Law and Aging
Washington DC, United States of
America

Joseph Sharit
University of Miami
Miami, FL, United States of America

Jo Anne Sirey
Weill Medical College of Cornell
University
New York, United States of America

Lachelle Smith
Salus University
Elkins Park, PA, United States of
America

Rani E. Snyder
The John A. Hartford Foundation
New York, United States of America

Joy Solomon
Weinberg Center for Elder Justice at the
Hebrew Home at Riverdale
Bronx, New York, United States of
America

Lauren Sponseller
Salus University
Elkins Park, PA, United States of
America

Allison Stark
Albert Einstein College of Medicine
Bronx, New York, United States of
America

Tristan Sullivan-Wilson
Weinberg Center for Elder Justice at the
Hebrew Home at Riverdale

Bronx, New York, United States of
  America

Mary Jo Thomas
Salus University
Elkins Park, PA, United States of
  America

Emily Vasile
Salus University
Elkins Park, PA, United States of
  America

Melissa A. Vitek
Salus University
Elkins Park, PA, United States of
  America

Kimberly Williams
Vibrant/Mental Health Association
New York, United States of America

Robert Wolf
Silverman Charitable Group
New York, United States of America

Megan M. Wolfe
Trust for America's Health
Washington DC, United States of
  America

Alexandra Woods
Weill Medical College of Cornell
  University
New York, United States of America

1 Introduction

Rani E. Snyder

CONTENTS

1.1   WHY AGING? WHY SAFETY? WHAT MATTERS

This book was conceived by a dear friend and colleague, Bob Wolf, whom I first met years ago at a Grantmakers in Aging conference. Like me, Bob has had a long career in aging-related philanthropy. Now, in my role as Chair of the board of Grantmakers in Aging, I have a bird's-eye view of the many approaches to funding in aging, and the ability to learn about the many excellent programs and initiatives in the field and to think long and hard about where programs are siloed, how to expand them and sometimes how to improve them. Through his years of scanning the field as a program officer, Bob saw the need for a resource that is different from what already exists; not a new wellness book, and not content focused on anti-aging, but rather a true guide to safety for older adults at the population level, a book focused on an approach to healthy aging from the perspective of safety.

Another privilege of working in philanthropy is the ability to make connections, to partner, and to cross-reference good work. The collaboration that created this book is an illustration of Bob's great strengths in engaging others and creating a strong partnership. As Bob thought through the content that would be necessary to make up this resource, he reached out to peers in the health and aging community, Barry Eckert the Provost and Vice President for Academic Affairs at Salus University with many years of experience in health professions education and another dear friend of mine, Amy Ehrlich, a geriatrician and professor at Albert Einstein College of Medicine, to collaborate with him as editors. Together they built up his vision for this resource and identified just the right mix of content and authors. This book is a tribute to Bob's collaborative nature and his deep expertise in the field of aging.

What if we could create a world where older adults are safe and free from harm? This book, written for the people who work within the systems of care that affect older people, will help get us there. If you are reading this, you already know about the need to improve the care of older adults. The demographic imperative has been

DOI: 10.1201/9781003197843-1

clear for decades. The population of those aged 65 and over has been on a steep growth curve. Whereas the current population of the group numbered 54.1 million in 2019, it is projected to reach 94.7 million in 2060. These broad numbers gloss over the details underneath: the older population itself is growing older; the proportion of older adults who are racial and ethnic minorities is growing; the proportion of older people living alone increases with age; and as our population ages, the education level of older adults has increased as well, from 28% high school completion in 1970 to 89% in 2020.(1) All of these facts affect older adults and their caregivers alike. In other words, these trends will affect all of us.

The chapters in this book fall under five themes. First, prevention is key. The more we can intervene upstream to get at root causes, the better we will be able to achieve the triple aim of improving care, improving population health and reducing costs of care. The relationship between what we spend on health care and the results of that health care is not linear, in part because too many people are actually harmed by the health care that is designed to help them.(2) This is especially true for older adults who are more likely to have chronic and multiple conditions and live with functional limitations that complicate their health, their lives, and their options. Patient safety is defined by the World Health Organization as "the prevention of errors and adverse effects to patients" (3) and therefore any effort that heads off the need for health care is an excellent place to start. Older adults who are healthy want to stay that way. The age-friendly practice of care in its many forms includes prevention first.

Second, all care needs to be age-friendly. Our quest for improved access, quality, and cost-containment strategies in the care of older adults must begin with an increase in both public and professional education as well as systematic approaches that put older adults and their families at the center of discussions about safety and autonomy. These are the tenets that underpin an approach to age-friendly care that began with support from the John A. Hartford Foundation to the Institute for Healthcare Improvement, partnered with the American Hospital Association and the Catholic Health Association of the United States called Age-Friendly Health Systems (AFHS). This social movement is based on a person-centered approach to the health of older adults. AFHS relies on evidence-based care that improves health outcomes and prevents avoidable harm. This *is* safety, on a systemic level. AFHS uses a framework called the 4Ms (what matters, medication, mentation, and mobility) to ensure that care is both reliable and evidence-based. The 4Ms operate together and are mutually reinforcing. When we refer to an AFHS, we are describing every place where older adults access care, ranging from the home to the emergency room, the hospital setting, the nursing home, or for example, a convenient care clinic.(4) Many of the chapters in this book discuss crucial needs in health care for older adults and specifically relate to the 4Ms of age-friendly care: mobility (falls prevention), mentation (chapters covering anxiety and depression and one on safety for older adults with cognitive decline) and medication management. All of the chapters, by necessity, take into account the first and overarching M of what matters.

AFHS is only one part – albeit an important, even outsized part – of the larger age-friendly ecosystem that includes age-friendly communities, age-friendly public health systems, and age-friendly universities, among other critical institutions.(5)

These other ecosystem elements make up a third theme in the universe of what affects the health and safety of older adults. There is a natural relationship between age-friendly care resulting from AFHS and public health, for which there is also important ongoing work to create an age-friendly *public* health system. Public health and emergency and disaster relief are two areas that have been neglected for older adults at our communities' peril. Other important topics fall outside the health care system but are nevertheless critical to both health and quality of life, such as home modifications so that people can do what we overwhelmingly all say we want to do: age in place in our own homes whenever possible. The topics of driver safety, travel, caregiver care, and legal tools come under this umbrella of supportive care topics, often thought of as instrumental activities of daily living. Workplace and financial safety dramatically affect older adults and their caregivers alike.

Fourth, the best programs and practices to improve the safety and care of older adults are interdisciplinary and multi-professional. Major efforts and millions of dollars of funding from private philanthropy and government sources have been invested in ensuring that the health care workforce is prepared to care for older adults.(6) This book is designed for all professionals and students serving older adults, including clinicians, occupational therapists, physical therapists, social workers, public health professionals, and others such as audiologists. Students will find this book central for their understanding of the many and multifaceted approaches to improving care for older adults. Anyone with an interest in mental and behavioral health issues will also find important content here, including key topics such as depression, anxiety, social isolation, and suicide prevention. An underlying theme, consistent throughout the book, is that anything that ensures professionals will focus on what matters in the lives and care of older adults and their families is based on the respect of the individual and family unit that is foundational for safety issues.

Fifth and finally, elder mistreatment in all its forms is epidemic and we must be dedicated to preventing and eradicating this malady in all its forms.(7) Mistreatment of older adults is prevalent, and it has consequences that are medically and socially serious, if not fatal.(8)

These are the most important topics in safety and aging; they are absolutely critical to the quality of life for older adults and their families. That's why it's particularly meaningful that this book deals with not only the challenges of safety for older adults, but also with a discussion on concepts of risk. Safety and independence are fundamental ideals that for older adults require thoughtful consideration. There may be compromises between what matters most to an individual and what professionals or family members prefer. Respect and understanding on all sides are key. The authors explain best practices in safety and how to infuse what matters most in a way that it may change the way you think about safety for older adults.

Finally, the ongoing coronavirus pandemic that began in 2020 laid bare not only the special needs of older adults, but also the vulnerabilities of those who care for them, from clinicians to community-based organizations to the paid and unpaid caregiver workforce. All have worked tirelessly on behalf of the older adults. Please join me in thanking the authors and all who have contributed to this book, which will improve the understanding of safety, care, and *what matters*.

REFERENCES

2020 PROFILE OF OLDER AMERICANS (ACL.GOV).

1. Administration for Community Living. *2020 PROFILE of OLDER AMERICANS.*; 2021.https://acl.gov/sites/default/files/Aging%20and%20Disability%20in%20America/2020ProfileOlderAmericans.Final_.pdf

2. Merry AF, Shuker C, Hamblin R. Patient safety and the Triple Aim. *Internal Medicine Journal.* 2017;47(10):1103–1106. doi:10.1111/imj.13563

3. Quality and Patient Safety. CooperHealth.org. www.cooperhealth.org/about-us/quality-and-patient-safety

4. Mate K, Fulmer T, Pelton L, et al. Evidence for the 4Ms: Interactions and Outcomes across the Care Continuum. *Journal of Aging and Health.* Published online February 8, 2021;33(7–8):469–481. doi:10.1177/0898264321991658

5. Dash K, Shue J, Driver T, Bonner A, Pelton L, et al. (2022) Developing a Shared Language to Describe the Age-Friendly Ecosystem: Technical Meeting Report. Int J Geriatr Gerontol 6: 134. DOI: http://doi.org/10.29011/2577-0748.100034

6. Snyder R, Lundebjerg NE. Show Us the Money: Investments That Support the Eldercare Workforce. *Generations: Journal of the American Society on Aging.* 2016;40(1):115–121. www.jstor.org/stable/26556188

7. Fulmer T. The Rosalie Wolf Memorial Lecture: Abuse-free Care in a World of Age-friendly Health Systems. *Journal of Elder Abuse & Neglect.* 2018;30(3):167–175. doi:10.1080/08946566.2018.1452658

8. Rosen T, Elman A, Dion S, et al. Review of Programs to Combat Elder Mistreatment: Focus on Hospitals and Level of Resources Needed. *Journal of the American Geriatrics Society.* Published online March 22, 2019;67(6): 1286–1294. doi:10.1111/jgs.15773

2 Hazards of Hospitalization

Michael Bogaisky and Randi Rothbaum

CONTENTS

2.1   INTRODUCTION

Older adults represent a large portion of the hospitalized population. While representing only 16% of the U.S. population, people aged 65 years and older, accounted for 37% of all hospitalizations in 2018. Older adults are more than twice as likely to require hospitalization compared with middle aged adults. Approximately 17% of adults older than 65 are hospitalized at least once in a calendar year. Older adults can develop a variety of complications in the hospital that are not related to their admitting diagnosis. These complications can lead to significant functional impacts and worsened prognosis. As a part of the aging process, older adults are at heightened susceptibility to various stressors. This increased vulnerability, or loss of physiologic reserves, in the setting of hospitalization and acute illness, can push older adults past their ability to compensate and can lead to functional decline and loss of independence. Even when the acute problems which led to hospitalization are resolved, a patient may never return to their pre-hospitalization functional status. Many of these so-called "hazards of hospitalization" can be avoided by identifying specific risk factors and modifying them or the existing hospital environment. Assessment of an older adult's functional and cognitive status is crucial to ascertaining areas of

DOI: 10.1201/9781003197843-2

5

vulnerability and interventions that can avoid these hazards. This chapter will review the most common hazards and appropriate interventions.

2.2 LOSS OF FUNCTION

Loss of function is a frequent consequence of hospitalization. Studies have shown that at the time of hospital discharge, between 35 and 46% of older adults have suffered a decline in ability to perform their basic activities of daily living (Cohen, 2019). Only a third of these people will regain their lost function in the months after hospital discharge. Loss of mobility is one of the more common functions lost. This loss of function is a major driver behind the transfer of previously independent community dwelling people to skilled nursing facilities following their hospital stay, rather than a return to their home.

2.3 FALLS

Falls are a not uncommon event in the hospital. On a 36-bed medicine ward, there will be between 3 and 7 falls per month. The combined effects of an altered physical status, an unfamiliar hospital environment, changes in medications, and caregivers who are unfamiliar with the patient all contribute to an increased risk of falls during hospitalization in vulnerable older adults. Falls can be a catastrophic event in older adults. Around 26% of falls in the hospital are injurious. Although most of these injuries are considered minor, 4% of falls result in a major injury, such as one requiring surgery or casting, and 0.2% result in death. Even a minor injury can cause significant functional impairment in a frail older adult with limited functional reserves. To prevent falls, many hospitals restrict older adults to bed rather than providing the supports necessary for them to ambulate safely. These restrictions, and the resultant effects of prolonged bedrest, contribute to the loss of function and mobility commonly seen in older adults during hospitalization.

2.4 INCONTINENCE

Incontinence is common during hospital stays. In community-dwelling older adults the prevalence of urinary incontinence ranges between 10 and 15%, while in the hospital it is present in 30–40%. Incident (new onset) urinary incontinence occurs in between 10 and 21% of hospitalized older adults. The forced immobility and functional decline commonly seen during hospitalization, as well as a lack of support necessary to help patients with toileting, contributes to the development of urinary incontinence. Incontinence can be psychologically distressing for previously continent older adults. It also increases the risk for development of urinary tract infections, skin ulcers, and falls. Incontinence can also lead to rejection by caregivers, who may not be willing to carry out the tasks needed for caring for an incontinent older adult.

2.5 DELIRIUM

Delirium is a sudden change in mental status characterized by poor levels of attention with a fluctuating mental status, and either disorganized thought processes or altered levels of arousal. Delirium occurs in 13–50% of older adults in the hospital with a particularly higher incidence after surgery. Delirium is often precipitated by an acute physical stressor such as infections, surgery, injury, pain, or can be caused by exposure to medications which affect the central nervous system. Risk factors for developing delirium include older age, dementia, poor vision, and hearing impairment. The hospital environment itself contributes to the development of delirium. Hospital rooms are often dimly lit and featureless. Lighting, noise, and activity levels in the hospital often do not follow normal circadian rhythms. Sleep patterns are often disrupted and patients are routinely cut off from cues to the normal passage of time. The sensory and sleep deprivation of the hospital plays a role in the occurrence of delirium.

People who are delirious are at increased risk for falls. Oral intake is often poor, contributing to the development of malnutrition. A depressed mental status often leads to prolonged periods of immobility increasing risk for loss of muscle strength and pressure ulcers. Agitation and refusal of care can become a significant issue. Agitation may lead to the prescription of sedating medications which further impairs participation in care and inhibits walking and eating. Delirium during hospitalization is associated with increased risk of functional decline during hospitalization, greater likelihood of discharge to a skilled nursing facility rather than back to the patient's home, and increased probability of dying during the hospital stay.

2.6 MALNUTRITION

Oral intake is often poor in the hospital. Hospital food at its best is usually not appealing. Older adults are frequently given "therapeutic" diets in the hospitals, such as low salt or low cholesterol content, which make food even less appetizing. For example, roughly 70% of older adults have hypertension and are thus given low salt diets. Many illnesses that lead to hospitalization can diminish the appetite. There may be further impacts on appetite from medications patients are given as well as from constipation. Patients with limited mobility may be unable to reach their tray tables and water pitchers. In addition, oral intake may suffer if they do not have access to their dentures. Other factors that contribute to poor appetite, or inadequate nutrition, may be difficulties with swallowing or lack of assistance with feeding. Studies consistently show that older adults in the hospital do not eat well (Bally, JAMA Intern Med, 2016). In one study, 92% of subjects did not eat enough to meet their estimated daily caloric requirements; 21% ate less than half of their caloric needs (Sullivan, JAMA, 1999).

2.7 PRESSURE INJURY

Pressure, or soft tissue injuries, are one of the most common complications that occur for older adults during hospitalization. The incidence of pressure ulcers in hospitalized patients varies from 5% to 30%. Pressure injuries occur when unrelieved pressure causes damage to skin and soft tissue areas of the body. There are multiple risk factors for pressure injuries. Traditionally risk factors are grouped into extrinsic and intrinsic factors. Extrinsic factors include prolonged pressure, shearing forces, and friction. The most common intrinsic factors include immobility, incontinence, sensory impairment, and malnutrition. Immobility, whether temporary or prolonged, is one of the greatest risk factors for pressure injuries. Patients may be placed on bedrest due to acute illness which in turn promotes immobilization. Pressure injuries can arise within a few hours of complete immobility. For example, an older adult who presents to the Emergency Department with an incapacitating illness may develop a pressure ulcer during their initial evaluation while lying on an emergency room stretcher.

Many hospitalized older adults suffer from dementia, delirium, or neuropathy. These conditions may impact the patient's ability to detect sensory discomfort from a pressure injury. Another important cause of pressure injuries is inadequate blood flow. This commonly occurs when hospitalized patients experience low blood pressure or shock, or have a history of peripheral arterial disease. Complications of pressure injury include increased risk of infections, reduction in quality of life due to pain, increased length of hospital stay, and delay in functional recovery. The presence of a pressure ulcer is a sign of poor prognosis and increased mortality.

2.8   VENOUS THROMBOEMBOLISM

Hospitalization is a significant risk factor for developing blood clots in veins, which are called venous thromboembolic events (VTE). VTE usually develop in the legs and can spread to the lungs. Blood clots in the lungs, called pulmonary emboli (PE), can be fatal. Risk factors include older age and immobility.

Both VTE and PE are treated with anticoagulant ("blood thinning") medications. Older adults who are hospitalized are at increased risk for VTE during and following hospital admission. The incidence of VTE is thought to be 300,000–600,00 cases per year in the United Status.

2.9   NOSOCOMIAL INFECTION

Nosocomial infections, also known as hospital-acquired or healthcare-associated infections, are infections that are acquired during hospitalization. Owing to declines in the function of the immune system with aging, older adults are more susceptible to infection, particularly in the hospital setting. Risk factors include poor nutritional status, severity of acute illness, and multiple underlying conditions. Common nosocomial infections include pneumonia, urinary tract infection, and gastrointestinal infections. Infections can be acquired from hospital staff, other patients, or the hospital facility itself. The risk of infection is elevated in older adults who reside in communal settings or participate in communal activities such as daycare programs or attend senior centers.

2.10   THE BIOLOGY OF NORMAL AGING

As people age, many changes occur in the body across most organ systems. Unfortunately, these expected and normal changes frequently interact negatively with the typical hospital environment. Adults reach their peak muscle mass between their late 20s and early 30s. Between their 40s and 50s, most people begin to slowly lose muscle mass and muscle strength. Those who survive to their 80s may have lost up to 30% of their muscle mass and 50% of their muscle strength. This progressive loss of muscle mass and strength is called sarcopenia.

In the hospital, many factors conspire towards rapid loss of muscle mass. Hospital patients are often immobile. Beyond the contribution of the illness which brought them to the hospital, the hospital environment itself encourages immobility. Medical equipment such as intravenous drips, nasal cannulas, and urinary catheters are frequently attached to patients. These can act as tethers, making it difficult for the elders to stand up and walk. Hospital gowns which gape in the back, exposing buttocks, discourage mobility. Older adults who use walkers or canes often do not bring them to the hospital. Many hospitals do not provide walkers or canes to people who need them. Shoes are often placed in storage bags with other possessions and patients are expected to walk in hospital socks. Older adults are often discouraged from walking, out of fear that they are at increased risk for falling, even when these fears are inaccurate. Much of hospital care is oriented around the hospital bed, providing little incentive to get out of bed. Therapies and meals are delivered at the bedside. The TV, which is often the only form of entertainment, is typically centered on the bed. There is often nowhere to go on a routine basis save the bathroom.

One study showed that patients who were previously ambulatory, spent 83% of their time in the hospital lying down and 14% of their time sitting, leaving only 30 minutes of the day spent standing or walking (Brown et al. 2009). Bedrest has a number of physiologic effects. Muscle strength declines rapidly with lack of exercise. In one study of elderly volunteers, who spent 10 days in bed except for walks to the bathroom, it was found that leg muscle strength declined by 15% (Kortebein et al. 2008). The lower muscle strength was accompanied by decreased physical function, including slower walking speed, and worsened ability to climb stairs.

The loss of muscle strength with bedrest occurs in both young and older adults. However, because older adults enter the hospital at a lower level of strength, they are more likely than younger adults to cross a threshold where loss of strength leads to significant changes in function. This can range from declines in gait speed or stair climbing ability to complete loss of the ability to transfer or walk independently.

Normally, upon standing, there is a shift of blood from the circulatory system in the torso to the legs. Receptors in our blood vessels, called baroreceptors, sense the drop in blood pressure which accompanies this shift and coordinate an increase in heart rate and constriction in blood vessels in the legs which raises the blood pressure.

With prolonged bedrest, the volume which usually shifts between our upper body and our legs with repeated postural changes during the day stays in our torso continuously. Our baroreceptors sense this as an increase in our blood volume and send a signal to our kidneys to urinate out this perceived excess fluid. After several days, our blood volume may decrease significantly, adapting us for the supine state. However, when the bed-adapted person does stand up, they have greater difficulty compensating for the usual shift in blood volume to their legs, as they are lacking overall blood volume. This makes them prone to sensations of dizziness and instability on standing. This increases their risk for falling and can lead to dizziness so severe that people feel unable to walk. Older adults are more vulnerable to this than younger adults.

The immobility of hospitalization and the ensuing loss of function also increases risk for other adverse events including blood clots, urinary incontinence, and skin ulcers.

With aging, there is frequently a blunting in the sense of taste and smell. These sensory losses decrease enjoyment of food. As a result, older adults may eat a more restricted range of foods, putting them at risk for micronutrient deficiencies prior to hospital admission. They also may eat less at every meal, increasing their risk for macronutrient deficiencies. In one study, the prevalence of malnutrition in hospitalized older adults was 41% (Covinsky et al. 1999). In another study eating less than 50% of caloric needs was associated with a 50% increase in the risk of dying during the hospital stay (Sullivan, JAMA, 1999).

Oral intake is often poor in the hospital, making malnutrition the norm. This leads to a catabolic (negative energy) state in which body fat and muscle may be lost. This catabolic state leads to a cascade of additional risks for the older adult including impaired wound healing, development of pressure ulcers, and loss of function. Immune function is also inhibited by malnutrition, raising risk for hospital-acquired infections.

Vision and hearing loss are common in older adults and rise rapidly with aging. Studies suggest that hearing loss is present in 45% of US adults aged 60–69, 68% of those aged 70–79, and 89 % of those aged 80 and above (Lin et al. 2011). Visual acuity worse than 20/40 in the better seeing eye is present in 1% of those aged 60–69, 3% aged 70–79 and 16% aged 80 and older. Only 44% of those aged 70–79 and 19 % of those aged 80 and older had no vision or hearing impairment (Swenor et al. 2013). These sensory impairments impact function in a variety of ways, including increasing risk for falls as well as increasing risk for delirium in the hospital.

2.11   INTERVENTIONS

A number of interventions have proved to decrease risk of adverse events in hospitalized older adults.

Acute Care for Elders (ACE) Units are specialized units within hospitals that treat the primary problems patients are admitted for with an interdisciplinary model of care that provides increased attention to patient function. Features of the environment in an ACE Unit are designed for the needs of adults with functional impairments, including elevated toilet seats, large format clocks and calendars, door levers instead of knobs, handrails, and open walkways. Nursing protocols are designed to prevent immobility, foster independence in self-care, promote continence, protect sleep, and support cognition. Interdisciplinary rounds occur routinely involving nursing, physicians, physical and occupational therapy, nutrition, and social work. Discharge planning starts soon after admission and prioritizes putting necessary supports in place so that patients can return home.

ACE Units have been evaluated in multiple studies. Patients cared for on ACE Units have a lower risk of functional decline than with conventional care and a higher likelihood of being discharged home rather than to a skilled nursing facility. They have a lower risk of delirium, fewer falls, a shorter length of hospital stay, and lower costs.

The Hospital Elder Life Program (HELP) focuses on delirium prevention. It uses an interdisciplinary team and trained volunteers to implement interventions in several domains: cognitive stimulation and reorientation, supports for patients with hearing and vision impairment, early mobilization with daily ambulation, support for oral nutrition and hydration, and non-pharmacologic sleep enhancement. HELP is deployed on a regular hospital unit as an additional service to all older adults admitted there.

HELP has been evaluated in multiple studies and has been shown to have a number of positive effects including a 44% reduction in delirium and a 64% reduction in the rate of falls. Importantly, the number of patients discharged with a significant decline in their activities of daily living (ADLs) was reduced from 33% to 14% (Hshieh et al. 2015).

A variety of interventions have focused on increasing mobility levels in the hospital including education and mobility protocols for both nurses and nursing aids. One such study showed lower rates of decline in ADLs at the time of hospital discharge.

Attempts to improve malnutrition in hospitalized patients have included customizing diets, increasing caloric content of meals, provision of snacks or protein-calorie supplements between meals, and assisted feeding. Results of many trials have been variable with no consistent evidence of a significant benefit (Bally 2016). However, one large trial examined an intervention for hospitalized adults at risk for malnutrition (Schuetz 2019). A nutritionist created an individualized nutritional support plan which could include food adjustments based on patient preferences, snacks, protein supplements, and oral nutritional supplements. Enteral and para-enteral nutrition were recommended when considered necessary. The intervention led to a significant increase in caloric intake which was accompanied by a reduction in all-cause mortality from 10% to 7% as well as a significant reduction in decline of ADLs.

There are numerous interventions aimed at preventing falls in the hospital. These include risk stratification protocols which attempt to identify those patients most at risk for falling and select them out for specialized care: interventions which focus on modifying a single risk factor for falls in all patients; multicomponent interventions which address several risk factors for falling in all patients; and multifactorial interventions which attempt to customize fall prevention interventions to patients' individual risk factors. Experts in the field have suggested that efforts to identify patients most at risk for falls are misguided, as existing risk prediction tools, such as the Morse Fall Scale or the St. Thomas's Risk Assessment Tool, have poor specificity. They incorrectly identify large numbers of older adults as at risk for falling, leading to needless restrictions on activity and waste of hospital resources. The literature on fall prevention interventions is rife with poorly designed studies which are difficult to interpret or replicate. One intervention, with a fair evidence base to support effectiveness, is Fall Tailoring Interventions for Patient Safety (TIPS). Fall TIPS is a multifactorial intervention in which a systematic screen for risk factors for falling leads to a nurse-generated personalized fall prevention plan with both a patient education plan and a bedside poster for hospital staff, so that all caregivers are aware of the fall prevention plan. In a multicenter randomized trial, FALL TIPS reduced falls by 25%.

For older adults admitted to a hospital with no specialized programs, the individual practitioner can provide care in a way which reduces risk of adverse events in older adults. Assessment of function is critical to all preventive efforts. Practitioners should determine the patient's baseline level of cognition and ADL function prior to their illness and then assess current functional levels in all patients starting at admission and periodically thereafter. A simple mobility assessment consists of asking the patient to turn in bed, sit up, stand, and walk. If declines in function are already apparent, plans should be made to address them. To combat immobility and prevent loss of function, an individualized mobility plan should be created for all patients. Patients should be given a realistic and safe mobility goal every day (e.g., one lap of the floor three times per day for more mobile patients, or three sets of ten leg and arm lifts for patients who are unable to ambulate). Barriers to mobility should be removed with elimination of tethers wherever possible. Families should be encouraged to bring in walkers and canes if these are not present in the hospital, and clear pathways should be maintained around the bedside for walking. For patients who require assistance with transferring and walking, a plan should be made for how to provide them with the support necessary to mobilize, be it from nursing staff, physical therapy, or family members.

For patients at increased risk of delirium due to advanced age, poor vision or hearing, or poor cognition, practitioners should test orientation on a daily basis and provide reorientation when needed. During the day, bed and window curtains should be open and lights kept on. Efforts should be made to ensure that people who use glasses and hearing aids have them in the hospital. Sleep deprivation can be avoided through minimizing routine nighttime interruptions. Unless necessary, orders can be changed to avoid early morning or nighttime medication administration, blood draws, and vital signs measurements. Use of ear plugs at night can also support sleep. For patients at increased risk for delirium, practitioners should avoid medications which act on the central nervous system wherever possible.

To promote nutrition, use of therapeutic diets (e.g., low salt, low cholesterol) should be avoided unless they are deemed essential by the clinical team. Snacks between meals and protein calorie supplements should be ordered on admission for patients at increased risk for malnutrition. Family should be encouraged to bring food from home if possible. Efforts should be made to avoid medications which are nauseating, constipating, or alter taste. All patients who need assistance with eating should be offered adequate support.

A focus on preserving mobility will help to prevent the development of incontinence in the hospital. For patients who are unable to ambulate to the bathroom but are able to transfer safely, a bedside commode should be ordered. Routine use of diapers should be avoided as these prevent normal toileting.

2.12 SUMMARY

Older adults are at increased risk for a variety of adverse events during hospitalization. Normal physiological changes which occur with aging can interact negatively with a hospital environment which is poorly designed for the needs of older adults, leading to increased risk for hospital-associated loss of function, falls, malnutrition, delirium, pressure ulcers, venous thromboembolic events, and nosocomial infections. There is a rich evidence base showing that certain systematic interventions, such as ACE Units and the HELP, can prevent a variety of hospital-associated adverse events and improve outcomes in older adults.

Practitioners in hospitals with no such specialized programs can assess individuals risk factors for hospital-associated adverse events and create individualized prevention plans.

BIBLIOGRAPHY

Anderson FA Jr, Zayaruzny M, Heit JA, et al. Estimated annual numbers of US acute-care hospital patients at risk for venous thromboembolism. *Am J Hematol.* 2007; 82(9):777.

Bally MR, Blaser Yildirim PZ, et al. Nutritional support and outcomes in malnourished medical inpatients: a systematic review and meta-analysis. *JAMA Intern Med.* 2016 Jan;176(1):43–53.

Bouldin EL, Andresen EM, Dunton NE, et al. Falls among adult patients hospitalized in the United States: prevalence and trends. *J Patient Saf.* 2013 Mar;9(1):13–17.

Brown CJ, Redden DT, Flood KL, Allman RM. The Underrecognized Epidemic of Low Mobility During Hospitalization of Older Adults. *J Am Geriatr Soc.* 2009 Sep;57(9):1660–5.

Centers for Disease Control and Prevention. Persons with hospital stays in the past year, by selected characteristics: United States, selected years 1997–2018. Available at: www.cdc.gov/nchs/data/hus/2019/040-508.pdf (Accessed on July 29, 2021).

Covinsky KE, Martin GE, Beyth RJ et al. The Relationship Between Clinical Assessments of Nutritional Status and Adverse Outcomes in Older Hospitalized Medical Patients. *J Am Geriatr Soc.* 1999 May;47(5):532–8.

Covinsky KE, Pierluissi E, Johnston CB. Hospitalization-associated disability: "She was probably able to ambulate, but I'm not sure." *JAMA.* 2011 Oct 26;306(16):1782–93.

Creditor MC. Hazards of hospitalization of the elderly. *Ann Intern Med.* 1993;118 (3):219–23.

Cohen Y, Zisberg A, Chayat Y, Gur-Yaish N, et al. Walking for better outcomes and recovery: the effect of walk-for in preventing hospital-associated functional decline among older adults. *J Gerontol A Biol Sci Med Sci.* 2019 Sep 15;74(10):1664–70.

Coll PP, Phu S, Hajjar SH, Kirk B, et al. The prevention of osteoporosis and sarcopenia in older adults. *J Am Geriatr Soc.* 2021 May;69(5):1388–98.

Dykes PC, Carroll DL, Hurley A, et al. Fall prevention in acute care hospitals: a randomized trial. *JAMA.* 2010 Nov 3;304(17):1912–18.

Hshieh TT, Yue J, Oh E et al. Effectiveness of Multicomponent Nonpharmacological Delirium Interventions: A Meta-Analysis. *JAMA Intern Med.* 2015 Apr;175(4):512–20.

Inouye SK, Westendorp RG, Saczynski JS. Delirium in elderly people. *Lancet.* 2014 Mar 8;383(9920):911–22.

Kortebein P, Symons TB, Ferrando A, Paddon-Jones D et al. Functional impact of 10 days of Bed Rest in Healthy Older Adults. *J Gerontol A Biol Sci Med Sci.* 2008 Oct;63(10):1076–81.

Lin FR, Niparko JK, Ferrucci L. Hearing Loss Prevalence in the United States. *Arch Intern Med.* 2011 Nov 14;171(20):1851–2.

Reddy M, Gill SS, Rochon PA. Preventing pressure ulcers: a systematic review. *JAMA.* 2006 Aug 23;296(8):974–84.

Schuetz P, Fehr R, Baechli V, et al. Individualised nutritional support in medical inpatients at nutritional risk: a randomised clinical trial. *Lancet.* 2019 Jun 8;393(10188):2312–21.

Sullivan DH, Sun S, Walls RC. Protein-energy undernutrition among elderly hospitalized patients: a prospective study. *JAMA.* 1999 Jun 2;281(21):2013–19.

Swenor BK, Ramulu PY, Willis JR, et al. The prevalence of concurrent hearing and vision impairment in the United States. *JAMA Intern Med.* 2013 Feb 25;173(4):312–13.

3 Atypical Presentations of Common Diseases

Rubina Malik and Alice Guo

CONTENTS

3.1 INTRODUCTION

Older adults have a longer life expectancy compared with previous generations. As adults age they are more likely to develop multiple chronic health conditions contributing to increased disability and illness. There is a complex interplay of aging, chronic illness, and functional and cognitive decline which lead to *atypical presentations* of common diseases in older adults.

A standard definition of an *atypical presentation* is when an older adult presents with clinical signs and symptoms of an illness or disease that is different from what would be expected in a younger adult. Sometimes this can translate to having no signs or symptoms (e.g., no fevers or cough with pneumonia), unusual or unrelated signs or symptoms (e.g., confusion with urinary tract infections), or even the opposite of what is expected (e.g., lethargy after a fall with a hip fracture).

Risk factors for atypical presentations include older age (especially aged > 85 years), multiple medical comorbidities, dementia, large number of medications, and increased functional dependence. Many of these risk factors are critical elements that geriatricians assess when caring for complex and frail older adults using the Age-Friendly "Geriatrics 4 Ms" paradigm which focuses on: Mentation,

DOI: 10.1201/9781003197843-3

Mobility, Medications, and what Matters Most. Some geriatricians also add a 5th M for Multimorbidity. Older adults may have multiple atypical presentations due to interactions of the domains of the 5Ms. Identifying high risk patients will help guide all clinicians to be more attentive to search for and recognize atypical presentations of common diseases.

Atypical presentations may lead to an increase in morbidity, functional decline, and even mortality; this is often attributed to a delay in diagnosis. Atypical presentations lead to increased emergency department visits and hospitalizations. Recognizing atypical presentations is important to improve healthcare outcomes in older adults.

3.2　COMMON ATYPICAL PRESENTATIONS

Those who work in the field of geriatrics are aware that older adults are more likely to present with *atypical presentations*, many of which are similar to *geriatric syndromes*. Common geriatric syndromes are confusion, unsteady gait, falls, incontinence, and failure to thrive. Unlike geriatric syndromes, which tend to be of a chronic nature, atypical presentations are acute in onset. Common atypical presentations are altered mental status, falls, poor oral intake, and weakness. Both atypical presentations and geriatric syndromes share similar risk factors and have multiple causes. The appropriate medical interventions for both atypical presentations and geriatric syndromes, in general, are multipronged with a variety of steps requiring input and actions from the entire interprofessional team as well as the caregivers. When atypical presentations occur frequently over a longer period of time, they become geriatric syndromes.

One of the initial steps in assessing older adults is to identify if there is a *change from the baseline* in function, mental status, or behaviors. This is especially important in older adults with cognitive or sensory impairments who are unable to describe symptoms. An astute caregiver or family member will often report that "something is not right."

A common symptom may be **altered mental status** leading to **confusion** or **delirium** (which is defined as an acute change in mental status associated with waxing and waning mental status) or **lethargy**. Delirium is underdiagnosed by clinicians in all settings including the office, emergency room (ER), hospital, and nursing homes. During the covid pandemic, confusion was one of the most common symptoms of the COVID-19 infection in older adults, compared to younger adults who presented with fever, cough, and muscle aches. It is also important to ask the patient and caregiver about substance use and underlying psychiatric illnesses, both of which are often not considered as potential sources of atypical presentations in older adults.

Patients may have altered mental status associated with extreme **behavioral symptoms.** These behavioral symptoms can be hyperactive or hypoactive and at times vary between the two extremes. The hyperactive behavior often manifests with agitation, restlessness, and constant motion. The hypoactivity behavior will present with decreased movement, diminished speech, increased sleepiness, and somnolence. Often this can be challenging to recognize in patients who also have dementia, sensory deficits, or who do not have collateral information from caregivers. In addition, patients may receive medications for the hyperactive behaviors making it even more difficult to identify the source of the problem.

Older adults may present with one or more atypical presentations. A common atypical presentation seen in the ER is *unsteady gait* with or without a *fall*. It is not unusual to see an older adult in the ER who has both altered mental status and a fall. In fact, altered mental status and falls accounted for 80% of the atypical presentations in older adults in the ER. Falls are both an atypical presentation and a common geriatric syndrome very prevalent in older adults.

Other atypical symptoms that can coexist include *dehydration, anorexia (poor oral intake), and weakness or generalized fatigue.* It is not surprising to see a variety of atypical presentations as older adults have multiple medical comorbidities and can rapidly become ill, with the involvement of many different critical organ systems which are all interrelated including the heart, lungs, kidney, gastrointestinal track, and central nervous system.

Urinary incontinence rises with increasing age in both men and women. Problems with the bladder in both men and women can be due to primary bladder issues or there may be an atypical presentation due to medical comorbidities. It is critical to determine what is the change from baseline and to perform a detailed examination and evaluation. Marked abnormalities in a person's electrolytes (critical chemicals and minerals) such as a very elevated blood glucose in poorly controlled diabetes or an elevated calcium in cancer are common findings, in addition to urinary tract infections, as a cause of new urinary incontinence.

3.3 MAKING A DIAGNOSIS

Determining a diagnosis for any of the atypical presentations may be challenging. The medical team needs to consider a broad range of possibilities including common infections, poorly controlled chronic diseases, or a new illness such as a heart attack or stroke. Often it will be one or more factors that contribute to the atypical presentation. For example, older adults can have both a urine infection and poorly controlled diabetes, and their atypical presentation could be a change in mental status, incontinence, unsteady gait, weakness, and poor appetite. Keeping this in mind will help identify the cause in older adults. A thorough investigation should always include atypical presentations of common illnesses.

3.4 ASSESSMENT

Engaging an attentive caregiver is critical to identify if there is a change from baseline in an older adult. Obtaining collaborative information focusing on the patient's baseline function including mobility, daily activities, use of assistive devices, presence of sensory impairments, underlying psychiatric, and a thorough medication review including listing over-the-counter (OTC) medications, identifying new changes in medications, and asking about substance use can help to narrow possible etiologies. Polypharmacy, described as taking five or more medications, has been associated with geriatric syndromes and atypical presentations including unsteady gait and falls, altered mental status, and anorexia (see Chapter 30).

Performing a thorough physical exam and utilizing laboratory and imaging data can often help to identify a possible cause for the atypical presentation. The physical exam should include assessment of orthostatic hypotension (blood pressure taken both standing and lying flat), vision, hearing, and a thorough neurological evaluation. Sometimes pain can be elusive to identify in older adults and can cause many of the atypical presentations. Evaluating uncommon sites, such as performing a rectal exam or abdominal X-ray to evaluate for accumulation of hard, retained fecal material (fecal impaction) or evaluating the mouth to find a tooth or gum infection should be considered. Obtaining appropriate imaging and electrocardiograms based on the history and physical exam may be required. It is not uncommon for older adults to present with altered mental status, and a thorough work-up would reveal a new abnormality in the heart rhythm such as atrial fibrillation, or an abdominal X-ray show fecal impaction, or an abdominal ultrasound demonstrate retained urine, all possible causes for the new onset confusion. Older adults may have multiple symptoms of atypical presentations and have multiple etiologies contributing to the presentation. Or, alternatively, a frail older adult may have a relatively simple reason, such as hard wax obstructing their hearing, as a cause for the atypical presentation.

3.5   ATYPICAL PRESENTATIONS OF COMMON ILLNESSES

3.5.1   CARDIOVASCULAR DISEASE

Older adults who have coronary artery disease (CAD) are less likely to complain of the classic symptoms of substernal chest pain or pressure. In fact, a significant portion of older adults with acute heart attacks (acute myocardial infarction [AMI]) present with atypical symptoms. The most common of these is shortness of breath, followed by weakness or fatigue, shoulder or arm pain, indigestion, or nausea. Altered mental status and worsening clinical signs of congestive heart failure are also common presentations of AMI in older adults and are frequently the first presenting symptoms. This can make the diagnosis of cardiac pain difficult to tease apart from other chronic comorbidities, thereby delaying time to treatment and increasing risk of mortality.

Aging affects the cardiovascular system in many ways and may also cause abnormal heart rhythms (arrhythmias). Older adults may experience a variety of symptoms associated with the change in the pump function of the heart (cardiac output) caused by the arrhythmias. Symptoms may include fatigue, weakness, and/or dizziness due to the associated low blood pressure. Some may have falls, lose consciousness, or experience shortness of breath due to the decreased blood flow. Some older adults may not even have any of the prior symptoms, presenting only with the end consequence of the arrhythmia instead, such as a stroke.

3.5.2   INFECTIONS

Typical signs and symptoms of infection are frequently absent in older adults. As our body ages, the baseline temperature typically decreases by up to 1° Fahrenheit. Therefore, the traditional definition of fever is often not seen in older adults, since they are starting at a lower temperature. In addition, many older adults do not have any increase in their baseline temperature in the setting of an acute infection. This means that the classic criteria we use to assess for infection in this population should be adjusted. Infection should be suspected with any new change in the functional status in the older adults. This change generally can include, but is not limited to, new or increasing confusion, change in baseline behavior, incontinence, falls, decline in mobility, and reduced food intake.

3.5.2.1 Urinary Tract Infections

Adults aged 65 years and older are less likely to present with traditional symptoms of urinary tract infections (UTI) such as fever, pain with urination (dysuria), increased urination frequency, or increased urge to urinate. It is estimated that up to two-thirds of older adults present instead with atypical symptoms of UTIs. Altered mental status is a common presenting symptom in this population. Other atypical presenting symptoms include a decline in activities of daily living, the development of new incontinence, a decline in oral intake, and/or generalized weakness.

3.5.2.2 Pneumonia

Older adults with pneumonia are also more likely to present with atypical symptoms, have more severe symptoms, and have higher mortality compared with younger adults. Fever has been shown to be absent in more than one-third of those with community-acquired pneumonia. Owing to poor oral intake, dehydration, and mental status changes, typical radiographic findings and physical findings such as cough or abnormal lung exam findings may be absent. Fernández-Sabé et al. (2003) showed that older adults are more likely to present with altered mental status instead of chest pain, headache, and muscle aches which are typical symptoms for younger adults.

3.5.2.3 Sepsis

Sepsis is a medical emergency indicating a severe infection in which bacteria are present in the blood stream. Younger adults typically develop fevers, chills, and rigors (shaking chills). Like other infections, older adults often have an atypical presentation and are less likely to present with fevers and chills. The septic older adult may present with low blood pressure, kidney failure, altered mental status, and difficulty breathing. Often, the source of the infection is not readily apparent, and clinicians need to perform broad investigations including physical exam, laboratory data, and imaging.

3.5.3 ABDOMINAL PAIN

There are various changes that occur within an aging gastrointestinal (GI) system that can make it more susceptible to various illnesses. These changes, together with an increased number of comorbidities and concurrent medication use, means the older adult is much more likely to present atypically. Atypical presentations of GI illnesses will mainly mean the *absence* of expected symptoms. This includes, but is not limited to, the absence of pain, fever, or expected laboratory value changes.

As a result of changes in the gut and nervous system, older adults tend to present with vague, nonspecific symptoms that pose a great barrier to diagnosis. Atypical presentations, such as poor oral intake, weakness with dehydration, and acute kidney injury can be the initial presentation for GI illnesses. Obtaining information from collateral sources can help to identify changes in baseline oral intake and bowel patterns. Asking about swallowing problems or changes in bowel movements can provide important information. Performing a thorough physical exam including a rectal exam and ordering additional testing including laboratory and imaging is also helpful.

When it comes to abdominal pain, the older adult population is prone to worse outcomes, higher rates of admission and surgical interventions, and prolonged emergency department and hospital stays compared with younger patients. Owing to diagnostic challenges, delayed clinical presentation, associated medical comorbidities, and more illness-related complications, the mortality rate in this population is much higher than in younger adults.

3.6 SUMMARY

The way illnesses present in older adults can be quite different from that of younger adults. Key diagnostic signs and symptoms such as fever and pain may be absent in older adults. Older adults may instead present with altered mental status, falls, poor oral intake, and weakness as the first manifestation of a new and potentially treatable medical illness including heart attack, pneumonia, or uncontrolled diabetes. The approach to diagnosis of diseases in older adults should, therefore, also be different. Obtaining information from key caregivers is essential, including identifying a change in the typical level of functioning, performing a detailed physical exam, and initiating appropriate further investigation (laboratory and imaging) which are necessary to ensure that potentially serious conditions are not overlooked or missed. The medical community, older adults, and their caretakers need to be vigilant in reporting and evaluating changes and recognizing atypical presentations in an older adult to improve outcomes.

BIBLIOGRAPHY

Carpenter CR, Shelton E, Fowler S et al. Risk factors and screening instruments to predict adverse outcomes for undifferentiated older emergency department patients: a systematic review and meta-analysis. *Acad Emerg Med.* 2015;22(1):1–21. doi:10.1111/acem.12569

Caterino JM, Ting SA, Sisbarro SG, Espinola JA, Camargo CA Jr. Age, nursing home residence, and presentation of urinary tract infection in U.S. emergency departments, 2001–2008. *Acad Emerg Med.* 2012;19(10):1173–1180. doi:10.1111/j.1553-2712.2012.01452

Clifford KM, Dy-Boarman EA, Haase KK, Maxvill K, Pass SE, Alvarez CA. Challenges with diagnosing and managing sepsis in older adults. *Expert Rev Anti Infect Ther.* 2016;14(2):231–241. doi:10.1586/14787210.2016.1135052

Faverio P, Aliberti S, Bellelli G et al. The management of community-acquired pneumonia in the elderly. *Eur J Intern Med.* 2014;25(4):312–319. doi:10.1016/j.ejim.2013.12.001

Fernández-Sabé N, Carratalà J, Rosón B, et al. Community-acquired pneumonia in very elderly patients: causative organisms, clinical characteristics, and outcomes. *Medicine (Baltimore).* 2003;82(3):159–169. doi:10.1097/01.md.0000076005.64510.87

Gbinigie OA, Ordóñez-Mena JM, Fanshawe TR, Plüddemann A, Heneghan C. Diagnostic value of symptoms and signs for identifying urinary tract infection in older adult outpatients: systematic review and meta-analysis. *J Infect.* 2018;77(5):379–390. doi:10.1016/j.jinf.2018.06.012

Goch A, Misiewicz P, Rysz J, Banach M. The clinical manifestation of myocardial infarction in elderly patients. *Clin Cardiol.* 2009;32(6):E46–E51. doi:10.1002/clc.20354

Gómez-Belda AB, Fernández-Garcés M, Mateo-Sanchis E et al. COVID-19 in older adults: what are the differences with younger patients? *Geriatr Gerontol Int.* 2021;21(1):60–65. doi:10.1111/ggi.14102

High KP, Bradley SF, Gravenstein S et al. Clinical practice guideline for the evaluation of fever and infection in older adult residents of long-term care facilities: 2008 update by the Infectious Diseases Society of America. *Clin Infect Dis.* 2009;48(2):149–171. doi:10.1086/595683

Hofman MR, van den Hanenberg F, Sierevelt IN, Tulner CR. Elderly patients with an atypical presentation of illness in the emergency department. *Neth J Med.* 2017;75(6):241–246.

Huggins HE, Brady M, Emma JP, Thaler DE, Leung LY. Differences in presenting symptoms of acute stroke among young and older adults. *J Stroke Cerebrovasc Dis.* 2020;29(8):104871. doi:10.1016/j.jstrokecerebrovasdis.2020.104871

Jung YJ, Yoon JL, Kim HS, Lee AY, Kim MY, Cho JJ. Atypical clinical presentation of geriatric syndrome in elderly patients with pneumonia or coronary artery disease. *Ann Geriatr Med Res.* 2017;21(4):158–163. doi:10.4235/agmr.2017.21.4.158

Leuthauser A, McVane B. Abdominal pain in the geriatric patient. *Emerg Med Clin North Am.* 2016;34(2):363–375. doi:10.1016/j.emc.2015.12.009

Liang SY. Sepsis and other infectious disease emergencies in the elderly. Emerg Med Clin *North Am.* 2016;34(3):501–522. doi:10.1016/j.emc.2016.04.005

Limpawattana P, Phungoen P, Mitsungnern T, Laosuangkoon W, Tansangworn N. Atypical presentations of older adults at the emergency department and associated factors. *Arch Gerontol Geriatr.* 2016;62A:97–102. doi:10.1016/j.archger.2015.08.016

Lyon C, Clark DC. Diagnosis of acute abdominal pain in older patients. *Am Fam Physician.* 2006;74(9):1537–1544.

Nicolle LE, Bradley S, Colgan R, Rice JC, Schaeffer A, Hooton TM; Infectious Diseases Society of America; American Society of Nephrology; American Geriatric Society. Infectious Diseases Society of America guidelines for the diagnosis and treatment of asymptomatic bacteriuria in adults. *Clin Infect Dis.* 2005 Mar 1;40(5):643–654. doi: 10.1086/427507

Ouellet GM, Geda M, Murphy TE, Tsang S, Tinetti ME, Chaudhry SI. Prehospital delay in older adults with acute myocardial infarction: the comprehenSIVe evaluation of risk factors in older patients with acute myocardial infarction study. *J Am Geriatr Soc.* 2017;65(11):2391–2396. doi:10.1111/jgs.15102

Perissinotto CM, Ritchie C. Atypical presentations of illness in older adults. In: Williams BA, Chang A, Ahalt C, Chen H, Conant R, Landefeld C, Ritchie C, Yukawa M. eds. *Current Diagnosis & Treatment: Geriatrics, Second Edition.* McGraw Hill; 2014. Accessed January 02, 2022. https://accessmedicine.mhmedical.com/content.aspx?bookid=953&sectionid=53375629.

Perry A, Macias Tejada J, Melady D. An approach to the older patient in the emergency department. *Clin Geriatr Med.* 2018;34(3):299–311. doi:10.1016/j.cger.2018.03.001

Shenvi C, Wilson MP, Aldai A, Pepper D, Gerardi M. A research agenda for the assessment and management of acute behavioral changes in elderly emergency department patients. *West J Emerg Med.* 2019;20(2):393–402. doi:10.5811/westjem.2019.1.39262

van Duin D. Diagnostic challenges and opportunities in older adults with infectious diseases. *Clin Infect Dis.* 2012;54(7):973–978. doi:10.1093/cid/cir927

Vonnes C, El-Rady R. When you hear hoof beats, look for the zebras: atypical presentation of illness in the older adult. *JNP*. 2021;17(4):458–461. doi:10.1016/j.nurpra.2020.10.017

Welsh TP, Yang AE, Makris UE. Musculoskeletal pain in older adults: a clinical review. *Med Clin North Am*. 2020;104(5):855–872. doi:10.1016/j.mcna.2020.05.002

Wester AL, Dunlop O, Melby KK, Dahle UR, Wyller TB. Age-related differences in symptoms, diagnosis and prognosis of bacteremia. *BMC Infect Dis*. 2013;13:346. doi:10.1186/1471-2334-13-346

Zazzara MB, Penfold RS, Roberts AL et al. Probable delirium is a presenting symptom of COVID-19 in frail, older adults: a cohort study of 322 hospitalised and 535 community-based older adults. *Age Ageing*. 2021;50(1):40–48. doi:10.1093/ageing/afaa223

4 Hip Fractures in Older Adults

Wanda Horn, MD

CONTENTS

Hip fractures in older adults constitute a life-defining sentinel event. Hip fractures routinely lead to a functional decline and frequently portend a rapid downward spiral in an older adult's overall health status. Overall, orthopedic surgeries are among the most common major surgeries in older adults. However, the most frequently performed orthopedic surgeries, joint replacement of the knee and hip, are elective surgeries in which the patients are carefully screened prior to surgery based on their medical history. Thus, in general, joint replacement surgery does not carry significant greater risk compared with other elective surgeries in older adults. This is in marked contrast to hip fracture surgery. A hip fracture is a surgical emergency; there is no time to prepare. The common wisdom is that the surgeons can generally safely operate and fix the broken bones in the operating room. However, the major morbidity and mortality occur in the days, weeks, and months after the hip fracture.

In 2020, there were close to 300,000 hospital admissions due to hip fractures and this number is predicted to continue to rise due to both increased longevity and the proportion of older adults in the population. The estimated cost associated with hip fractures is around $15 billion per year with huge burden on the healthcare system (Malik et al. 2020).

Over 95% of hip fractures in older adults occur from falls, with the highest incidence in those aged 70 years and older. Women account for two-thirds of all hip fractures due to higher incidence of osteoporosis (a medical illness in which bones become fragile) (Iskrant 2010). Sometimes even a simple twisting or tripping injury may lead to a fracture. In some cases, the bone may be so weak that the fracture occurs spontaneously while someone is walking or standing. In this instance, it is often said that "the break occurs before the fall."

DOI: 10.1201/9781003197843-4

While most falls result in no injury, one-third result in an injury requiring medical attention or restriction of activities for at least one day. Most of these are minor soft tissue injuries, but 10–15% of falls result in a fracture.

Hip fractures should be treated in a hospital and the majority require surgery within one or two days after admission. The exception would be if an older adult is at the end of life and a decision is made to manage the hip fracture as a terminal event or as part of a palliative approach to care. The goals of the surgery are to allow the patient to return to their prior level of function, control pain, reduce morbidity, and improve quality of life. Unfortunately, approximately 30–50% of patients with a hip fracture do not regain their previous level of ambulation and up to 20% become nonambulatory.

Patients who are older, have cognitive deficits, and have used an assistive device prior to surgery are more likely to suffer a loss in functional status and have a higher risk of becoming nonambulatory.

At least 30% of patients sustaining a hip fracture require care in a skilled nursing facility after hospital discharge, and the associated mortality is 20% within one year. Higher mortality is associated with older age, male sex, multiple under-lying medical problems, psychiatric illness, and the presence of postoperative complications.

The rates of a person being able to return home permanently after a hip fracture vary widely between 50% and 90%, based on regional differences, the availability of services in the community, and individual preferences. Factors associated with nursing home placement are age over 80, presence of delirium, poor prior functional status, and lack of social support.

4.1   RISK FACTORS

There are multiple risk factors associated with hip fractures, the most important one being osteoporosis with the development of low bone mass and deterioration of the bone tissue. More than 70% of women in America aged 80 years and older are living with osteoporosis. However, despite widely available noninvasive tests for osteoporosis and multiple treatment options, many women are not diagnosed or treated for osteoporosis. Even after a wrist or vertebral fracture, only 10% of women are started on treatment for osteoporosis (Andrade et al. 2003).

Risk factors for osteoporosis include genetic factors (family history, white Caucasian), gender (female), diet (calcium balance), exercise, smoking, alcohol abuse, immobility, and chronic medical diseases which affect the bone. Hip fractures account for 72% of osteoporotic fractures, followed by other types such as fractures of the vertebra and wrists which are generally less serious.

4.2   TYPES OF HIP FRACTURE

There are different types of hip fractures based on anatomical differences, each of which requires a different type of surgical management. Intertrochanteric and fem-oral neck fractures are the most common types of hip fracture. Femoral head fractures are extremely rare and are usually the result of a high-velocity event such as a motor vehicle accident. There are also isolated fractures of the greater trochanter that usually come from a low-energy household fall. Although they are often painful, they usually heal without surgery. These fractures are stable and can be treated with protected weight bearing with either crutches or a walker.

4.3 DIAGNOSIS

The diagnosis of a hip fracture is usually made in an emergency room or urgent care, where the older adult presents after a fall. On exam, the hip is usually very painful, often with local bruising and limited motion. The affected limb is usually shorter and may be rotated abnormally from the torso. The fracture is generally diagnosed by X-rays of the hip. If a patient experiences significant hip pain and the hip X-ray is negative then a fracture may still be present and additional testing including a CAT scan can be performed.

Research findings prove that the last decade has witnessed significant effort to reduce morbidity and mortality among older adults with hip fractures, resulting in a heightened evidence-based approach. Early and continued comprehensive medical care, rapid time to surgery, early and aggressive physical therapy, and osteoporosis management have been demonstrated as effective strategies for reducing the rate of possible complications after a hip fracture.

The medical consultant, often an internist or geriatrician, plays an important role in the assessment and treatment of patients before surgery. Most research shows that early surgery, within 24–48 hours after injury, is associated with a reduction in one-year mortality and fewer postoperative complications (Seong et al. 2020). The medical consultant will ensure that key issues are addressed quickly so that surgery can proceed. Common problems that require attention include stabilization of fluid and electrolyte imbalances, identification of potential cardiovascular risk factors, and treatment of underlying infections. Important areas requiring attention in older adults are serial cognitive assessments with screening for delirium and underlying cognitive impairment, nutritional and functional status, polypharmacy, screening for depression, and discussion with patient and family regarding expectations, postoperative goals, and caregiver support. Geriatrics orthopedics co-management is a new model of care in which a geriatrician works side by side with the orthopedics team to deliver comprehensive, integrated care for older adults with a hip fracture. The team works together starting from the time the older adult is evaluated in the emergency room until they are discharged from the hospital. This successful model of care has demonstrated enhanced treatment preoperatively with a prevention of further complications postoperatively and after discharge.

4.4 TREATMENT

Surgical treatment remains the gold standard for the majority of hip fractures with very few exceptions. If the patient has been nonambulatory with serious comorbidities and imminent risk of death prior to the fracture, after a rigorous and multidisciplinary discussion between the patient, family, and medical team, the decision may be to treat the patient without surgery.

There are two major types of hip fractures: femoral neck fractures and intertrochanteric fractures. Intertrochanteric hip fractures are the most common type of hip fractures; they are usually associated with significant bleeding and routinely require blood transfusions. Treatment is generally internal fixation in which pins, rods, nails, and plates are used to stabilize the fracture. The recovery is generally longer than with other types of fractures.

Femoral neck fractures may be either displaced, where the bone has moved out of position, or nondisplaced where the bone remains in position. For nondisplaced femoral neck fractures, the most common treatment is internal fixation with multiple screws. In this procedure, surgical pins are placed across the fracture site to hold the bone in place while the fracture heals. Pinning prevents the femoral head from dislodging or slipping off of the femoral neck, a situation that would require a total joint replacement. Displaced fractures are more complex and require a partial (hemiarthroplasty) or total hip replacement surgery based on age and prior functional status. Both internal fixation and joint replacement surgery permit immediate postoperative mobilization with ambulation and provide complete functional recovery.

Postoperative care consists of aggressive physical therapy that starts the day of surgery and pain management together with medical management of comorbidities and treatment for osteoporosis. All surgical patients receive blood thinning medications to prevent blood clots in the legs, so called deep vein thrombosis (DVTs). These are generally injections with a low molecular weight heparin or oral medications depending on the specific clinical setting.

Early mobilization is defined as having a patient get out of bed shortly after hip fracture surgery. This generally requires the evaluation and assistance of a physical therapist. Early mobilization following hip fracture repair is imperative. Patients and families should be instructed that early mobilization, sometimes even the same day of the surgical intervention, should be the goal for reducing complications. Bedrest has been associated with multiple serious side effects including problems with the heart, lung, urinary systems as well as decrease in muscle strength and tone, loss of bone strength, and negative psychological effects. Specifically, early physical therapy has been associated with a reduction in postoperative complications such as clots in the legs (thromboembolism), pneumonia, wound breakdown, pressure ulcers, and worsening confusion (delirium).

Studies have shown that patients who are mobilized within the first 48 hours of surgery were able to walk further, required less assistance to transfer, and were more likely to be discharged directly home, when compared to patients whose ambulation was postponed.

Sometimes there is reluctance by hospital staff as well as family to encourage early mobilization, due to concerns about increased risk of falls and pain. Early mobilization may require increased pain medications, which has its own potential for side effects. As is often the case in medical care, there is a fine line between risks and benefits that need to be taken into consideration and fully disclosed to patients and caregivers.

4.5 POSTOPERATIVE COMPLICATIONS

The most common postoperative complications are delirium and DVT. Older adults with underlying cognitive and functional status are at higher risk for developing delirium during their hospital stay. More than a quarter of all patients admitted with a hip fracture suffer from an episode of delirium during their hospitalization. It is often underrecognized and difficult to treat. Delirium increases mortality and morbidity, leads to lasting cognitive and functional decline, and increases both length of stay and institutionalization at discharge. Risk of delirium should be discussed prior to surgery and the team should institute multiple measures to prevent and decrease incidence. Environmental factors (noise, light, hearing aids, eyeglasses, and family at bedside), frequent orientation (clocks and calendars), pain control, and avoidance of specific known medications with anticholinergic effects, electrolyte, and fluid management are usually the most effective measures to prevent delirium.

Deep vein thrombosis prophylaxis is instituted in all patients after surgery and is usually well tolerated. In spite of multiple protocols for prophylaxis treatment, studies have reported that the incidence of perioperative DVT due to hip fractures ranges from 11% to 35%. Some studies have also reported that the prevalence of DVT is as high as 62% in patients with hip fractures when the time to surgery is delayed by more than two days. Several risk factors were associated with the rate of DVT after hip fractures, including age, female sex, cardiovascular disease, pulmonary disease, cancer, previous hospitalization for DVT, and type of anaesthesia.

Other common postoperative complications include pneumonia, atelectasis, which is a minor collapse of the lung tissue, urinary retention, constipation, urinary tract infection, anemia from acute blood loss, and malnutrition. Many of these complications are potentially preventable and strategies should be implemented in appropriate older adults. Postoperative urinary tract infection can be decreased by not routinely placing urinary catheters prior to surgery and recognizing and treating urinary retention when present.

All older adults should be instructed preoperatively on how to perform incentive spirometry (a hand-held medical device used to help patients improve their lung functioning), in order to prevent potential lung complications such as atelectasis.

Anemia should be treated with blood transfusions only when patients have symptoms from the acute blood loss (low blood pressure, rapid heart rate, dizziness, and inability to engage in physical therapy) or at the advice of the consultant due to other medical conditions

Older adults are at high risk of developing constipation after surgery and should be placed on a bowel regimen and have their bowel movements recorded and tracked.

Older adults are also at risk of malnutrition and a nutritional consult should be obtained when decreased oral intake is present or patient shows significant weight loss prior to surgery. Oral supplements should be prescribed under these circumstances, and weight should be monitored.

Surgical wound infections occur in 5% of patients and are treated with antibiotics, but sometimes requires a second surgical intervention and possible revision surgery. Problems with fracture healing can be found in up to 15 % of older adults and almost always require additional surgery.

Unfortunately, recognition and treatment of the patient's underlying osteoporosis is often not started by the inpatient medical team in spite of significant data that shows higher risk of a second hip fracture after the first fracture. Vitamin D levels should be measured and supplementation should be initiated during the hospitalization. In addition, discussions with the patient, family, and primary care provider regarding starting medication for osteoporosis after hospital discharge should be initiated. Multiple medical institutions have introduced a Fracture Liaison Service as a multidisciplinary team to reduce the risk of subsequent fractures in patient with recent osteoporotic fractures.

Discharge planning after an admission for hip fracture can be a very difficult and emotionally challenging process for an older adult faced with a new functional impairment, and the transfer to a short rehabilitation facility is the recommended practice for the vast majority of older adults with a hip fracture. Some younger patients, with excellent social support, can return to the community if appropriate services can be provided. All older adults with risk factors for a fall should have a home safety evaluation to institute measures to prevent future falls.

BIBLIOGRAPHY

American Academy of Orthopaedic Surgeons Management of Hip Fractures in Older Adults Evidence-Based Clinical Practice Guideline. www.aaos.org/hipfxcpg.pdf Published December 3, 2021.

Andrade SE, Majumdar SR, Chan KA, Buist DS, Go AS, Goodman M, Smith DH, Platt R, Gurwitz JH. Low frequency of treatment of osteoporosis among postmenopausal women following a fracture. Arch Intern Med. 2003 Sep 22;163(17):2052–7.

Iskrant AP. The classic: the etiology of fractured hips in females. 1968. Clin Orthop Relat Res. 2010 Jul;468(7):1731–5.

Judd KT, Christianson E. Expedited Operative Care of Hip Fractures Results in Significantly Lower Cost of Treatment. Iowa Orthop J. 2015;35:62–4. PMID: 26361446; PMCID: PMC4492154

Kates SL, Mendelson DA, Friedman SM. Co-managed care for fragility hip fractures (Rochester model). Osteoporos Int. 2010 Dec;21(Suppl 4):S621–5. doi: 10.1007/s00198-010-1417-9. Epub 2010 Nov 6. PMID: 21058002.

Koizia LJ, Wilson F, Reilly P, Fertleman MB. Delirium after emergency hip surgery - common and serious, but rarely consented for. World J Orthop. 2019 Jun 18;10(6):228–234

Malik AT, Khan SN, Ly TV, Phieffer L, Quatman CE. The "Hip Fracture" Bundle Experiences, Challenges, and Opportunities. Geriatr Orthop Surg Rehabil. 2020 Mar 5;11:2151459320910846.

Marcantonio ER, Flacker JM, Michaels M, Resnick NM. Delirium is independently associated with poor functional recovery after hip fracture. J Am Geriatr Soc. 2000 Jun;48(6):618-24

Mary Ann Forciea, Risa Lavizzo-Mourey, Edna P. Schwab, Donna Brady Raziano Elsevier Health Sciences, Feb 24, 2004

Roe J. Montefiore comprehensive model of care for elderly hip fracture patients, Montefiore J. Musculoskeletal Med Surg 2016

Ryan DJ, Yoshihara H, Yoneoka D, Egol KA, Zuckerman JD. Delay in Hip Fracture Surgery: An Analysis of Patient-Specific and Hospital-Specific Risk Factors. J Orthop Trauma. 2015 Aug;29(8):343–8.

Seong YJ, Shin WC, Moon NH, Suh KT. Timing of Hip-fracture Surgery in Elderly Patients: Literature Review and Recommendations. Hip Pelvis. 2020 Mar;32(1):11–16.
Swenning T, Leighton J, Nentwig M, Dart B. Hip fracture care and national systems: The United States and Canada. OTA Int. 2020 Mar 23;3(1):e073.

5 Functional Fitness for Optimal Aging

James F. Konopack

CONTENTS

5.1 INTRODUCTION

Older adults comprise an increasingly large segment of the population in the United States, with individuals ages 65 and above projected to increase from just over 50 million at this writing to a total of 80 million in the year 2040, with an even more pronounced growth among those ages 85 and older. With increasing chronological age comes an associated increase in the risk of various chronic health conditions, which coupled with the nation's shifting aging demographics will continue to create a proportional demand for healthcare professionals' services. Fortunately, many of the adverse health conditions associated with human aging are amenable to change through regular physical activity, yet the costs of physical inactivity on individual health and society remain significant. The maintenance of functional fitness through regular physical activity – or its initiation at any point among those who are insufficiently active – is an achievable aim with wide-ranging benefits and remains a critical component to optimal health and well-being during older adulthood.

5.2 BENEFITS OF PHYSICAL ACTIVITY FOR OLDER ADULTS

Physical activity is an essential component of healthy living for adults and especially so for older adults. The benefits of aerobic physical activity are many. These include the reduction in risk for the development of several chronic diseases including, among others, coronary artery disease, stroke, certain types of cancer, and non-insulin-dependent diabetes mellitus.

DOI: 10.1201/9781003197843-5

Most research and most public health recommendations focus on aerobic modes of physical activity, such as walking, running, and bicycling. There is ample literature demonstrating a causal relationship between these activities and many of the aforementioned conditions as well as weight reduction and the associated reduction in obesity-linked sequelae. However, non-aerobic activities (e.g., weight lifting, yoga, tai chi) have received increased attention by researchers and have demonstrated positive outcomes of their own. Strengthening skeletal muscle results in improved functional fitness (i.e., ability to successfully complete activities of daily living), particularly among less active older adults. Strength-building movements and balance-training activities alike have been shown to reduce the risk of falls in older adults as well as reduce individuals' fear of falling, which itself is a psychological barrier to getting even more physical activity (e.g., not walking due to concerns of slipping or tripping).

The metabolic changes that occur with aging can disrupt health but respond to interventions such as resistance training. The aging process in addition to an accumulation of visceral body fat and a depletion of bone mineral density on the rate of 0.5–1.0% per year after age 40 is also associated with a graduate decrease in lean body mass, which can be measured as a decrease in muscle cell cross-sectional area and a disproportionate loss of type II (i.e., fast-twitch) muscle fibers compared with type I (i.e., slow-twitch) fibers and worsening muscle quality. This process of age-related, progressive muscle loss is known as sarcopenia and can frequently result in frailty among older adults. In the average person, sarcopenia results in a reduction of approximately 40% of one's skeletal muscle mass between roughly the ages of 20 and 70 (Nair, 2005). Altogether, the age-related loss of muscle and bone with concomitant increase in body fat has been dubbed by some as "osteosarcopenic obesity," the risks and effects of which are markedly diminished with regular strength-building exercise. Fortunately, skeletal muscle generally retains its plasticity throughout the lifespan, responding to strength-building stimuli at any age, so there is no point in an individual's life after which the effects of resistance training diminish. Strength training, even among the oldest old, can improve functional fitness and the performance of activities of daily living.

5.3   PSYCHOLOGICAL BENEFITS

Of course, physical activity of all types has also been widely demonstrated to elicit positive psychological effects, and the psychological health challenges in older adulthood are amenable to change through physical activity. These include reduction of negative psychosocial health outcomes such as depression and dysthymia, anxiety and stress reactivity, and fatigue as well as increases in positive outcomes such as self-esteem, life satisfaction, mood, perceived energy, sleep quality, and social support when participating in group activities. And a robust literature supports the neurocognitive benefits of regular exercise (see US Department of Health and Human Services, 2021), with widely demonstrated improvements in cognitive function, particularly executive control processes that involve response inhibition, switching, or other "mental gymnastics" or thinking on the spot. Despite the many benefits of resistance exercise training mentioned thus far and the convincing evidence that high-intensity strength training is an effective tool in treating clinical depression, the preponderance of data on the psychological benefits of exercise pertains to aerobic forms of activity. Thus, although resistance training confers tremendous and multi-faceted functional benefits to older adults, it should not be done at the expense of aerobic activity but rather in conjunction with it.

5.4   PSYCHOSOCIAL DETERMINANTS OF EXERCISE

While there is a voluminous literature on the mechanisms of exercise's benefits, of particular relevance to healthcare workers are the salient mechanisms of exercise's psychosocial benefits. Self-efficacy, or an individual's beliefs in their ability to exe-cute a particular task (or overcome barriers to same), is perhaps the most important of these mediators in the exercise-psychosocial health relationship. Self-efficacy is both a mediator of this relationship and reciprocally predictive of future physical activity. That is, the more a newly active person walks, the greater that individual's belief in his/her ability to execute that behavior under increasingly challenging conditions, such as walking for 30 minutes without stopping. These increases in self-efficacy for walking are associated with improvements in several psychosocial outcomes and lead to further increases in walking, as greater self-efficacy makes one more resilient in the face of adversity when it comes to a specific task.

Those who develop and provide exercise plans to others may benefit from an exploration of the sources of self-efficacy information, as maximizing this infor-mation can boost self-efficacy and therefore make exercise plans more efficacious. As first described by the preeminent psychologist Albert Bandura and subsequently validated by empirical research, the four basic sources of self-efficacy information are: (1) **enactive mastery experiences;** (2) **verbal persuasion and social influence;** (3) **vicarious experiences**; and (4) **interpretation of physiological and affective states**.

Enactive mastery experiences are the most powerful source of efficacy infor-mation and can be thought of simply as an individual's direct experience with the behavior along with the feedback they receive from engaging in it. For example, in the case of a sedentary older adult who is beginning a walking-based exercise program, they may not be especially confident in their ability to walk uninterrupted for 60 minutes at a brisk pace, unsurprisingly. To increase walking self-efficacy through enactive mastery, the individual may be asked to walk uninterrupted for 5 minutes at a brisk pace, or a similar goal that is somewhat of a challenge but very likely to be achieved. In this way, "baby steps" in exercise progression can lead to eventual fitness gains by way of self-efficacy gains along the way.

The other major sources of efficacy information are less significant but still worthy of consideration. **Vicarious experiences** are comparisons an individual makes with others, sometimes simply referred to as modeling. If an individual can observe someone with whom they identify completing the target task, they experience a vic-ariously derived boost in efficacy; it is an experience of, "If they can do it, I can, too." **Verbal persuasion and social influence** are simply the means by which efficacy is increased by a coach, mentor, peer, or significant other's encouragement. Though a less significant source than the others, an individual's **interpretation of physiological**

and affective states (e.g., heart rate, respiration, muscle discomfort, sweating) are cues to exercise intensity, the accurate and behavior-promoting interpretation of which may be informed by an experienced exerciser, trainer, or instructor's knowledge and feedback, particularly for uninitiated exercisers.

Another important consideration in the promotion of physical activity among older adults is the recognition of self-reported barriers to activity. Among the most common of these are: lack of time, illness or injury, self-reported fatigue, and limited access to a facility or other exercise resources.

5.5 PHYSICAL INACTIVITY

Despite the well-known benefits of physical activity for nearly every dimension of health and wellness, physical inactivity remains an ongoing public health challenge. Recent data suggest that only approximately 25% of adults in the United States maintain minimum physical activity levels sufficient to accrue health benefits (US Department of Health and Human Services, 2021). Of the remainder, over 31 million are sedentary, engaging in little to no physical activity beyond that required to complete basic activities of daily life. These figures are even more pronounced among older adults, with leisure time physical activity levels generally diminishing with advancing age. This creates a threat to the security and well-being of older adults not merely because of the absence of cognitive, emotional, cardiovascular, metabolic, and other benefits derived from exercise-induced fitness, but because the risk for those responsive conditions increases throughout the course of normal aging, and approximately half of the adult population in the United States currently has one or more preventable diseases. The Centers for Disease Control and Prevention estimate the annual healthcare costs in the United States attributed to low levels of physical activity to be $117 billion, leading to 10% of premature mortality nationwide. For individuals and society alike, physical inactivity has its costs that can, by definition, be remedied by individual behavior.

5.6 RECOMMENDATIONS FOR PRACTITIONERS

1. Promote the adoption and maintenance of physical activity levels currently recommended for older adults (from USDHHS, 2018):
    a. Adults should move more and sit less throughout the day. Some physical activity is better than none. Adults who sit less and do any amount of moderate-to-vigorous physical activity gain some health benefits.
    b. For substantial health benefits, adults should do between 150 minutes (2 hours and 30 minutes) and 300 minutes (5 hours) a week of moderate-intensity, or 75 minutes (1 hour and 15 minutes) to 150 minutes (2 hours and 30 minutes) a week of vigorous-intensity aerobic physical activity, or an equivalent combination of moderate- and vigorous-intensity aerobic activity. Preferably, aerobic activity should be spread throughout the week.
    c. Additional health benefits are gained by engaging in physical activity beyond the equivalent of 300 minutes (5 hours) of moderate-intensity physical activity a week.

    d. Adults should also do muscle-strengthening activities of moderate or greater intensity and that involve all major muscle groups on two or more days a week, as these activities provide additional health benefits. Practitioners and public health advocates should promote a pattern of weekly physical activity for older adults that includes balance training as well as aerobic and muscle-strengthening activities.

    e. Older adults should determine their level of effort for physical activity relative to their level of fitness.

    f. Older adults with chronic conditions should understand whether and how their conditions affect their ability to safely perform regular physical activity.

    g. When older adults cannot do 150 minutes of moderate-intensity aerobic activity per week (i.e., the recommended minimum) because of chronic conditions, they should be as physically active as their abilities and conditions allow.

2. Develop exercise programs with a foundation of functional movements performed with correct form and appropriate frequency and intensity (from Konopack, 2016):

    a. **Select Functionally Relevant Activities:** When designing a resistance training program, exercises should encompass all major muscle groups across approximately 8–10 exercises. Complex, multi-joint movements performed to their full range of motion are preferred over isolation exercises or movements with restricted range of motion when possible. The resultant strength gains from such movements should have obvious functional benefits (e.g., improved stair-climbing, sit-to-stand transitions, lifting and carrying objects around the home, etc.) Sample movements include squats, lunges, or leg presses; bench presses, shoulder raises, or push-ups; and upright or seated rows, among many alternatives. For a free and accessible guide to strength-focused activities that can be done at home, see the web resource provided by the Centers for Disease Control and Prevention which is provided at the end of this chapter.

    b. **Train with Appropriate Frequency and Intensity:** For many older adults, a resistance training regimen that includes major muscle groups of both the upper and lower body can be well tolerated when performed twice weekly. Several sources recommend a frequency of two or more times per week, but two weekly training sessions of the same muscle group can be effective for improving muscle quality in older adults. In addition, adults of all ages will often experience delayed-onset muscle soreness (DOMS) shortly after undertaking a new exercise program, particularly following movements that involve longer eccentric muscle action (i.e., lengthening of the muscle, usually associated with the lowering of the weight). DOMS results from the body's natural inflammatory response to minute damage incurred during exercise stimulus and can last up to four days (McCall, 2015, p. 422). For this reason as well as the fact that muscular growth occurs during the rest *between* rather than *during* training sessions, strength training should not exceed three times per week after the initial few weeks for most older adults, if ever. During a resistance training workout, each movement should be easy enough to permit at least 6–8 repetitions but challenging enough to bring about failure before reaching a maximum of 15–20 repetitions. With a high degree of effort, sets of strength movements for older adults are ideally in the range of 10–15 repetitions.

c. **Learn and Practice Correct Form:** Physical activities such as walking and bicycling require little to no formal instruction in order to perform them safely and effectively throughout one's lifetime, but resistance training exercises, when improperly executed, can result in sprains, dislocations, or impact injuries from the weights themselves. Learning correct form under the supervision of an expert is an important safety measure.

d. **Emphasize Additional Physical Activity Modes:** In addition to (but not necessarily in place of) weight training, the activities that follow are recommended for their potential to increase muscular strength in older adults when performed with regularity. Many of these also directly help improve balance:

   i. Stair-climbing
   ii. Strenuous gardening (i.e., involving carrying, digging, and similar activities)
   iii. Pilates
   iv. Brisk walking or walking on hills
   v. Yoga
   vi. Tai Chi

3. Encourage older adults to safely and gradually progress to minimum activity levels. To do so, older adults should do the following (also from USDHHS, 2018):

a. Understand the risks, yet be confident that physical activity can be safe for almost everyone. In determining a person's risk stratification and readiness for exercise, the reader is strongly encouraged to first consult the second chapter of the *ACSM's Guidelines for Exercise Testing and Prescription* (11th edition, 2021) and have prospective exercise participants confer with their primary healthcare providers as needed.

b. Choose types of physical activity that are appropriate for the older adult's current fitness level and health goals, because some activities are safer than others.

c. Increase physical activity gradually over time to meet key guidelines or health goals. Inactive people should "start low and go slow," that is, they should start with lower intensity activities and gradually increase how often and how long activities are done. One need not exercise at minimum levels at the onset. The most significant improvements come from a transition from no activity or complete sedentariness to some activity. Patients or clients can work their way up to meeting minimum recommended levels of activity over time.

    d. Protect themselves by using appropriate gear and sports equipment, choosing safe environments, following rules and policies, and making sensible choices about when, where, and how to be active.

    e. Be under the care of a healthcare provider if they have chronic conditions or symptoms. People with chronic conditions and symptoms can consult a healthcare professional or physical activity specialist about the types and amounts of activity appropriate for them.

4. By extension of the previous recommendation, providers may also wish to consult with certified specialists, such as a Senior Fitness Specialist (SFS) or Certified Strength and Conditioning Specialist (CSCS) credentialled by the National Strength and Conditioning Association (NSCA), a Clinical Exercise Physiologist certified by the American College of Sports Medicine (ACSM), or a Medical Exercise Specialist certified by the American College on Exercise (ACE). Healthcare providers may also be interested in exploring the ACE's specialty certifications for themselves, such as ACE Senior Fitness Certification, Functional Aging Institute Functional Aging Certification, and others (see the link provided in this chapter's bibliography).

5. Structure physical activity programs with psychosocial variables in mind, such as social support and self-efficacy. To maximize the latter,

    a. Create conditions for older adults to develop **enactive mastery experiences**: small, progressive achievements directly related to the target behavior that will increase their beliefs in their ability to do similar and even more challenging behaviors in the future.

    b. Provide behavior-promoting **verbal persuasion and social support** through coaching, encouragement, instruction, or other forms of social reinforcement. The quality of feedback provided to exercisers by instructors or clinicians can augment the exerciser's feelings of mastery.

    c. Provide direct feedback or educational resources or references to help participants accurately **interpret physiological and affective responses** to exercise, which may be unfamiliar and potentially worrisome to those who have been sedentary or active but with limited intensity for some time.

    d. Pair participants with peer leaders or instructors with whom they can identify (i.e., active older adults) to maximize participants' **vicarious experiences**, or use instructional videos that feature older adults engaging in the behavior.

    e. Increase opportunities for socialization before, during, or after bouts of physical activity, as positive **social support** is a means by which to organically augment self-efficacy through encouragement and accountability and confers its own benefits to the individual.

    f. Having exercisers develop their own reasonably difficult, specific, measurable goals that represent accurate expectations of future performance can guide efforts and provide ready-made opportunities to create enactive mastery experiences.

5.7 CONCLUSION

In summary, the safety of older adults – expressed as reduced risk from injurious falls, several diseases, and chronic conditions, and improvements in mental health – can be improved through regular physical activity. Though lifelong physical activity is ideal, there is never a point at which it is "too late" to reap the health and safety benefits of regular physical activity, including those derived from strength gains by way of resistance training, which improves several key aspects of functional fitness.

Interestingly, there is also a growing body of literature linking higher amounts of skeletal muscle with better immune function in older adults (e.g., Nelke et al., 2019), with data suggesting that the health of older adults may be improved not merely by increasing cardiorespiratory fitness but by restoring the body to a more robust composition characterized by increased lean body mass. Indeed, researchers have recently shown muscle strength to be independently associated with COVID-19 hospitalization among adults aged 50 and above, such that those with greater grip strength were less likely to be hospitalized than their weaker peers (Cheval et al., 2021), adding more evidence to the literature supporting strength as an important element of safety for older adults.

BIBLIOGRAPHY

ACSM's Guidelines for Exercise Testing and Prescription (11th ed.). Baltimore, MD: American College of Sports Medicine; 2021.

The American Council on Exercise. Fitness Certifications. www.acefitness.org/fitness-certifi cations/default.aspx Accessed October 2021.

Bandura A. Self-efficacy: Toward a unifying theory of behavioral change. *Psychological Review,* 1977; 84(2), 191–215.

Centers for Disease Control and Prevention. Growing Stronger: Strength Training for Older Adults. www.cdc.gov/physicalactivity/growingstronger/ 2011. Accessed August 2021.

Cheval B, Sieber S., Maltagliati S, Millet GP, Formanek T, Chalabaev A, Cullati S, Boisgontier MP. Muscle strength is associated with COVID-19 hospitalization in adults 50 years of age and older. *Journal of Cachexia, Sarcopenia and Muscle,* 2021; 12(5), 1136–1143. doi: https://doi.org/10.1002/jcsm.12738

Demontiero O, Vidal C, Duque G. Aging and bone loss: New insights for the clinician. *Therapeutic Advances in Musculoskeletal Disease,* 2012; 4(2), 61–76. doi: 10.1177/ 1759720X11430858

Fleck SJ, Kraemer WJ. *Designing Resistance Training Programs* (4th ed.). Champaign, IL: Human Kinetics; 2014.

Konopack JF. Resistance training: Recommendations for age-relevant benefits. *Annual Review of Gerontology and Geriatrics,* 2016; 36(1), 193–204. doi: 10.1891/0198-8794.36.193

McCall P. Resistance training. In: Porcari J, Bryant C, Comana F (Eds.), *Exercise Physiology.* Philadelphia: F.A. Davis, 2015; 412–438.

Nair KS. Aging muscle. *American Journal of Clinical Nutrition,* 2005; 81(5), 953–963.

Nelke C, Dziewas R, Minnerup J, Meuth SG, Ruck T. Skeletal muscle as potential central link between sarcopenia and immune senescence. eBioMedicine, 2018; 49, 381–388. doi: https://doi.org/10.1016/j.ebiom.2019.10.034

National Strength and Conditioning Association. www.nsca.com/ Accessed October 2021.

Office of Disease Prevention and Health Promotion. Healthy People 2030. https://health.gov/healthypeople Accessed December 2021.

Ormsbee MJ, Prado CM, Ilich JZ, Purcell S, Siervo M, Folsom A, Panton L. Osteosarcopenic obesity: The role of bone, muscle, and fat on health. *Journal of Cachexia, Sarcopenia, and Muscle*, 2014; 5(3), 183–192. doi: 10.1007/s13539-014-0146-x

Signorile JF. *Bending the Aging Curve: The Complete Exercise Guide for Older Adults.* Champaign, IL: Human Kinetics, 2011.

US Department of Health and Human Services. *Physical Activity Guidelines for Americans* (2nd ed.). Washington, DC. https://health.gov/our-work/nutrition-physical-activity/physical-activity-guidelines 2018. Accessed August 2021.

6 Care Transitions

Allison Stark, MD, MBA and Janet Kasoff, EdD, RN, NEA-BC, BC, CPHQ, CPHIMS

CONTENTS

6.1 OVERVIEW

Patient Collins is a 72-year-old retired male. He returns to the emergency department from his home after he was discharged from the hospital two days ago. His symptoms include nausea and vomiting, shortness of breath, fatigue, headache, and his blood glucose level is very elevated. The discharge medication list was different from the medication list he was taking pre-hospitalization. He did not take the medications post-discharge in hopes of clarifying the medication list with his primary care provider during their next scheduled appointment. There was no telephone call by a care team member after discharge to evaluate these post-acute gaps in care. This is one of many cases where effective communication, collaboration, coordination, effective planning, patient education, and health literacy intersect in ways that can impact the patient's experience and outcomes. What went wrong here? Could this readmission be prevented? Was there an adequate post-acute care transition? What are care transitions?

Care transitions is a term used broadly to refer to a movement of patients between care settings. Examples of this include a patient transitioning from home to the hospital, hospital to home, or hospital to a skilled nursing facility. Care transitions are necessary as a patient's condition and care needs change during an acute or chronic illness.

Care transitions fall under various types. These include within settings, between settings, across health states, and between providers. The care transitions process has several key components, such as proper patient identification, provider notification, support through the transition, coordination, communication, collaboration, tracking, ongoing reassessment of the plan of care, and medication reconciliation, which is defined as the process of identifying the most accurate list of all medications that the patient is taking. There can be many care transitions in a patient's life when moving from and to home. Important in this transition process is coordinated communication among and across caregivers within the settings. The chart here describes the different types of care transitions with examples provided.

Transition Type	Examples
Within Settings	Primary Care to Specialty Care, Recovery Room to a Surgical Unit
Between Settings	Hospital to Home, Hospital to Skilled Nursing Facility, Home Care to Self-Care
Across Health States	Curative Care to Palliative Care
Between Providers	Generalist to Specialist

6.2 SIGNIFICANCE OF CARE TRANSITIONS AND IMPACT ON PATIENT OUTCOMES

When patients transition from one care setting to another, they are at risk to receive care that is not well coordinated. If information exchange is not adequate, the risk of medical errors, poor quality of care, and patient harm all emerge as major issues. Quality, financial, and patient satisfaction outcomes can also all be impacted. Cross-setting communication can have an impact on multiple domains including the length of a hospitalization or skilled nursing facility admission, the acuity of a patient's illness, and medical finances.

6.3 SUCCESSFUL CARE TRANSITIONS

There are multiple elements to consider in evaluating whether a care transition was successful. It is important to differentiate what success means to the various stakeholders involved in the care transition and to understand that each element of a care transition can be evaluated for success. These elements include discharge planning, determination of the optimal post-acute care setting, ambulatory follow-up appointments and any needed social supports, care coordination, communication, medication reconciliation and safety, promotion of patient self-management and empowerment, advance care planning, symptom management, and monitoring post-discharge.

From the patient and caregiver perspective, key considerations include whether the patient and any identified caregivers felt prepared and cared for by the healthcare professionals they interacted with during the transition period. Well-executed communication is at the heart of successful care transitions. Patients and caregivers desire communication that is timely, well organized, and at the appropriate health literacy level. Patients and caregivers want to fully understand and feel prepared to follow the plan of care, which should be developed with their input to ensure optimal feasibility. Did they know what medications and treatments were needed, what they were for, and whether there have been any changes from prior regimens? Were the services that were described as part of the plan of care initiated without interruption? Were follow-up appointments scheduled at times that were feasible? Did the patient and caregiver feel that the healthcare providers were committed to their recovery? Did they know who they could turn to for advice or medical care through the transition period? Success for patients also includes that their post-acute care was delivered at the right time, in the optimal care setting or location, and that they were able to follow the plan of care.

From the provider and health system perspective the goals of a successful care transition revolve around health outcomes, patient safety, and healthcare quality. Discharging providers seek to avoid adverse outcomes for patients including medication errors, unexpected progression of illness, insufficient post-discharge follow-up, and avoidable emergency department visits or readmissions. In addition, providers care about the satisfaction of patients, families, and staff in the transition process.

At the hospital and health system level the goals of successful care transitions include improving patient satisfaction and Hospital-Consumer Assessment of Healthcare Providers & Systems (H-CAHPS) survey scores related to discharge and care transitions as well as the avoidance of financial penalties through reductions in reimbursement from the Centers for Medicare and Medicaid Services (CMS) or other health plan payers. These financial penalties are tied to multiple metrics including, but not limited to, H-CAHPS scores and 30-day readmissions and excess days in acute care settings for certain clinical conditions. In addition, for hospitals and health systems involved in value-based arrangements with health plans, where the hospital or health system is financially at risk for a specified population of patients, successful transitions of care tie directly to quality incentives that can be realized by reducing readmissions to achieve quality metric and any total cost of care targets. Finally, providers, hospitals and health systems that are in value-based arrangements that involve higher levels of financial risk can achieve savings when healthcare utilization is appropriate and potentially preventable admissions and readmissions are avoided.

Health plans and government payers are focused on member or beneficiary satisfaction and optimal health outcomes with care transitions as well as on appropriate utilization of healthcare services and overall healthcare costs. These goals are well aligned with those of providers, hospitals, and health systems.

6.4 CARE TRANSITION MODELS

The care during transitions requires interventions that enhance coordination and continuity. There are various models that exist to manage care transitions. The Care Transition Model by Dr. Eric Coleman is an effective one. Dr. Coleman postulated four pillars that influence the success of a care transition. These pillars are medication self-management which includes medication reconciliation, follow-up with provider post-discharge, understanding the red flag symptoms that influence disease exacerbation, and access to a comprehensive personal discharge record. Patients and/or caregivers obtain a complete list of the facility discharge medications. This discharge medication list must be compared with the medications the patient has at home. Any concerns or discrepancies should be discussed with the patient's provider. It is important that the patient and/or caregiver understand each medications' dosage, uses, frequency of administration, and the reason for taking the medications.

Dr. Coleman's model uses a four-week program, during which patients with complex care needs receive specific tools, are supported by a Transition Coach, and learn self-management skills to ensure their needs are met during the transition from hospital to home. The goals of the Care Transitions Program are to support patients and families, increase skills among healthcare providers, enhance the ability of health information technology to promote health information exchange across care settings, implement system level interventions to improve quality and safety, develop performance measures and public reporting mechanisms, and influence health policy at the national level.

Patients who received Dr. Coleman's Care Transition Program interventions were significantly less likely to be readmitted to the hospital and more likely to achieve self-identified personal goals around symptom management and functional recovery. The findings were sustained for as long as six months after the program ended.

What is significant about Dr. Coleman's model is the identification of a practitioner or team dependent on setting, to facilitate and coordinate the patient's transitions plan. The model uses a Transitional Care Nurse whose role is to perform a comprehensive assessment of the patient's needs and then collaborate with patient/caregiver and the care team to develop the transition's plan. The post-acute needs of the patient are taken into account in the formation of this plan and follow-up post-discharge is provided.

Better Outcomes by Optimizing Safe Transitions (BOOST) is another care transitions model. Developed by the Society of Hospital Medicine in 2010, BOOST is an example of an evidence-based decision support program in care coordination. The purpose of the program is to provide safe care transitions and prevent unnecessary rehospitalizations.

Project BOOST goals are to reduce 30-day readmission rates for general medicine patients (with particular focus on older adults), improve facility patient satisfaction scores and the institution's H-CAHPS scores related to discharge, improve the flow of information between hospital and outpatient physicians, identify high-risk patients and target specific interventions to mitigate their risks for adverse events, and improve patient and family preparation for discharge.

The BOOST model uses interdisciplinary coaches who work with patients as they transfer from hospital to home. The model identifies eight indicators that are predictive of an unsuccessful care transition and are used as a guideline to format information for sharing. The eight Ps are: problem medications, psychological concerns, principal diagnosis, polypharmacy, patient support, poor health literacy, prior hospitalization within the last six months, and palliative care. The project outcomes included reduced readmission rates.

Additional care transition models include Project Re-Engineered Discharge (RED) which utilizes a standardized approach to patient discharge. The State Action on Avoidable Rehospitalizations (STAAR) is prescriptive in its approach and employs a risk stratification process that categorizes patients daily into high, medium, and low risk discharges and post-discharge care is predicated on risk.

Timely access to care post-hospitalization is key in all models. Post-discharge calls focusing on medication management, self-care, red flag symptoms, and post-acute appointment validation are part of the model as well as discussion of end-of-life wishes.

6.5   CARE TRANSITION TOOLS

There are various tools spearheaded by experts in the field that address safe care transitions' best practices for the Interdisciplinary Care Team and the patient and/or caregiver. The Care Transitions Program website https://caretransitions.org/ provides tools and resources for clinicians, patients, and caregivers. The Family Caregiver Activation in Transitions Tool (FCAT) assists with identifying gaps in preparatory readiness for the discharged patient based on the responses of the family caregiver. Care Transitions Measure Tool (CTM-15) assesses the quality of the care transition. The Family Caregiver Tool – DECAF- measures the collaborative role of the family/caregiver in various aspects of the patient's care. The Care Transitions Measure (CTM-3) reports the quality of preparation for self-care among adult patients discharged from acute care hospitals within 30 days. The Tips for Managing Care at Home provides information on how to adapt to the home environment post-discharge. The Tips for Effective Medication Management provides information on how to manage medications at home. Identification of red flags focuses on the indicators that demonstrate early warning signs of disease exacerbation. Universal discharge or transition checklists prepare the patient for home. There are numerous standardized templates developed for health professionals which provide key information to be shared with the next level of care. These are: standard plan of care, transfer tool, transition record, and transition summary. The Situation, Background, Assessment, Recommendation (SBAR) communication template uses a standardized method to organize communication effectively.

6.6   IMPACT ON PATIENT SAFETY

Communication challenges present a major risk in the care transition process. The Center for Transforming Healthcare categorized risk factors related to ineffective communication which includes the sender and receiver having different expectations, lack of teamwork and respect, inadequate time allotted for an appropriate patient hand-off, and lack of a standardized communication template. The use of the SBAR tool can be an effective one for standardizing verbal patient hand-offs. The management of the patient in the hospital by specialists and lack of involvement or communication back to their primary care doctor is another issue which is compounded by the lack of a team-based approach and accountability for care across the continuum. It is important to develop and utilize a timely discharge and transition of care document to communicate to all post-acute providers and care teams. The lack of access to the electronic medical record is another barrier. In addition, frequently there are challenges to complete additional diagnostic testing, specialty visits, and therapy after hospital discharge.

When these gaps in care exist, they put patient at risk for adverse outcomes and rehospitalization. Nearly one in four patients with heart failure or renal disease are readmitted within 30 days of patient discharge. It is estimated that 20% of patients hospitalized in the United States are readmitted within 30 days of discharge.

6.7   CARE TRANSITIONS TEAM MEMBERS

Many stakeholders and interdisciplinary team members are involved in the transition of care process, and coordination among them is an essential component of success. Key team members include the attending physician, registered nurse, and physical therapist who determine the patient's functional status and care needs after the hospital stay, the social worker and registered nurse who place referrals for post-discharge services and the floor-registered nurse who is typically responsible for communicating the discharge plan and reviewing medications. The floor nurse plays a key role in assisting the entire interdisciplinary team and ensuring the patient's physical and psychological readiness for hospital discharge. Others who may be involved include care managers who follow the patient for complex or chronic condition case management, pharmacists, primary and specialty care providers, and clinicians in any post-acute setting that the patient may experience, including nurses, therapists, and social workers from home care agencies or skilled nursing facilities.

6.8   PATIENT EDUCATION AND TEACH-BACK ROLE IN PATIENT SAFETY AND CARE TRANSITIONS

The 2007 Med Pac Report found that "patient adherence with discharge instructions had a significant impact on the rate of hospital readmissions." Discharge instructions can be complicated, not written for the right literacy level or language, and not take into account cultural preferences. Effective patient education is a key driver in patient adherence. As part of effective patient education strategy, the Teach-Back Method is recommended. Teach-Back is a strategy used to validate that the educator explained relevant content in a way that the patient/caregiver can understand. Utilizing Teach-Back, the patient/caregiver explains the content in their own words back to the educator, and the educator then can evaluate their level of understanding.

It is important that any learning gaps in education are addressed to maximize effective outcomes. The process for Teach-Back is a feedback loop. In the first step, the educator shares and explains a new concept to the patient and/or caregiver. The educator then asks the patient and/or caregiver to explain the concept again, listens to the patient's/caregiver's explanation and then shares again, focusing on reviewing any missing points. Finally, the educator asks for additional information if they feel the process needs to repeat again. Using this process is a systematic way of ensuring patient/caregiver discharge instructions are clearly communicated and understood.

According to the Agency for Healthcare Research and Quality (AHRQ), Teach-Back should be utilized whenever new health information is presented. Some examples include prescribing a new medication or making changes to the current medication regimen, explaining home care instructions, or describing the next steps in the follow-up of care. It is important when using the Teach-Back method that one employs basic principles of health literacy to ensure comprehension. Communication needs to be simple and at the right literacy level, so that the risk of miscommunication is low. Only 12% of the adults in the United States have health literacy skills required to manage our healthcare system, and stress and illness may further lower comprehension capability. The Institute for Healthcare Improvement (IHI) recommends using the "Ask Me 3" questions to help patients and their families understand their health conditions. This method outlines three key questions patients should ask their providers: What's my main problem? What should I do for that problem? Why is that important?

6.9   IMPROVEMENTS IN CARE TRANSITIONS AND FUTURE DIRECTIONS

Efforts to improve care transitions can focus on a number of interventions. These may include prospectively identifying patients at high risk for rehospitalization and developing specific interventions to mitigate potential adverse events and outcomes. Interventions for these high-risk patient may include condition-specific discharge pathways that are evidence based, including use of checklists and specific educational tools and interventions, as well as standardized documentation tools and communication strategies. In addition, identifying and utilizing key clinical team members, such as nurse navigators or pharmacists, who may follow patients through a care transition and interact in both the inpatient and outpatient setting to ensure that the plan of care is followed and patients are connected back to ambulatory care providers may have multiple benefits. Another care transition improvement strategy is to systematically engage in early discharge planning, including determination of the optimal discharge care setting and destination based on the identified patient needs and barriers. Early discharge planning should begin at the time of admission and can be highly effective in terms of reducing both the length of stay for the current admission and the risk of readmission. Warm hand-off protocols are also being used as a care transition improvement strategy to enhance communication between the hospital-based discharging team and the receiving post-acute care team. Warm hand-offs focus on real-time sharing of key information relevant to the safe and effective ongoing care of the patient and are typically conducted between peer clinicians at the sending and receiving care setting (i.e., nurse-to-nurse or physician-to-physician).

Significant innovation is taking place in the area of transitions of care, with many health systems, providers, and vendors identifying new opportunities to impact outcomes. Digital health interventions, including remote patient monitoring, is one of the most exciting areas of innovation, with use increasing dramatically in the setting of the COVID-19 pandemic. Patients may be discharged with devices to monitor their vital signs including blood pressure, pulse, blood oxygenation, weight, or glucose. The devices transmit data back to a clinical team that is actively monitoring the patients and intervening when clinical measurements hit defined thresholds. Another innovation includes the use of community paramedicine providers who follow a patient in the community after discharge from another care setting such as the emergency department, hospital, or skilled nursing facility. These teams coordinate with usual care providers and are available to respond real-time to patients with clinical concerns. Home-based models of care, including home-based primary care and hospital-at-home, are also increasing in scope and use as a means of improving patient care by bringing the care to where patients need it most, and minimizing the number of care transitions experienced by patients.

6.10   SUMMARY

Patient and caregiver self-management is key to successful care transitions. In the case study provided at the beginning of this chapter, Patient Collins care transition was fraught with suboptimal experiences. He did not participate in a thorough medication reconciliation. His knowledge of red flag symptoms was not verified, and he was not assessed for health literacy. There was not close coordination with the care team, and he did not have an understanding of the need to follow up timely with post-acute medical management. This resulted in Patient Collins experiencing adverse events and ultimately being rehospitalized. Using the best practices, interventions, and workflows from the models identified has the potential to improve quality, patient satisfaction, health outcomes, and healthcare costs.

BIBLIOGRAPHY

AHRQ Health Literacy Universal Precautions Toolkit. www.ahrq.gov/sites/default/files/wysi wyg/professionals/quality-patient-safety/quality-resources/tools/literacy-toolkit/health literacytoolkit.pdf. Accessed August 18, 2021.

AHRQ Patient Guide to Teach Back. September 2020. www.ahrq.gov/health- literacy/improve/ precautions/toolkit.html. Accessed August 18, 2021.

AHRQ Readmission Data. August 2018. www.ahrq.gov/data/infographics/readmission-rates. html. Accessed August 18, 2021.

Auerbach AD, Kripalani S, Vasilevskis EE et al. Preventability and causes of readmissions in a national cohort of general medicine patients. *JAMA Intern Med.* 2016;176(4):484–493. doi:10.1001/jamainternmed.2015.7863

Campagna V, Nelson SA, Krsnak J. Improving care transitions to drive patient outcomes: the triple aim meets the four pillars. *Prof Case Manag.* 2019 Nov/Dec;24(6):297–305. doi:10.1097/NCM.0000000000000387. PMID: 31580296.

Campbell BM, Hodshon B, Chaudhry SI. Implementing a warm handoff between hospital and skilled nursing facility clinicians. *J Patient Saf.* 2019 Sep;15(3):198–204. doi:10.1097/ PTS.0000000000000529. PMID: 30095538.

Cibulskis CC, Giardino AP, Moyer, VA. Care transitions from inpatient to outpatient settings: ongoing challenges and emerging best practices. *Hospital Practice.* 2011;39(3):128–139.

Dusek B, Pearce N, Harripaul A, Lloyd. Care transitions. *J Nurs Care Q.* 2015;Jul/Sep 30(3):233–239. doi:10.1097/NCQ.0000000000000097

Forster AJ, Clark HD, Menard A, Dupuis N, Chernish R, Chandok N, Khan A, Van Walraven C. Adverse events among medical patients after discharge from hospital. CMAJ. 2004 Feb 3;170(3):345–9. Erratum in: CMAJ. 2004 Mar 2;170(5):771.

Fox MT, Persaud M, Maimets I, Brooks D, O'Brien K, Tregunno D.. Effectiveness of early discharge planning in acutely ill or injured hospitalized older adults: a systematic review and meta-analysis. *BMC Geriatr.* 2013 Jul; 6;13:70. doi: 10.1186/1471-2318-13-70.

Gupta S, Perry JA, Kozar R. Transitions of care in geriatric medicine. *Clin Geriatr Med.* 2019 Feb;35(1):45–52. doi:10.1016/j.cger.2018.08.005. Epub 2018 Oct 3. PMID: 30390983.

Hass SA, Swan BA, Haynes TS. *Care coordination and transition management core curriculum.* Pitman, NJ: American Academy of Ambulatory Nursing; 2014.

Kansagara D, Chiovaro JC, Kagen D, Jencks S, Rhyne K, O'Neil M, Kondo K, Relevo R, Motu'apuaka M, Freeman M, Englander H. So many options, where do we start? An overview of the care transitions literature. *J Hosp Med.* 2016 Mar;11(3):221-30. doi: 10.1002/jhm.2502. Epub 2015 Nov 9.

Kripalani S, Theobald CN, Anctil B, Vasilevskis EE. Reducing hospital readmission rates: current strategies and future directions. Annu Rev Med. 2014;65:471-85. doi: 10.1146/annurev-med-022613-090415. Epub 2013 Oct 21.

Labson M. Innovative and successful approaches to improving care transitions from hospital to home. Home Healthcare Now. 2015 Feb;33(2):88–95. doi:10.1097/NHH.0000000000000182

Maimon HR, Zimlichman E. Does care transition matter? Exploring the newly published HCAHPS measure. *Am J Med Quality.* 2020;35(5):380–387. doi:10.1177/1062860620905310

Mitchell SE, Laurens V, Weigel, Hirschman KB, Scott AM, Nguyen HQ, Howard JM, Laird L, Levine C, Davis TC, Gass B, Shaid E, Jing L, Williams MV, Jack BW. Care transitions from patient and caregiver perspectives. *Ann Fam Med.* 2018 May;16(3):225–231; doi:https://doi.org/10.1370/afm.2222

Morkisch N, Upegui-Arango LD, Cardona MI et al. Components of the transitional care model (TCM) to reduce readmission in geriatric patients: a systematic review. *BMC Geriatr.* 2020;20(1):345. Published 2020 September 11. doi:10.1186/s12877-020-01747-w

National transitions of care coalition. www.ntocc.org. Accessed June 23, 2021.

Powell SK, Tahan, HA. *Case management a practical guide for education and practice*, 4th ed. Philadelphia, PA: Wolters Kluwer; 2019.

Project BOOST (Better outcomes for older adults through safe transitions). www.hospitalmedicine.org/globalassets/clinical-topics/clinical-pdf/8ps_riskassess-1.pdf. Accessed June 4, 2021.

Project RED (Re-engineered discharge). www.bu.edu/fammed/projectred/index.html. Accessed June 3, 2021.

Project Boost® implementation guide, 2nd ed. Coffey C, Greenwald J, Budnitz T, Williams MV. Eds. Society of Hospital Medicine. www.hospitalmedicine.org/BOOST

Rutherford P, Nielsen GA, Taylor J, Bradke, P, Coleman E. *How to guide: improving transitions from the hospital to community settings to reduce avoidable re-hospitalizations.* Cambridge, MA: Institute for Healthcare Improvement, 2013.

Shah MN, Hollander MM, Jones CM, Caprio TV, Conwell Y, Cushman JT, DuGoff EH, Kind AJH, Lohmeier M, Mi R, Coleman EA. Improving the ED-to-home transition: the community paramedic-delivered care transitions intervention-preliminary findings. *J Am Geriatr Soc.* 2018 Nov;66(11):2213–2220. doi:10.1111/jgs.15475. Epub 2018 Aug 10. PMID: 30094809; PMCID: PMC6235696.

Strategy 4: Care transitions from hospital to home: IDEAL discharge planning. Content last reviewed Dec 2017. Agency for Healthcare Research and Quality, Rockville MD. www. ahrq.gov/patient-safety/patients-families/engagingfamilies/strategy4/index.html

The care transitions program leaving the hospital: what you must know. https://caretransitions. org/leaving-the-hospital-what-you-must-know/. Accessed August 11, 2021.

The care transitions program® – transitional care and intervention. www.caretransitions.org. Accessed June 16, 2021.

Transitions of care coalition: care transitions bundle seven essential intervention categories. https://static1.squarespace.com/static/5d48b6eb75823b00016db708/t/5e837a30f7518 a6872e34876/1585674803444/SevenEssentialElements_NTOCC+logo.pdf. Accessed August 11, 2021.

Zwart DLM, Schnipper JL, Vermond D, Bates DW. How Do Care Transitions Work?: Unraveling the Working Mechanisms of Care Transition Interventions. Med Care. 2021 Aug 1;59(Suppl 4):S387–S397. doi: 10.1097/MLR.0000000000001581. PMID: 34228021; PMCID: PMC8263132.

7 Medication Management Issues in the Older Adult Patient

Tracy Offerdahl-McGowan and Marvin Pathrose

CONTENTS

7.1 TERMINOLOGY PEARLS

- "Medication" and "drug" can be used interchangeably.
- "Adverse Drug Event" (ADE), "Adverse Effect," and "Side Effect" can be used interchangeably.
- "Nonprescription" refers to Over-the-Counter (OTC) Medications.
- "Co-Morbidities" refers to more than one disease or condition in a person. For example, Type 2 diabetes and hypertension.
- "Multi-Morbidities" refers to multiple diseases or conditions in a person. For example, Type 2 diabetes, hypertension, dyslipidemia, heart failure, etc.

As people age, there is generally an increase in the number of health problems that older patients experience. This often results in co-morbidities or even multi-morbidities. It is not surprising, then, that our older adults are the largest per capita consumers of prescription medications. This presents a number of challenges for the older adults, their caregivers, and their healthcare and medical practitioners. In addition to the prescription medications, many older adults also take nonprescription medications (over-the-counter or OTC) and dietary supplements (e.g., vitamins, herbs, etc.; see Table 7.1 and Figure 7.2 for common OTC medications and dietary supplements used in older adult patients). This combination may result in a huge financial burden, a large daily "pill burden," potential drug interactions, and potential side effects or adverse drug effects.

DOI: 10.1201/9781003197843-7

Medication use in our older adult population is complicated. When a prescription or nonprescription drug is studied in clinical trials prior to approval, it is most commonly evaluated in younger adult patients. Dietary supplements do not go through clinical trials under the Food and Drug Administration (FDA) at all (www.fda.gov/food/dietary-supplements/information-consumers-using-dietary-supplements#:~:text=Under%20DSHEA%2C%20FDA%20is%20not,market%20without%20even%20notifying%20FDA). So, when we use a drug or dietary supplement in older adults, we are left to try and predict how they will be tolerated by the older adult's body. This further complicates the treatment of multi-morbidities in our older patients, as they frequently have multiple medications that need to be taken for the individual diseases or conditions. When multiple medications are taken, the chance that one or more "drug-drug interactions" can occur is drastically increased.

In the last couple of decades, there has been a large "push" by pharmacy and medical organizations to address this issue of a large pill burden or "polypharmacy." Traditionally, "polypharmacy" can be defined as ≥ 5 medications taken daily or on most days, and it is estimated that 30% of people ≥ 65 years old take 5 or more drugs. Others define polypharmacy as the use of more medications than clinically indicated or use of unnecessary/harmful medication(s). Polypharmacy is problematic for several reasons, with the most significant being that the number of medications a person is taking is the single most important predictor of harm. This may result in overall frailty, falling, disability due to a fall, hospitalization, and even death. In addition, the risk of adverse drug events (ADEs) is correlated to very old age, multi-morbidities, dementia, frailty, short life expectancy, and polypharmacy.

One of the reasons that polypharmacy is riskier in older adult patients is because this group of people have a number of physiologic changes that can alter the pharmacokinetics and pharmacodynamics of drugs. Generally, "pharmacokinetics" refers to what the body does to a drug as the drug is moving through the body (i.e., how the body is working to metabolize and clear the drug), whereas "pharmacodynamics" refers to what the drug does to the body (i.e., mechanism of action and adverse effects). Where a drug will go once it is in the bloodstream is called "distribution." It is a pharmacokinetic principle used to predict what receptors, organ systems, and compartments a drug might travel to, and this helps us to predict some of the potential adverse drug effects that we might see in patients. Distribution is different in older adults when compared to younger adults, as there tends to be more body fat and less lean body mass. Drugs that are more fat (lipid) soluble will have a larger distribution and a prolonged clearance rate in older adult patients. The liver's oxidative metabolic enzymes known as the cytochrome P450 system are also decreased in older adult patients when compared to younger patients, and the renal clearance of drugs tends to be less efficient in older adult patients. Inefficient clearance or elimination of a drug or drugs means that more drug than anticipated is left behind to potentially cause problems. Finally, the "good" that comes from medications cannot be separated from the "bad" that can potentially come from them.

Patients take medications to prevent or treat a condition or a disease, but these medications distribute to "extra" receptors or compartments or organ systems that are in addition to the area that is needed for treatment. In other words, these additional distribution sites are what provide opportunities for harm from the medication. We cannot separate the target drug sites (why a patient is taking a drug) from the "extra" drug distribution sites (potential adverse drug effects from the drug), nor can we stop the drug from targeting all of them. The significance of this cannot be overstated. This is why practitioners must always monitor for efficacy and toxicity.

7.2 CLINICAL PEARLS

Older adult patients are at risk for experiencing a medication "triumvirate of doom."

- They are more likely to have multi-morbidities that may result in several medications being taken.
- They are physiologically different from younger adult patients resulting in differences in how their bodies metabolize and eliminate medications.
- They are at risk for unrecognized drug-drug interactions as well as multiple ADEs that increase their risk of falling.

It is estimated that one-third of community-dwelling people over the age of 65 fall every year, and medication use can contribute to this. Older adults who have fallen or have balance and gait disturbances have an even higher chance of a falling again. Individual medications with ADEs that cause cognitive impairment or frailty can result in a fall, and when multiple medications are taken by an older adult the chances of cognitive changes and fall risk increases even more. Typically, any patient who has had a fall within the past year should be medically evaluated with an assessment that checks multiple factors for gait, balance, and lower extremity strength. Evaluating a patient's drug regimen should be added to this list, as well. By regularly evaluating our older patients' drug regimens, we lessen the chance that avoidable adverse drug effects may occur.

The Office of Disease Prevention and Health Promotion (ODPHP) as part of the US Department of Health and Human Services Adverse Drug Event Action Plan (ADE Action Plan) provides the following statistics about the general population:

In inpatient settings, adverse drug events

- Account for an estimated 1 in 3 of all hospital adverse events
- Affect about 2 million hospital stays each year
- Prolong hospital stays by 1.7–4.6 days

In outpatient settings, adverse drug events

- Account for > 3.5 million physician office visits
- Account for an estimated 1 million emergency department visits
- Account for approximately 125,000 hospital admissions

When we look at outpatient older adult patients, they are three times more likely to experience adverse drug effects that requires a trip to the emergency department and they are seven times more likely to experience an adverse drug effect that requires a hospital admission. The numbers for inpatient older adults and those in long-term care facilities are even higher. This shows us that safe prescribing practices in older adult patients such as evaluation of kidney function that may lead to an adjusted dosing regimen, medication allergies, and drug doses and interactions is an underutilized practice.

No practitioner prescribes a drug or recommends a nonprescription drug or supplement with the intention to make a person sicker; however, many older adult patients have multiple prescribers for their various health issues and diseases. In addition, many patients have more than one pharmacy, as they may get some drugs from mail-order pharmacies and other drugs from their local drug store (see Table 7.1).

More prescribers and pharmacies on board will increase the likelihood that another drug or remedy will be prescribed or recommended and simply added to the patient's drug list. It is easy for a practitioner to look at a patient's drug list and think, "well,

---

TABLE 7.1
Common Nonprescription Medications and the Potential Risks of Use

Nonprescription Medications (OTC)	Potential Risks
Nonsteroidal Anti-inflammatory Medications (NSAIDS) • Ibuprofen (Advil™, Motrin™) • Naproxen sodium (Aleve™) • Uses: pain, fever, inflammation • **Aspirin (Bayer)** • Uses: Low-dose for anti-platelet effect to prevent heart attack and ischemic stroke; higher doses for treatment of pain	• Will interact with anti-platelet and anti-coagulant medications, which may lead to bleeding • May ↑ blood pressure • May cause stomach irritation, heartburn, reflux, gross bleeding, peptic ulcer disease, and perforation after long-term use. • May prolong bleeding time • All NSAIDs (with the exception of aspirin) ↑ risk of acute myocardial infarction in short-term and long-term use • May cause vasoconstriction of the kidneys in patients who are dehydrated or who are taking some anti-hypertensive agents • All NSAIDs may cause tinnitus, but aspirin is the most likely agent
Miscellaneous Anti-Pain Agent • Acetaminophen (Tylenol™) • Uses: pain, fever	• May cause liver inflammation and liver damage • Use with caution in patients who have liver disease or regularly drink more than ≥ 3 alcoholic beverages daily
Laxative Agents **Bulk-Forming Agents** • Psyllium (Metamucil™) • Methylcellulose (Citrucel™) • Polycarbophil (FiberLax™) • Wheat dextrin (Benefiber™)	• All laxatives may cause some degree of dehydration, which may lead to an ↑ risk of falling. • Bulk-forming agents will ↓ the absorption of other drugs if given within 2–4 hours of each other

TABLE 7.1 (Continued)
Common Nonprescription Medications and the Potential Risks of Use

Nonprescription Medications (OTC)	Potential Risks
Stool Softener • Docusate sodium (Colace™)	
Stool Softener + Stimulant • Docusate sodium + senna (Colace Plus™) **Anti-Acid medications** • Aluminum+Magnesium chewable tabs/ liquid (Mylanta™, Maalox™) • Calcium Carbonate (TUMS™) **Histamine-2 Antagonists** • Famotidine (Pepcid™, Zantac 360™) **Proton-Pump Inhibitors** • Omeprazole (Prilosec™) • Esomeprazole (Nexium™) • Lansoprazole (Prevacid™) • Uses: Heartburn, gastro-esophageal reflux disease (GERD); some use TUMS™ as a calcium supplement	• All anti-acid medications may ↓ the absorption of other drugs; patients and caregivers must carefully follow instructions for individual agents • Proton pump inhibitors ↑ risk of fractures, ↑ risk of gastrointestinal and lower-respiratory infections (e.g., pneumonia), and they may also contribute to dementia
Anti-Allergy Medications First generation agents: • Diphenhydramine (Benadryl™) Second generation agents: • Loratidine (Claritin™) • Fexofenadine (Allegra™) • Cetirizine (Zyrtec™) • Levocetirizine (Xyzal™) • Uses: seasonal and perennial allergies Diphenhydramine only: anaphylaxis, insomnia, motion-sickness and vertigo, Parkinson's disease	• Generally, second generation agents are well-tolerated • Diphenhydramine is most likely to have anti-cholinergic adverse drug effects that are typically more pronounced in older adults; these may include dry eyes, dry mouth, urinary retention, constipation, and change in mental status which may lead to an ↑ risk of falling • These agents may be particularly detrimental in older adult males with enlarged prostates, as the agents may cause significant urinary retention in these patients; Diphenhydramine is most likely to cause this problem. • Concomitant use of alcohol will increase sedative properties of these agents and further ↑ risk of falling • Diphenhydramine is also found in some over-the-counter sleep aids (Unisom™, Simply Sleep™, etc.)

the other prescriber(s) probably added this drug or drugs for a good reason, so I will just leave that one alone!" Before the patient knows it, they have multiple medications on their daily drug list. Patients or their caregivers should be able to tell medical practitioners why they take every drug on their list. If they say "I don't know! The doctor told me to take it!" then a thorough drug review with their pharmacist or general practitioner is warranted. This dynamic may also result in the "prescribing cascade" whereupon a medication's adverse drug effect is inappropriately thought to be due to a new or worsening medical condition. Instead of a thorough reevaluation of a patient's medication list for possible discontinuation of one or more meds, another medication is added to the ever-growing list of drugs. With this in mind, when an older adult presents with any new sign or symptom, we should always look to the medication list first. If there is a temporal relationship between adding a new medication and a new problem or symptom, then the new drug (or potential drug-drug interaction) should be "guilty until proven innocent" (see Figure 7.1).

Several scholarly endeavors have produced guidelines that help to differentiate between appropriate polypharmacy and inappropriate/harmful polypharmacy. This differentiation is generally based on ensuring that older adult patients are not given any medication that is deemed inappropriate or dangerous by specific sets of guidelines. The American Geriatric Society (AGS) expert panel has produced, perhaps, the most well-known set of guidelines for practicing clinicians called the "Beers Criteria®," with an update in 2019. These guidelines are for use in adults ≥ 65 years in ambulatory, acute care, and in institutionalized settings of care (long-term care facilities). These guidelines are not for use in hospice and palliative care settings. This set of guidelines is updated every three years and includes dozens of medication classes and individual medications that are potentially inappropriate medications (PIM) in older adult populations. These recommendations are meant to "guide" medication choices in older adult patients and are not meant to be "absolute" in every patient or situation. See Table 7.2 for dangerous medications in older adults.

The 2019 Beers Criteria® has five lists/tables of potentially dangerous medications in older adult patients:

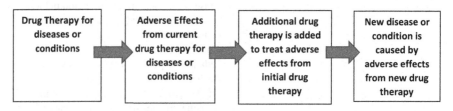

| Drug Therapy for diseases or conditions | Adverse Effects from current drug therapy for diseases or conditions | Additional drug therapy is added to treat adverse effects from initial drug therapy | New disease or condition is caused by adverse effects from new drug therapy |

FIGURE 7.1 The Prescribing Cascade: This concept portrays how drug therapy produces adverse effects and why a new drug therapy is recommended to treat the adverse effects from the initial drug therapy. The newly added drug therapy may now cause a new set adverse effects that produce a new set of problems (disease or condition) that may now need to be treated with a third addition of drug therapy, and so on. This is one of the ways that polypharmacy occurs, and it is a vicious cycle.

TABLE 7.2
American Geriatric Society Beers Criteria® Partial List of PIM and Miscellaneous Precautions and Recommendations

Drug or Drug Class	Potential Risks
Benzodiazepines (BZDPs) Alprazolam (Xanax™) Lorazepam (Ativan™) Diazepam (Valium™) Clonazepam (Klonopin™) Uses: anxiety, insomnia, seizures, alcohol withdrawal	Older adults are more sensitive to BZDPs and most agents have a longer length of effect. The BZDPs cause cognitive impairment, delirium, falls, fractures, and may cause motor vehicle accidents in older adults. Recommendation: AVOID
Sleep Agents ("Z-Hypnotics") Zolpidem (Ambien™) Zaleplon (Sonata™) Eszopiclone (Lunesta™) Uses: Insomnia	These are called "non-benzodiazepine benzodiazepine receptor agonist hypnotics." These have adverse effects similar to traditional BZDPs (alprazolam, etc.) including cognitive impairment, delirium, falls, fractures, and potential for motor vehicle accidents. Additional adverse effects include ↑ in emergency room visits and hospitalizations, and minimal improvement in sleep patterns. Recommendation: AVOID
Opioid Agents Morphine (MSIR™, Avinza™, MSContin™) Hydromorphone (Dilaudid™) Long-Acting Oxycodone (OxyContin™, XTampa™) Short-Acting Oxycodone (Roxicodone™) Oxycodone + Acetaminophen (Percocet™, Endocet™) Hydrocodone + Acetaminophen (Vicodin™, Norco™, Lortab™) Codeine + acetaminophen (Tylenol #3™) Fentanyl (Duragesic™) Methadone Uses: Acute and chronic pain management	These agents are well known for causing sedation and central nervous system (CNS) impairment. When used in combination with BZDPs and/or Z-Hypnotics, the risk of fractures and falls is substantially ↑. The use of two or more CNS depressants (including alcohol) should be avoided. These agents may also cause constipation, blurred vision, confusion, and syncope (drop in blood pressure resulting in "passing out") in patients and they have the potential for death due to respiratory depression. All agents have the potential for addiction and abuse. Recommendation: Use opioids at the lowest dose for the shortest period of time. AVOID combination with other CNS depressant medications.
Opioid-5HT, NE Agent Tramadol (Ultram™) Uses: Acute and chronic pain management	Tramadol is considered to be an opioid analgesic; however, it is unique because it also increases serotonin (5-HT) and norepinephrine (NE).

(continued)

TABLE 7.2 (Continued)
American Geriatric Society Beers Criteria® Partial List of PIM and
Miscellaneous Precautions and Recommendations

Drug or Drug Class	Potential Risks
	Owing to the mixed mechanism of action, tramadol has slightly different adverse effects which includes sleep problems, sedation (not as likely as traditional opioids like morphine, etc.), dizziness, central nervous system stimulation, and diarrhea.
	Tramadol has addiction and abuse potential (believed to be less than the risk from traditional opioids like morphine, etc.)
	Tramadol can also cause serotonin syndrome which is potentially fatal, particularly when mixed with other serotonergic agents like antidepressants and migraine medications such as
	• Selective-Serotonin-Reuptake Inhibitors (SSRIs):
	Fluoxetine (Prozac™) Sertraline (Zoloft™) Citalopram (Celexa™) Escitalopram (Lexapro™) Paroxetine (Paxil™)
	• Serotonin-Receptor Agonists ("Triptans"):
	Sumatriptan (Imitrex™) Eletriptan (Relpax™) Rizatriptan (Maxalt™)
	Recommendation: Use tramadol at the lowest dose for the shortest duration of time. Avoid combining tramadol with other serotonergic agents for prolonged periods of time.

1. Potentially inappropriate medication used in older adults (by drug or drug class)
2. Potentially inappropriate medication use in older adults due to drug-disease or drug-syndrome interactions that may exacerbate the disease or syndrome
3. PIM: drugs to be used with caution in older adults
4. Potentially clinically important drug-drug interactions that should be avoided in older adults
5. Medications that should be avoided or have their dosage reduced with varying levels of kidney function in older adults

The AGS Beers Criteria® states that the intention of the criteria is to

- improve medication selection
- educate clinicians and patients
- reduce ADE
- serve as a tool for evaluating quality of care, costs, and patterns of drug use in the older adult population

Another respected expert panel of guidelines and screening tools come from the United Kingdom, and is called the screening tool of older people's prescriptions (STOPP) and screening tool to alert to right treatment (START) Criteria (i.e., STOPP/START Criteria for potentially inappropriate prescribing in older people: version 2). The original iteration of this set of guidelines was published in 2008, and version 2 is from 2015. The main differences between the Beers Criteria® and the STOPP/START Criteria are that the STOPP/START guidelines list PIMs that should absolutely be avoided (rather than degrees of inappropriateness), and the STOPP/START guidelines also include potential prescribing omissions (PPOs). Like the Beers Criteria®, the STOPP/START screening tools aim to provide adverse drug reaction/adverse drug event prevention benefits in our older adult patients in a variety of clinical settings.

There is most certainly room to use both sets of guidelines for prescribers, practitioners, and pharmacists. Finally, another tool to help with polypharmacy is reinforcing the importance of "deprescribing" medications to prescribers. Deprescribing entails discontinuing or reducing the dose of medications that are no longer appropriate or in line with a patient's care plan. This is hugely significant in our older patient population who have multi-morbidities and long medication lists.

There are published protocols for deprescribing that involve several steps that are a crucial part of improving patient outcomes:

1. Evaluate the list of all prescription medications, OTC medications, and supplements that a patient is taking and the reasons for use for each one.
2. Evaluate potential for IPMs and risks for drug-induced harm for each patient to help determine the intensity of the deprescribing intervention.
3. Evaluate and assess each individual agent in terms of current or future benefit versus current or future harm.
4. List and prioritize drugs for decrease in dose or discontinuation, by assessing the benefit-to-harm ratio.
5. Implement a decrease in dose and/or discontinuation plan.
6. Monitor patients closely for improvement and onset of adverse drug effects.

7.3 WHEN TO SEEK HELP FROM YOUR PRESCRIBER OR PHARMACIST

Older adult patients should seek a medication review from their doctor or medical practitioner and/or pharmacist if any of the following scenarios describe them.

1. They get prescriptions from more than one doctor or prescriber and/or pharmacy.
   - Many people see more than one doctor or prescriber; when medication prescriptions come from two or more doctors or prescribers then it is prudent to have a primary care practitioner or a pharmacist completely organize and review the medication and supplement regimen.
   - The same is true for patients who use more than one pharmacy (e.g., mail order and local pharmacy).
2. They regularly use OTC and/or dietary supplement. See Table 7.1
   - OTC medications and dietary supplements have risks associated with them, particularly when mixed with prescription medications, or when multiple OTC medications or supplements are taken.
   - Patients should always include these in their medications lists along with prescription medications.
3. They have been taking one or more medications for a long time.
   - Some medications may require use for a lifetime. Common examples include thyroid replacements, diabetes, and heart disease medications. Even these types of medications need to be periodically reviewed for appropriateness of use, correct dosing, and potential drug interactions.
   - There are some medications that patients may take longer than necessary, including topical steroids, medications for osteoporosis prevention, hormone replacement, or medications used for "reflux" or "heartburn."
   - Some medications are no longer appropriate as a patient ages.
4. They are taking more than one medication to treat a specific disease or condition.
   - Some disease states or conditions do require more than one medication to treat it. This may also be a situation where the patient is taking more medications than necessary.
5. They recently started a new prescription medication, OTC medication, or dietary supplement that is causing a notable adverse effect.
   - It is sometimes necessary for a patient's body to "re-equilibrate" when starting a new medication, and usually the prescriber and the pharmacist will counsel the patient regarding "typical" or "predictable" adverse effects.
   - If a patient feels very poorly or is experiencing something that is different from what they were told to expect, they should contact the prescriber and the pharmacist to let them know.
6. They must take additional prescription or OTC medications to treat adverse effects that come from another medication.
   - This is part of the "prescribing cascade" and may result in a vicious cycle of "new problem = new drug" over and over. Sometimes adverse effects can be lessened or completely mitigated by switching to another medication, lowering the dose of a drug or taking a drug less often.
7. They have a difficult time affording one or more of their medications.
   - Prescribers typically do not know the cost of a medication they recommend.

- Patients need to ask appropriate questions regarding less expensive alternatives like a generic formulation or even a completely different medication in some cases.
- Patients can solicit the help of their pharmacist to look for manufacturer coupons or to contact the prescriber on their behalf to explain the situation.
8. They feel as though they misuse or abuse one or more substances, such as alcohol, benzodiazepines, opioids (prescription or illicit).
   - It is imperative that the older adult is able to seek help and support for their substance use disorder in a timely manner. This can decrease significant morbidities and mortalities associated with the use of these medications alone or along with other medications for multi-morbidities.

7.4 MEDICATION MANAGEMENT TOOLS

Sometimes, polypharmacy results in a reduced adherence of medications or inappropriate use of medications. This may happen because the older adult thinks that they are spending too much money on their medications or they believe that they just take too many medications so they may take them less often (e.g., every other day or one week on and one week off). This, too, may result in significant adverse drug effects including increased risk of falling. Even if there are no visible drug adverse effects or drug-drug interactions, patients and their caregivers often times have a difficult time taking all of their prescription, nonprescription, and supplements at the correct time each day. Taking an organized approach is a great way to minimize errors and lighten the burden, and there are various tools available to make medication schedules easier.

The very first thing that patients and/or their caregivers should do is to make sure that there is an accurate log of everything the patient takes, including prescription, OTC, and dietary supplements. The list should include both brand and generic names, the name of the prescriber, drug allergies, and how the patient takes the medications and supplements. This list or drug card (for the wallet) should have the patient's name and date of birth, the phone numbers for the patient's primary care practitioner and pharmacy, as well as the information of an emergency contact. These types of templates are easily downloadable from the FDA (www.fda.gov/drugs/resources-you-drugs/my-medicine-record) and from the Institute of Safe Medication Practices (https://consumermedsafety.org/tools-and-resources/medication-safety-tools-and-resources/taking-your-medicine-safely/keep-track-of-your-medicine). A copy should be kept in the patient's and caregiver's wallet or purse, and it should be updated and reviewed often (Figure 7.2).

The simplest solution involves using a pill box, which is available online and at most local pharmacies. These pill boxes help organize the medications and supplements into individual days or into days with multiple dosing intervals (i.e., some medications and supplements need to be taken more than once per day). This method can be further enhanced when patients set reminder alarms on their cell phones or on other "smart" devices like watches, Amazon Alexa, or Google Assistant.

Another option that your local pharmacist may offer is blister packs. A blister pack is a card with punch-out compartments for each day of the week and time (e.g.,

Calcium and Vitamin D – to treat or prevent osteoporosis
ARED™/AREDS 2™ – prevent and treat age-related macular degeneration (AMD); general eye health
Coenzyme Q10 – decrease musculoskeletal pain, systemic inflammation; decrease or prevent muscle aches from using "statins" (eg. atorvastatin (Lipitor), simvastatin (Zocor), etc.)
Tumeric/Curcumin – to decrease systemic inflammation and aches and pains
Fish Oil – "good" fat replacement; may help lower cholesterol, cardiac benefits, eye health
Melatonin – helps to induce and maintain sleep
B-Complex – to improve cognition, eye health, etc.
Multi-Vitamins – general health

FIGURE 7.2   Common Dietary Supplements in Older Adults.

TABLE 7.3
Medication Management Tools

Medication Management Tool	Availability	Average Price Range
Pill organizers	Local pharmacy Amazon.com Smart pill organizers EllieGrid.com Herohealth.com MedQpillbox.com	$3–$10 $20–$60 (some have monthly membership)
Reminder phone APP	Google or apple app store	Free

Blister packs	Upon special request at some
	   local pharmacies
	Forgettingthepill.com
	SimpleDose
	PillPack.com
	Amazon.com
		0$–$40
		Some with monthly
		   subscriptions

Automatic tablet dispensers	Herohealth.com Amazon.com	Wide range from 7$ to > $200
Smart tablet organizers	Elliegrid.com Amazon.com	10$–150$
Timer caps	Amazon.com Or local pharmacy	7$–10$

morning, afternoon, and evening). The blister pack compartment is prefilled with the medications needed at that corresponding time.

For the more technology-savvy patients there are a few new more advanced options on the market. Modern automatic tablet dispensers can secure medications, dispense the correct amount throughout the day, and some offer subscription refill services for a monthly fee. Another option is a smart tablet organizer. These gadgets provide feedback information such as time taken and adherence levels while also serving as a place to keep medicines organized. In addition, timer caps for medication bottles are used to track the time elapsed since the last dose. This helps solve any questions of "when did I last take my medicine?" (see Table 7.3).

7.5   MEDICARE PART D (DRUG) PLAN INFORMATION

Medicare is difficult to understand for older adult patients, their caregivers, and, frankly, the prescribers and pharmacists who must work within its parameters and guidelines. Having Medicare plans that may be inadequate to meet the medical and drug coverage for a patient may result in poor outcomes, as older adult patients may not be able to afford the out-of-pocket expenses. One "plan" does not fit every older adult patient in every situation. In general, "Original Medicare" includes Medicare Part A which is hospital insurance (e.g., inpatient care, skilled nursing facilities, hospice, home healthcare) and Part B which is medical insurance (e.g., outpatient care, durable medical equipment, many preventive services/screenings). Older adult patients are eligible for Medicare when they turn 65. Each year from mid-October to mid-December (Open enrollment), patients can change their Medicare health or drug coverage for the following year. Coverage begins on January 1 of each year for already enrolled patients, or on the first day of the month you turn 65 for new enrollees. Original Medicare plans can be used by any doctor or hospital that takes Medicare anywhere in the United States.

Older adults may also choose to purchase Part D (drug coverage) plans and Medigap coverage. Part D coverage includes prescription drugs and vaccines. The main difference with Part D drug coverage is that it is run by private insurance companies (rather than the government) that follow the rules set by Medicare. Medigap coverage is supplemental insurance that helps pay for out-of-pocket costs associated with Original Medicare, like coinsurance and deductibles. Some older adults have other coverage from a former employer or union, or even Medicaid, and these benefits would be used to supplement Medicare plans for Parts A, B, and D.

Medicare Advantage (Part C) is a newer Medicare plan that is an "all-in-one" plan that bundles Parts A and B and usually Part D. This newer plan boasts lower out-of-pocket expenses as compared to Original Medicare; however, in most cases, patients will need to use doctors who are in the plan's network. In addition, Medicare Advantage plans will generally offer benefits like vision, hearing, and dental, which are not typically covered under Original Medicare plans. With this plan, older adults can make changes to their plans or switch to Original Medicare one time from January 1 to March 31 (general enrollment period), and the plan goes into effect the first month after the plan gets the patient's request.

The costs for each Medicare plan vary, sometimes greatly and change yearly, so older adults are encouraged to research and educate themselves on each of the plans. Also, because these plans cost money, it may place a financial strain on older adult patients, which may result in noncompliance.

7.6   IMPORTANT MEDICARE PART D TIDBITS

Monthly premium: The monthly fee that an older adult will pay for their coverage.

Yearly Deductible: The once yearly amount that an enrollee must pay before the plan begins to pay its share of covered medications. It is typically a bit more than $400. Some plans do not have a deductible.

Coinsurance (Drug "Copays"): The amount of money that an enrollee will pay for covered medications AFTER the deductible has been met. Basically, the enrollee pays their share (e.g., 20%) and the plan plays its share. These amounts may vary during the year due to changes in medication total costs.

Here is where things get complicated (on the basis of 2022 Medicare costs):

- Once you AND your plan spend $4,130 COMBINED on drugs (this also includes the deductible), you will now pay NO MORE than 25% of the cost of the prescription drugs.
- This will continue UNTIL your out-of-pocket spending is $6,550.
- The time difference between when you and your plan spend $4,130 and when you pay out-of-pocket for your drugs UP TO $6,550 ($2,420 in this example) is known colloquially as the dreaded "**Donut Hole.**" People dread it because they will have to personally pay the $2,420 (in this example).
- Once the $6550 maximum is fulfilled, then "**catastrophic coverage**" begins for the rest of the calendar year. With catastrophic coverage, you will pay no more than 5% of the cost of your covered drugs for the rest of the year.
- Medigap supplemental coverage, which is an additional yearly fee, can help cover the out-of-pocket costs like the deductible and the "donut hole."
- When January 1st rolls around, it starts all over again, and generally the costs as mentioned above increase.

Helpful Information:
Call 1-800-MEDICARE (1-800-633-4227) for assistance, or go to their website at Medicare.gov/plan-compare.

7.7   CONCLUSION

Compared to younger adult patients, older adults are at a significantly higher risk of ADEs. Polypharmacy is an urgent issue for older adult patients, and a coordinated, multidisciplinary effort is needed in order to better serve older patients with multi-morbidities. Regular medication review, multidisciplinary and interprofessional collaboration, and patient and caregiver involvement and education are effective ways to keep patients safe.

BIBLIOGRAPHY

American Geriatrics Society Beers Criteria Update Expert Panel. American Geriatrics Society 2019 updated AGS Beers Criteria for potentially inappropriate medication use in older adults. *J Am Geriatr Soc.* 2019;67(4):674–694. doi:10.1111/jgs.15767. Accessed October 21, 2021.

Bally M, Dendukuri N, Rich B, et al. Risk of acute myocardial infarction with NSAIDs in real world use: Bayesian met-analysis of individual patient data. BMJ. 2017;357(9):j1909. doi: https://doi.org/10.1136/bmj.j1909. Accessed October 28, 2021.

Benadryl Allergy (diphenhydramine) [prescribing information]. McNeil Consumer Healthcare; received March 4, 2014. Accessed October 24, 2021.

Brighton TA, Eikelboom JW, Mann K, et al. ASPIRE investigators. Low-dose aspirin for preventing recurrent venous thromboembolism. *N Engl J Med.* 2012;367(21):1979–1987. doi:10.1056/NEJMoa1210384. Accessed October 10, 2021.

Boult C, Reider L, Leff B, et al. The effect of guided care teams on the use of health services: Results from a cluster-randomized controlled trial. *Arch Intern Med.* 2011;171(5):46–466. doi: 10.1093/gerona/63.3.321. Accessed September 2, 2021.

Budnitz DS, Lovegrove MD, Shehab N, et al. Emergency hospitalizations for adverse drug events in older Americans. *N Engl J Med.* 2011;365(21):2002–2012. doi: 10.1056/NEJMsa1103053. Accessed August 27, 2021.

Celexa (citalopram) [prescribing information]. Irvine, CA: Allergan USA Inc; December 2018. Accessed October 24, 2021.

Chou J, Tong M, Brandt NJ. Combating polypharmacy through deprescribing potentially inappropriate medications. *J Gerontol Nurs.* 2019:45(1):9–15. doi: 10.3928/00989134-20190102-01. Accessed October 2, 2021.

Cordell CB, Borson S, Boustani M, et al. Alzheimer's Association recommendations for operationalizing the detection of cognitive impairment during the Medicare Annual Wellness Visit in a primary care setting. *Alzheimers Dement.* 2013;9(2):141–150. doi: 10.1016/j.jalz.2012.09.011. Accessed August 22, 2021.

Councell SR, Callahan CM, Clark DO, et al. Geriatric care management for low-income seniors: A randomized controlled trial. *JAMA.* 2007;298(22):2623–2633. doi: 10.1001/jama.298.22.2623. Accessed September 25, 2021.

Cuckler GA, Sisko AM, Poisal JA, et al. National health expenditure projections, 2017–26: Despite uncertainty, fundamentals primarily drive spending growth. *Health Affairs.* 2018;37(3):482–492. doi:10.1377/hlthaff.2017.1655. Accessed August 21, 2021.

Djatche L, Lee S, Singer D, et al. How confident are physicians in deprescribing for the elderly and what barriers prevent deprescribing? *J Clin Pharm Ther.* 2018;43:550–555. doi: 10.1111/jcpt.12688. Accessed October 23, 2021.

Evans RW, Tepper SJ, Shapiro RE, et al. The FDA alert on serotonin syndrome with use of triptans combined with selective serotonin reuptake inhibitors or selective serotonin-norepinephrine reuptake inhibitors: American Headache Society position paper. *Headache.* 2010;50(6):1089–1099. doi: 10.1111/j.1526-4610.2010.01691.x.Accessed October 26, 2021.

Facts and Comparisons. https://fco.factsandcomparisons.com/lco/action/doc/retrieve/docid/fc_dfc/5549562. Accessed October 15, 2021.

Fox A. The best tools for medication management. *Caring Village*, July 16, 2021. caringvillage.com/2018/03/30/best-tools-medication-management. Accessed September 24, 2021.

Ganz DA, Bao Y, Shekelle PG, et al. Will my patient fall? *JAMA.* 2007;297(1):77–86. doi: 10.1001/jama.297.1.77. Accessed October 2, 2021.

Haastrup PF, Thompson W, Sondergaard J, Jarbol DE. Side effects of long-term proton pump inhibitor use: A review. *Basic Clin Pharmacol Toxicol.* 2018;123(2):114–121. doi: 10.1111/bcpt.13023. Accessed October 15, 2021.

Hill-Taylor B, Walsh KA, Stewart S. Effectiveness of the STOPP/START (Screening tool of older persons' potentially inappropriate prescriptions/screening tool to alert doctors to the right treatment) criteria: systematic review and meta-analysis of randomized controlled studies. *J Clin Pharm Ther.* 2016;41(2):158–169. doi: 10.1111/jcpt.12372. Accessed October 23, 2021.

Ibuprofen tablet [prescribing information]. East Brunswick, NJ: Strides Pharma Inc; April 2021. Accessed October 24, 2021.

Institute for Safe Medication Practices. https://consumermedsafety.org/tools-and-resources/medication-safety-tools-and-resources/taking-your-medicine-safely/keep-track-of-your-medicine. 2021. Accessed October 24, 2021.

Kuhn-Thiel AM, Weiss C, Wehling M. Consensus validation of the FORTA (Fit for the Aged) list: A clinical tool for increasing the appropriateness of pharmacotherapy in the elderly. *Drugs Aging.* 2014;31(2):131–140. doi:10.1007/s40266-013-0146-0. Accessed September 22, 2021.

Lafrance JP, Miller DR. Selective and non-selective non-steroidal anti-inflammatory drugs and the risk of acute kidney injury. *Pharmacoepidemiol Drug Saf.* 2009 Oct;18(10):923–931. doi: 10.1002/pds.1798. Accessed September 14, 2021.

Levy HB. Polypharmacy reduction strategies: Tips on incorporating American Geriatrics Society Beers and screening tool of older people's prescriptions criteria. *Clin Geriatr Med.* 2017;33(2):177–187. doi: 10.1016/j.cger.2017.01.007. Accessed October 17, 2021.

Lexapro (escitalopram) [prescribing information]. Irvine, CA: Allergan USA Inc; January 2019. Accessed October 24, 2021.

Mangin D, Bahat G, Golomb BA, et al. International group for reducing inappropriate medication use & polypharmacy (IGRIMUP): Position statement and 10 recommendations for action. *Drugs Ageing.* 2018;35(7):575–587. doi: 10.1007/s40266-018-0554-2. Accessed July 27, 2021.

Marek KD, Antle L. Medication management of the community-dwelling older adult. In: Hughes RG (Ed.). *Patient safety and quality: An evidence-based handbook for nurses.* Rockville (MD): Agency for Healthcare Research and Quality (US); 2008 Apr. Chapter 18. Available from: www.ncbi.nlm.nih.gov/books/NBK2670/. Accessed October 29, 2021.

Medicare & You handbook. National Medicare Handbook. www.medicare.gov/pub/medicare-you-handbook. September 2021. Accessed October 21, 2021.

Ogawa R, Echizen H. Clinically significant drug interactions with antacids: An update. *Drugs.* 2011;71(14):1839–1864. doi: 10.2165/11593990-000000000-00000. Accessed October 15, 2021.

O'Mahoney D, O'Sullivan D, Byrne S, et al. STOPP/START criteria for potentially inappropriate prescribing in older people: Version 2. *Age Ageing.* 201(2):213–218. doi: 10.1093/ageing/afu145. Accessed October 11, 2021.

Paxil (paroxetine hydrochloride) immediate release [prescribing information]. Weston, FL: Apotex Corp; August 2017. Accessed October 24, 2021.

Philpot EE. Safety of second generation antihistamines. *Allergy Asthma Proc.* 2000 Jan–Feb;21(1):15–20. doi: 10.2500/108854100778249033. Accessed October 27, 2021.

Prozac (fluoxetine) [prescribing information]. Eli Lilly and Company; March 2017. Accessed October 24, 2021.

Qato DM, Alexander GC, Conti RM, et al. Use of prescription and over-the-counter medications and dietary supplements among older adults in the United States. *JAMA.* 2008;300(24):2867–2878. doi: 10.1001/jama.2008.892. Accessed July 15, 2021.

Rankin A, Cadogan CA, Patterson SM, et al. Interventions to improve the appropriate use of polypharmacy for older people. *Cochrane Database Syst Rev.* 2018;9. doi: 10.1022/14651858.CD008165.pub4. Accessed July 30, 2021.

Scott IA, Hilmer SN, Reeve E, et al. Reducing inappropriate polypharmacy: The process of deprescribing. *JAMA Intern Med.* 2015;175(5):827–834. doi: 10.1001/jamainternmed.2015.0324. Accessed July 30, 2021.

Steinman MA, Hanlon JT. Managing medications in clinically complex elders: "There's got to be a happy medium." *JAMA.* 2010;304(14):1592–1601. doi: 10.1001/jama.2010.1482. Accessed July 30, 2021.

Studenski S, Perera S, Patel K, et al. Gait speed and survival in older adults. *JAMA*. 2011;305(1):50–58. doi: 10.1001/jama.2010.1923. Accessed October 21, 2021.

Tylenol Regular Strength (acetaminophen) [prescribing information]. Ft. Washington, PA: McNeil Consumer Healthcare Division; received June 2020. Accessed October 24, 2021.

Ultram (tramadol) [prescribing information]. Titusville, NJ: Janssen Pharmaceuticals Inc; September 2021. Accessed October 24, 2021.

US Department of Health and Human Services, Office of Disease Prevention and Health Promotion. National Action Plan for Adverse Event Prevention. https://health.gov/our-work/national-health-initiatives/health-care-quality/adverse-drug-events/national-ade-action-plan. Washington, DC: U.S. Department of Health and Human Services, Office of Disease Prevention and Health Promotion: 2014 (Updated August 24, 2021). Accessed September 12, 2021.

US preventive services taskforce recommendation statement: Interventions to prevent falls in community-dwelling older adults. *JAMA*. 2018;319(16):1696–1704. doi: 10.1001/jama.20183097. Accessed October 21, 2021.

Vazalore (aspirin) [prescribing information]. Sparta, NJ: PLx Pharma Inc; April 2021. Accessed October 24, 2021.

Wagner AK, Zhang F, Soumerai SB, et al. Benzodiazepine use and hip fractures in the elderly: Who is at greatest risk? *Arch Intern Med.* 2004;164(14):1567–1572. doi: 10.1001/archinte.164.14.1567. Accessed September 30, 2021.

Wastesson JW, Morin L, Tan ECK, et al. An update on the clinical consequences of polypharmacy in older adults: A narrative. *Expert Opin Drug Saf.* 2018;17(12):1185–1196. doi:10.1080/14740338.2018.1546841. Accessed September 2, 2021.

Wise J. Polypharmacy: A necessary evil. *BMJ*. 2013;347(28):f7033. doi: 10.1136/bmj.f7033. Accessed August 11, 2021.

Zoloft (sertraline) [prescribing information]. New York: Pfizer; January 2018. Accessed October 24, 2021.

8 Oral Health and Older Adults

Leonard Brennan and Jennifer A. Crittenden

CONTENTS

> The mouth reflects general health and well-being … as the gateway of the body, the mouth senses and responds to the external world and at the same time reflects what is happening deep inside the body.[1]

Oral health is a key building block of overall health for older adults and an often-neglected source of risk for this population. The silent epidemic of oral diseases that present in older adults is described in the 2021 "Oral Health in America: Advances and Challenges," a follow-up to the 2000 "Surgeon General's Report on Oral Health."[1] Older adults have more oral health problems than any other age cohort.[1,2,3] The greatest obstacle to older adult care is financial as there is no current Medicare coverage for most dental care and few have access to comprehensive care under Medicaid and private insurance programs. In 2021, most adults over 65 were unable

DOI: 10.1201/9781003197843-8

69

to afford preventive and restorative dental care and entered the end-of-life phase unmanaged and unstable dentally.

Neglected oral conditions affect the ability to chew, to speak, and to swallow and can rapidly progress to affect the overall safety of the patient and place them at significant health risk. The rates for cavities and periodontal disease are overwhelming. Over 75% of older adults have gum disease and nearly 35% have caries.[1,3,4] In addition, 12% have pain, 20% have mouth sores, 25% have broken and sensitive teeth, 30% have dry mouth, and another 12% have candidiasis and dental abscesses.[1,3] These diseases and conditions cause severe mouth odor, pain, oral swelling, active bacterial, viral, and fungal infections, and compromised esthetics.[1,3,5] At a time when the quality of life and the ability to communicate and function are particularly important, older patients are often embarrassed with their appearance, suffer in pain, and are unable to eat. Not surprisingly, many enter self-imposed isolation as a result.

8.1   ORAL HEALTH AND SYSTEMIC HEALTH

The well-being of the aging mouth is tied to the health of the rest of the body. A person cannot have good general health without good oral health.[4] There is mounting evidence of a strong association between gum inflammation and chronic conditions such as diabetes, heart disease, stroke, Alzheimer's Disease, and respiratory problems. Bacteria and inflammatory components from gum infections can travel through the bloodstream to trigger inflammation in distant organs and tissues and vice versa systemic disease can travel from distant organs to the mouth and influence the oral disease process.[4] This bidirectional relationship between oral and systemic health is inevitable. In 2021, Dr. Thomas Van Dyke and colleagues demonstrated an intimate and predictive link between periodontal disease and arterial inflammation, which can cause heart attacks, strokes, and other dangerous manifestations of cardiovascular disease.[6]

One of the most effective ways to reduce inflammation and protect health is through excellent oral hygiene and regular dental visits. When adults learn to brush effectively, floss, and rinse, they can often avoid mouth problems. Good oral hygiene is fundamental for oral integrity.

8.2   ORAL HEALTH AND LONG-TERM CARE

Two care contexts that present significant safety risks in relation to oral health include long-term care (LTC) and palliative care settings as they provide high acuity care to patients and residents with significant health challenges and declines.

About half of adults 65 and older will develop a health condition requiring some form of long-term supports and services whether that care be provided in-home or in a specialized nursing care facility.[7] While the majority of adults will age in their own homes and communities, over half of adults 57 and older will spend some time during their lives receiving care in a nursing home facility either through a short skilled nursing stay or longer-term care.[8] Oral healthcare within LTC is a critical component of care due to the convergence of a variety of risk factors within this setting as residents with multiple health conditions are cared for in homes where there is often little staff time and financial reimbursement for extensive oral care provision.

Oral health status has been found to be linked to quality-of-life impacts for nursing home residents, regardless of whether or not they have natural teeth or dentures. Commonly cited impacts of poor oral health status within nursing home residents include difficulty with eating, speaking, and smiling as well as the emotional impacts of the stigma of poor oral health and its effect on socialization. In this setting, nursing home staff and family members are often the first to identify and address problems with loose-fitting dentures, gum disease, and poor food intake and nutrition due to oral health pain and function. Staff face numerous barriers while providing proper oral healthcare to residents including high resident care loads which reduces the time available for oral healthcare, resident behavior that makes it difficult to provide daily care, and a lack of local access to dental health providers.[9]

Several strategies can improve care for residents within the nursing home setting. A foundational strategy to support oral health in this setting is the use of an oral health screening assessment. The Oral Health Assessment Tool (OHAT) developed by Chalmers et al. is designed specifically for the long-term setting and for use by non-dental care providers. The OHAT assesses the following:[10]

- Moisture or cracking present in the lips
- Presence of any cracks, fissures, or swelling of the tongue
- State of the gums and tissues
- Saliva production
- Assessment of any natural teeth, fissures, or dentures for cavities, breaks, and fitting
- Assessment of overall cleanliness of the mouth and the presence of plaque or food
- Dental pain as evidenced by verbal, physical, or behavioral signs
- Documentation of follow-up care plans for dental care

Communication strategies used by staff and caregivers can improve the ability of residents to perform their own oral healthcare and engage those with memory loss or challenging behaviors in such care. These strategies include using task-focused communication with simple one- or two-step instructions, repeating simple care instructions, demonstrating the care task for the resident, using encouraging words, using touch to reassure and direct attention during oral care, conversing with the residents to put them at ease, and providing hands-on assistance when necessary.[11]

In addition to knowing the foundations of daily oral healthcare, LTC staff training is critical for maintaining a high standard of care. Along with training on daily care techniques, those on the frontlines of nursing home care need to concentrate on these:[12]

- Identifying oral health problems early
- Providing oral healthcare to residents with challenging behaviors

- Quick-fix strategies for addressing oral health problems
- Assessing when and how to act on an oral health issue
- The role of medication in oral health status

8.3   ORAL HEALTH AND PALLIATIVE CARE

Older adults frequently face serious oral health challenges that arise during palliative and end-of-life care that can affect both health and quality-of-life issues. The US Census Bureau estimates that the number of adults 65 and older will reach 95 million by 2060, representing 25% of the nation's overall population.[2] Estimates suggest that 75% of people approaching the end-of-life phase will benefit from palliative care.[2,13]

Oral palliative care dentistry is the management of patients with active, progressive, and advanced disease in whom the oral cavity has been compromised by the primary disease or by its treatment and presents the patients with unique risk factors for health and safety.[13,14] End-of-life oral care is the term used to describe the support and care given during the time surrounding death.[14] Wiseman further describes palliative dentistry as "an approach, through early identification and assessment to best manage oral health and treat pain."[5] Oral palliative care can be valuable to older adults living with a progressive illness and can be delivered concurrently with other treatment protocols and is focused on providing support.

Oral health factors that affect the safety and quality of palliative and end-of-life patients are:

- Status of oral health on entering palliative care
- Lack of oral health education for caregivers
- Quality of oral health during care
- Chronic health conditions
- Polypharmacy
- Radiation and chemotherapy
- Nausea and vomiting
- Smoking and alcohol
- Belief that dental disease is a normal part of the aging process

Many older adults entering palliative care and long-term care face these preexisting dental health challenges.

8.4   STRATEGIES TO CONSIDER BEFORE ENTERING CARE

Many retirees do not realize that Medicare does not cover routine dental care. It is crucial for older adults to have a strategy for a healthy mouth and a plan for dental expenses in advance of retirement. Older adults should attempt to complete recommended dental treatment while physically healthy and financially able to avoid the complications associated with an unstable mouth.

The first step toward a healthy mouth is a complete oral exam and a thorough assessment to develop a treatment plan. A well-thought-out treatment strategy can

TABLE 8.1
Develop an Action Plan for Entering Palliative Care

- Identify the oral health needs of a patient before they enter palliative care
- Complete a thorough patient assessment
- Develop and implement a comprehensive preventative and restorative strategy
- Identify the dental healthcare member of the palliative care team; the importance of dental care is often overlooked
- Complete recommended dental treatment
- Understand that most critically ill people are usually dependent on care providers for their oral care needs

Source: 5

prevent future dental problems developing during the normal aging process and should account for polypharmacy, systemic diseases, and the side effects of chemotherapy and radiation therapy (Table 8.1). A stable mouth can alleviate mouth problems and preserve cosmetics.

Monjon and MacEntee recommended "to consider a patient's general condition, life expectancy and overall propensity for oral health when considering a treatment plan."[5] Treatment plans should have the goals to provide comfort and maintain dental restorations and periodontal health, and surgery "should be performed only after thoughtful consideration."[5]

8.5 CARE STRATEGIES FOR IMPROVING ORAL HEALTH OF PALLIATIVE CARE PATIENTS

Older patients with physical limitations, lack of hand coordination, or grasping problems should be encouraged to brush their teeth two or more times daily with an electric or battery-operated toothbrush. Removable prosthetic devices, dentures, and partials should be removed and cleaned before bed and returned to the mouth in the morning after rinsing. For tips on maintaining a healthy mouth, see Table 8.2.

8.6 DIET AND SUGAR

Sugar is a critical factor that influences the rate and severity of oral disease. Patients, healthcare providers, and family members need to be aware that the types of sugars ingested and their frequency of usage affect the extent and progression of oral diseases. Bacteria utilize sugars for their energy metabolism and produce acids as a by-product. The more sugar that a patient eats, the more acid produced, and the greater risk for cavities, periodontal disease, and oral lesions. Important dietary considerations for the caregiver and patients are:

- Simple sugars are the easiest for bacteria to digest and efficiently produce acid. Examples are candies and soda. More complex sugars are not as efficient for

TABLE 8.2
Tips for Maintaining a Healthy Mouth

Action	Protocol	Rational
Brushing	• Ensure a minimum of twice a day for 2–3 minutes in addition to before bed	• Soft toothbrushes are recommended to avoid abrasion, recession, and tooth sensitivity • Electric toothbrush is preferred for everyone • For those lacking hand coordination and have difficulty grasping, electric toothbrushes have a greater impact

Flossing	• Floss a minimum of once per day, preferably before bedtime	• Flossing removes food particles and bacteria trapped between the teeth.
		  • Floss
		  • Floss aides
		  • Water jet removal systems

Fluoride Use	• Use fluoride as recommended by provider	• Fluoride remineralizes teeth, reduces decay, and reduces sensitivity • Use fluoride toothpaste, gels, and rinses to strengthen teeth and stop the decay process

Mouth Rinse	• Use mouth rinse as recommended by provider	• Use of appropriate mouth rinses to
		  • Manage dry mouth
		  • Prevent caries
		  • Kill bacteria, viruses, and fungi
		  • Manage plaque
		  • Remove stains

Care for Dentures and Partials	• Thoroughly clean appliances every day with products recommended by the American Dental Association	• Routine care for dental prostheses, dentures, and partials removes stain, calculus, bacteria, and food particles

Source: 5

bacteria to utilize, and acid production and damage are less. For every exposure to sugar, acid is produced for 20 minutes.

• Frequency of sugar intake plays an important role in oral disease. Candy that sticks to the teeth increases exposure time and increases the rates of oral disease rate. Lozenges used for dry mouth can result in prolonged exposure to sugar and acid.

• The buffering mechanism in saliva neutralizes these bacterial acids. Any medication or condition that reduces saliva will affect the ability of the body to protect itself.

• Saliva production normally decreases at night resulting in an inability to neutralize the bacterial acids that cause oral diseases. Minimize sugar intake before bedtime.

8.7   CHRONIC HEALTH AND MEDICATION EFFECTS ON ORAL HEALTH

Oral health status is inextricably linked to comorbid conditions and polypharmacy among older adults. The majority of older adults (80%) have at least one chronic condition, and 68% have two or more such conditions putting them at high risk for poor oral health outcomes.[15] Many are taking two or three over-the-counter medications for other ailments. The most frequently occurring chronic conditions among older adults are: hypertension (58%), high cholesterol (47%), arthritis (31%), heart disease (29%), and diabetes (27%).[15] Cancer is also a chronic disease of concern with two-thirds of new cancer diagnoses being made among those aged 60 and older.[16] In addition, changes in physical and cognitive function precipitated by conditions such as dementia, arthritis, and age-related vision loss can significantly reduce the ability to engage in oral health self-care.[17,18,19]

It has been estimated that more than 100 systemic diseases and over 500 medications prescribed have oral manifestations and frequently cause dry mouth while the top 200 medications often cause taste disorders.[20] Having a dry mouth affects the ability of an older person to digest food, taste, speak, swallow, and fight off serious oral diseases.[21] Prescribers must be aware of the relationship between medications and dry mouth. This understanding allows the provider to react quickly to alleviate symptoms and stabilize the mouth before damage occurs in the oral cavity (Table 8.3).[12,5]

Palliative and end-of-life patients frequently have additional risks for dry mouth because of chemotherapy, radiation therapy, and increased medications. These threats aggravate existing mouth pathology and increase the distress for the older patient.[21,22,23] One of the most debilitating effects of therapy can be severe tissue ulceration, mucositis causing considerable morbidity and diminished physical and psychological well-being (see Table 8.4).[23]

TABLE 8.3
Effects of Dry Mouth (Xerostomia)

Oral Symptoms of Dry Mouth	Clinical Findings of Dry Mouth
• Inability to wear dentures	• Thick saliva
• Difficulty chewing and tasting	• Dry tongue, lips, and mouth
• Burning sensations	• Cracked lips
• Thick mucous	• No saliva
• Difficulty speaking	• Oral candidiasis
• Halitosis	• Angular cheilitis
• Candidiasis, angular cheilitis, and bacterial infections	• Increased plaque, stain, and calculus
• Increased risk of tooth loss	• Increased caries
• Loss of buffering capacity	
• Cracked lips	

Source: 12,5

TABLE 8.4
Oral Effects of Polypharmacy, Disease, and Treatment

Physical Impact	• Difficulties in eating and drinking
	• Taste disorders
	• Denture instability
	• Dry mouth
	• Difficulty swallowing, chewing, speaking
	• Fungal and viral infections
	• Bacterial infections
	• Mouth ulcers, mucositis, and stomatitis
	• Pain
	• Nutritional
Social Impact	• Difficulties to communicate
	• Self-conscious of the effects of cancer, smell, and disfigurement
	• Embarrassed of inability to talk, eat, and smile
	• Difficult to socialize with pain and effect of caries, periodontal disease
	• Self-conscious of appearance
	• Social isolation because of bad breath
	• Uncomfortable showing physical emotions, hug, and kisses
	• Social withdrawal because of pain
Emotional Impact	• Emotional pain and embarrassment
	• Fear of dying
	• Fear for family
	• Depressed
	• Loss of self-value

Source: 5

8.8   EFFECTS OF SMOKING AND ALCOHOL ON THE ORAL TISSUES

Smoking and alcohol can irritate and dry out the mouth often aggregating sore mouth lesions, increasing tooth and mouth sensitivity and bad breath, delay healing, and causing oral pain. Chemotherapy and radiation therapy both can exacerbate the side effects of treatment.[23,24,25]

8.8.1   SMOKING

The Centers for Disease Control and Prevention estimates that 40 million adults, 14% of all adults, smoke cigarettes.[26] Smoking during palliative care can cause an increase in plaque and staining that can lead to irritation of the delicate gum tissues. It can result in decreased efficiency in swallowing and chewing and aggravate the symptoms of dry mouth.[24] Patients should be counseled to stop smoking totally, before radiotherapy, to minimize the duration of mucositis ulcerations.

Summary of Smoking Effects in Mouth[24]

Oral Symptoms	Clinical Manifestations
• Lips, gums, tongue, and throat irritation and dryness • Mouth discomfort • Impaired tissue healing • Increased plaque and staining • Poor oral hygiene	• Aggravated tooth decay • Increased gum disease • Burning sensation in the oral cavity • Decreased ability to swallow and chew • Plaque and staining increases • Poor oral hygiene

8.8.2 ALCOHOL

It is estimated that one-third of older adults suffer from alcohol use disorder. Patients in palliative care and LTC need to understand that alcohol consumption can affect the mouth.[26,27] Patients should be counseled to avoid oral health products that contain alcohol, for example, mouth rinses.

Summary of Alcohol Effects in Mouth[26, 27]

Oral Symptoms	Clinical Manifestations
• Dehydration of gum, tongue, and oral tissues causes irritation, dryness, and burning sensations • Tissue healing is impaired • Oral health habits tend to be poor	• Increased tooth decay • Increased gum disease with the continuous use of alcohol

8.9 MANAGING THE RISKS ASSOCIATED WITH NAUSEA AND VOMITING

According to the Cleveland Clinic, nausea and vomiting are symptoms of many different conditions or diseases during palliative care, such as infection, migraines, chemotherapy, medication, and radiation therapy.[28] The acidity from the vomit causes damage to the teeth, the gums, the inside of the lips, and the throat.[23,24,25] Tips to minimize damage from vomiting:

- Neutralize stomach acid before brushing teeth with a weak solution of baking soda; wait for 40–60 minutes before brushing. Brushing does not remove this acid; it spreads it around
- Maintain good oral hygiene
- Brush with a soft toothbrush; acid can weaken the tooth structure
- Brush below the gum line and between the teeth
- Avoid foods or drinks that are high in acid and sugar
- Minimize dietary exposure to sugary and acidic foods

8.10   RADIATION AND CHEMOTHERAPY AND ORAL HEALTH

"The mouth is one of the most distressed parts of the body in palliative care."[20,26] It is challenged during cancer therapy and has an extremely substantial risk of serious oral complications that affect, chewing and swallowing, pain, mucositis, and infection change in taste.[24,25] According to the National Cancer Institute,

- Chemotherapy and radiation therapy can slow or stop the growth of cancer cells and normal cells in the lining of the mouth and slow down its ability to repair itself.[24,25] The recovery from mouth disease and oral complications can take much longer to heal.
- Radiation therapy can affect the supporting tissues in the mouth and damage and break down oral tissue, bone, and the glands that produce saliva.[23,24,25]
- Chemotherapy and radiation therapy can also affect the balance of healthy normal flora in the mouth. This imbalance can encourage the growth of pathogens that result in mouth problems.

TABLE 8.5
Common Mouth Complaints Associated with Chemotherapy and Radiation Therapies

Oral Problems	Clinical finding and Consequences
• Mucositis	• Difficulty swallowing, talking, and eating; severe pain
• Caries and Periodontal Diseases	• Broken and sharp teeth • Caries and gum disease increase • Sensitive teeth • Compromise existing crowns, partials, and implants
• Intolerance to Dentures and Partials	• Dry mouth decreases denture retention • Lack of saliva irritates tissues under appliances • Patient motivation to wear dentures decrease • Loss of muscular control in mouth makes it difficult to support dentures for function
• Dry Mouth	• Difficulty swallowing, talking, and eating • Common side effects of medication radiation and chemotherapy
• Candidiasis	• Mouth odor • Removable plaque in the mouth • Burning sensations • Painful red or white lesions • Incidence: 90% of patients with cancer • Always treat oral appliances: denture, partials, mouthguards, and toothbrushes • Difficulty wearing dentures and partials
• Taste Disorders and Dysphagia	• Lack of saliva causes taste distortion and difficulty swallowing • Over 500 prescribed medications can cause dry mouth • Over 200 of the most prescribed medications can cause taste disturbances

Source: 25

TABLE 8.6
Managing Mucositis

Daily Habits	• Serve warm and not hot foods
	• Avoid substances that irritate dry mucosa
	• Avoid alcohol and tobacco
	• Moisten foods, use sauces to wet-dry foods or avoid dry foods
	• Serve chilled foods like yogurt or ice-creams
	• Remove and clean the denture and leave out at night (this applies to partial dentures too)
	• Change to soft toothbrushes
	• Avoid spicy foods
Pain Management	• Rinse with hydrogen peroxide diluted 1:4
	• Use analgesics and topical morphine
	• Rinse with a 50/50 mixture of Benadryl and Kao pectate
	• Rinse with Magic Mouthwash: xylocaine, Benadryl, topical steroid, antifungal
Comfort Measures	• Moisten mouth every 2–4 hours around the clock
	• Cryotherapy: ice chips
	• Soft reline of denture
	• Routine application of lip balm

Source: 23,24,25

For some patients, the complications can be so debilitating that they may tolerate only lower doses of therapy and need to postpone or reschedule treatments or discontinue treatments entirely (Table 8.5). Oral issues can also lead to serious systemic infections. Medically necessary oral care before, during, and after cancer treatment can prevent or reduce the incidence and severity of problems, enhancing both patient survival and quality of life. [23,24,25]

A common side effect of chemotherapy and radiation treatment is mucositis, extremely painful sores, and inflammation in the mouth. About 5–15% of cancer patients develop mucositis and severe ulcers and pain.[24,25]

About 50% of cancer patients receiving radiotherapy to the head and neck develop mucositis and it occurs within 32 weeks and lasts 6–8weeks.[24,25] Mucositis induced from chemotherapy has a more acute onset, may appear within 3–4 days. The management of mucositis is essential for the quality of life of these patient. (Table 8.6). The strategy to deal with these lesions should begin before treatment is initiated.[24,25]

8.11 ORAL INTEGRITY IN COMMUNICATION AND SOCIAL INTERACTIONS

As patients become progressively unwell, they will need assistance with oral hygiene. The inability of a patient to maintain good oral health causes mouth breakdown and a decrease in facial cosmetics and appearance. This affects their psychological quality of life.[28]

- Social interaction decreases with poor appearance and odor, loss of muscular function, and chewing ability.[28]
- There is a widespread belief that healthy teeth are a sign of attractiveness and intelligence.[28]
- Loneliness and depression double the likeliness of early death in older adults.[28]
- Oral health and disease influence social interactions in all age groups.[28]
- Bad breath creates a barrier between the patient and those around them and may worsen social isolation at the end of life.[28]
- Bad breath can often be managed through brushing, flossing, and rinsing.

Ultimately, the communication that occurs at the end of life between the terminally ill, family members, and healthcare specialists are critical for a "good death," because it is only through communication where peoples' true wishes are heard, understood, and followed that their loved ones are left without regret.[29]

Oral health has a pivotal role in verbal and nonverbal communications.

8.12   EDUCATING CAREGIVERS AND STAFF

Caregivers, when providing mouth care for palliative and long-term care patients, face a particularly challenging task. Even for the most experienced dentist, to comfortably enter the mouth and provide care in such a very intimate and personal space, when individuals may be confused and in discomfort, can be difficult.

Caregivers face the challenge of stepping in, often with limited training. There are many programs to educate members of the interprofessional team to minimize risk and promote dental health. A few of these programs are listed here:

- Smiles for Life (Continuing Education|Smiles for Life Oral Health)
- MOTIVATE Oral Health Training Modules (MOTIVATE Oral Health Resources|Lunder-Dineen (lunderdineen.org))
- Chemotherapy and your Mouth Chemotherapy and Your Mouth (nih.gov)
- CDC, Oral Health Program, learning to provide care (Oral Health|CDC)

8.13   THE INTERSECTION OF ORAL HEALTH AND SOCIAL DETERMINANTS OF HEALTH

Access to adequate oral healthcare is an important factor in risk reduction for older adults. Like many health issues, oral health is influenced by a variety of health equity factors including access to care and insurance coverage across the lifespan. For example, there are marked racial and ethnic differences in access to care with Black adults being less likely to have had a routine dental visit in the past year and 68% more likely to have unmet dental needs as compared to their White counterparts.[30] Throat cancer survival rates and the diagnosis of gum disease demonstrate similar disparities for Black and Hispanic older adults.[31]

Overall, as many as 20% of older adults report delaying needed dental care. The top barrier to accessing this much needed care is the significant out-of-pocket financial cost of care.[32] While about half of adults ages 65–80 report having some form of dental coverage, insurance coverage – whether through public or private options – is reported at lower rates among Black and Hispanic older adults when compared to other racial groups.[30,32] In addition to insurance, other social determinants of health that have been found to impact access to oral healthcare include income, cultural and language considerations, formal education, and mental health.[33,34]

Strategies for increasing access to care that help to mitigate some of these disparities in access include:

- Creating patient education materials with accessible, culturally appropriate, and jargon-free language
- Providing access to transportation
- Creating physically accessible spaces for the wheelchair-bound and those with mobility aids
- Conducting community-based outreach using peer champions[34]
- Increasing social supports through connections with family and friends who can assist with access and care
- Exploring options for free or reduced-price care through:[35]
  - Aging and Disability Networks that are facilitated by Area Agencies on Aging
  - Schools of dental hygiene
  - Federally qualified health centers and community health centers
  - Medicaid
  - Veterans' coverage through either that state Veterans' Administration (VA) or Federal VA system
- Advocating for the inclusion of a dental health benefit via Medicare and Medicaid to address systemic lack of access

8.14   MENTAL HEALTH AND ORAL DISEASE

Many patients in long-term care, palliative, and end-of-care situations develop emotional and mental health problems from the stress of disease and disability. Most medications used to treat behavioral and mental health issues cause a severe and dramatic decrease in saliva production and dry mouth.[21,22,27]

Behavioral medications can also decrease the motivation to continue with oral hygiene and to understand its value. Psychological distress results in behaviors that damage both the teeth and gums, such as grinding and clenching. Distress can also cause older adults to brush their teeth less, brush too aggressively, and eat foods high in sugar which results in tooth loss, enamel erosion, gingivitis, tooth abrasion, oral pain disorder, dry mouth, jaw and muscle pain, decay, and fractured teeth. Close monitoring of patient motivation, compliance to dental care, and evaluations of the oral effects of the disease and treatments are especially important[36] (Table 8.7).

TABLE 8.7
Factors Affecting the Oral Health of Individuals Living with Mental Health Issues

- Severity and stage of mental illness and ability to self-care
- Socioeconomic factors and access to a dentist
- Poor diet, high in sugar
- Lack of perception of oral health problems
- Smoking and alcohol
- Side effect of medication, dryness, and taste disturbances

Source: 36

8.15   ORAL HEALTH IS THE RESPONSIBILITY OF THE INTERDISCIPLINARY TEAM

The interdisciplinary care team and family members may be the only individuals available to assist an older adult in their oral management. Geriatricians, family physicians, nurse practitioners, certified nurse assistants (CNAs), occupational therapists, dieticians, social workers, and others all need to be enlisted to monitor oral palliative and end-of-life patients. Professional cross-training becomes critically important to meet these dental challenges for the older adults in palliative care and long-term care settings.

Prescribers need to be aware of how medications affect the oral environment; dieticians should know the link between diet and sugar and mouth diseases; mental healthcare providers should be aware of the fact that behavioral medication frequently can have effects that lead to serious oral disease; social workers can assist family members in accessing care coverage and programs as well as assist them in communicating with the care team; occupational and physical therapist need to understand how to best assist a patient with brushing and flossing techniques; and family members need to understand the connection between oral health and social isolation. Oral health is truly a team effort.

Working as part of an interdisciplinary team provides a comprehensive strategy to promote and provide the best clinical outcomes and delivery of care.

8.16   SUMMARY

Many oral health challenges are present in palliative, end-of-life, and long-term care patients. Careful and early assessment, prevention, treatment, and management can promote an improved quality of life for these patients. Key principles to minimize risks and promote patient safety are:

- Understand that dental disease is not a normal part of the aging process
- Include dental providers on the palliative care team and integrated within long-term care discussions

- Strategize a management protocol before the need for LTC, palliative, or end-of-life care
- Assess and plan for regular effective mouth care for all patients
- Monitor response to interventions
- Identify serious oral problems that require referral
- Manage oral health problems as soon as possible
- Connect with resources and programs to increase oral healthcare access
- Utilize an interdisciplinary team for the most effective and efficient oral healthcare across settings and care contexts

REFERENCES

1. U.S. Department of Health and Human Services. *Oral Health in America: A Report of the Surgeon General.* Rockville, MD: National Institute of Dental and Craniofacial Research, National Institutes of Health; 2000.
2. United States Census Bureau. Older and Growing-Percent Change among the 65 and Older Population: 2010 to 2019. www.census.gov/library/visualizations/2020/comm/map-popest-65-and-older.html. June 25, 2020. Accessed December 2021.
3. National Institute of Dental and Craniofacial Research. Dental Caries (Tooth Decay) in Seniors (Age 65 and Over). www.nidcr.nih.gov/research/data-statistics/dental-caries/seniors. July 2018. Accessed December 2021.
4. American Dental Association. Oral-Systemic Health. www.ada.org/resources/research/science-and-research-institute/oral-health-topics/oral-systemic-health. September 23, 2019. Accessed December 2021.
5. Wiseman M. Geriatric dentistry: caring for our aging population dentistry. In: *Palliative care dentistry* (pp. 17–28). John Wiley; 2014.
6. Van Dyke TE, Kholy KE, Ishai A, et al. Inflammation of the periodontium associates with risk of future cardiovascular events. *J Periodont.* 2021;92(3):348. https://pubmed.ncbi.nlm.nih.gov/33512014.
7. Favreault M., Dey J. Long-Term Services and Supports for Older Americans: Risks and Financing Research Brief. U.S. Department of Health and Human Services Office of the Assistant Secretary for Planning and Evaluation. https://aspe.hhs.gov/reports/long-term-services-supports-older-americans-risks-financing-research-brief-0. February 2016. Accessed December 16, 2021.
8. Hurd MD, Michaud P-C, Rohwedder S. Distribution of lifetime nursing home use and of out-of-pocket spending. *National Academy of Sciences.* www.pnas.org/content/114/37/9838/tab-figures-data. September 12, 2017. Accessed December 16, 2021.
9. Porter J, Ntouva A, Read A, Murdoch M, Ola D, Tsakos G. The impact of oral health on the quality of life of nursing home residents. *Health and Quality of Life Outcomes.* 2015;13(1):102. https://doi.org/10.1186/s12955-015-0300-y.
10. Chalmers J, King P, Spencer A, Wright F, Carter K. The oral health assessment tool – validity and reliability. *Austral Dent J.* 2005;50(3):191–199. https://doi.org/10.1111/j.1834-7819.2005.tb00360.x
11. Wilson R, Rochon E, Mihailidis A, Leonard C. Quantitative analysis of formal caregivers' use of communication strategies while assisting individuals with moderate and severe Alzheimer's disease during oral care. *J Comm Dis.* 2013;46(3):249–263. https://doi.org/10.1016/j.jcomdis.2013.01.004

12. Crittenden J, Gugliucci M. MOTIVATE Project Needs Assessment Executive Summary. Lunder-Dineen Health Education Alliance of Maine. https://lunderdineen. org/sites/default/files/MOTIVATE-Executive-Summary-FINAL-7-24-17.pdf. June 2016. Accessed December 16, 2021.

13. World Health Organization. How Many People Will Need Palliative Care in 2040? Past Trends, Future Projections, and Implications for Services. www.who.int/news-room/fact-sheets/detail/palliative-care. August 20, 2020. Accessed November 2021.

14. National Institute on Aging. Providing Care and Comfort at the End of Life. www. nia.nih.gov/health/providing-comfort-end-life#what. May 17, 2017. Accessed November 2021.

15. National Council on Aging. The Top 10 Most Common Chronic Conditions in Older Adults. www.ncoa.org/article/the-top-10-most-common-chronic-conditions-in-older-adults. April 23, 2021. Accessed December 16, 2021.

16. White MC, Holman DM, Goodman RA, Richardson LC. Cancer risk among older adults: time for cancer prevention to go silver. *Gerontol.* 2019;59(Supplement 1):S1–S6. https://doi.org/10.1093/geront/gnz038.

17. American Dental Association. Aging and Dental Health. www.ada.org/resources/research/science-and-research-institute/oral-health-topics/aging-and-dental-health. November 9, 2021. Accessed December 16, 2021.

18. Center for Disease Control. Percent of U.S. Adults 55 and Over with Chronic Conditions. www.cdc.gov/nchs/health_policy/adult_chronic_conditions.htm. September 2009. Accessed November 2021.

19. National Institute of Health. The Dangers of Polypharmacy and the Case for Deprescribing in Older Adults. www.nia.nih.gov/news/dangers-polypharmacy-and-case-deprescribing-older-adults. August 24, 2021. Accessed December 2021.

20. American Geriatric Society. Updated Beers Criteria for potentially inappropriate medication use in older adults. *J Am Geriatr Soc.* 2019; 67(4):674–694.

21. American Dental Association. Dental Health Before Cancer Treatment. www.mouth healthy.org/en/az-topics/c/cancer-before-treatment. November 2021. Accessed December 2021.

22. Cleveland Clinic. Nausea and Vomiting: Treatment and Care. https://my.clevelandcli nic.org/health/symptoms/8106-nausea--vomiting. 2021. Accessed December 2021.

23. Boyd CM, Darer J, Boult C, Fried LP, Boult L, Wu AW. Clinical practice guidelines and quality of care for older patients with multiple comorbid diseases: Implications for pay for performance. *JAMA.* 2005;294(6):716–724.

24. U.S. Department of Health and Human Services National Institutes of Health. Oral Complications of Cancer Treatment: What the Dental Team Can do. www.nidcr.nih.gov/sites/default/files/2017-09/oral-complications-cancer-dental-team.pdf. Accessed December 2021.

25. National Cancer Institute. Comprehensive Cancer Information. www.cancer.gov/. 2021. Accessed December 2021.

26. Center for Disease Control and Prevention. Health Effects, Smoking, and Tobacco. www.cdc.gov/tobacco/basic_information/health_effects/index.htm. April 28, 2020. Accessed November 2021.

27. Rigler SK. Alcoholism in the elderly. *American Family Physician.* 2000;61(6):1710.

28. Donnelly L, Clarke L, Phinney A, McEntee M. Oral healthcare in the frail elder. In: MacEntee MI, Müller F, Wyatt C, eds. *Contexts of body image and social interactions among frail elders.* Wiley-Blackwell; 2010:161–172.

29. Keeley MP. Family communication at the end of life. *Behavioral Sciences.* 2017;7(3):45.

30. CareQuest Institute for Oral Health. New Oral Health Data Reflect Inequities, Barriers. https://doi.org/10.35565/CQI.2020.4001. May 2020. Accessed December 16, 2021.
31. Centers of for Disease Control and Prevention. Disparities in Oral Health. www. cdc.gov/oralhealth/oral_health_disparities/index.htm. February 5, 2021. Accessed December 16, 2021.
32. Kramarow E. Dental Care Among Adults Aged 65 and Older 2017. National Center for Health Statistics. www.cdc.gov/nchs/products/databriefs/db337.htm. May 2019. Accessed December 16, 2021.
33. Sweier D. Dental Care and Coverage after 65: Experiences & Perspectives. National Poll on Healthy Aging. www.healthyagingpoll.org/reports-more/report/dental-care-coverage-after-65-experiences-perspectives. March 23, 2020. Accessed December 16, 2021.
34. Garcia RI, Cadoret CA, Henshaw M. Multicultural issues in oral health. *The Dental Clinics of North America*. 2008;52(2):319. https://doi.org/10.1016/j.cden.2007.12.006.
35. Administration for Community Living. Oral Health Overview. https://acl.gov/programs/health-wellness/oral-health. November 30, 2021. Accessed December 16, 2021.
36. Social Determinants of Health. Healthy People 2020. www.healthypeople.gov/2020/topics-objectives/topic/social-determinants-of-health June 2, 2020. Accessed December 16, 2021.

9 Preventing Falls and Fall-Related Injuries among Older Adults

Kathleen A. Cameron

CONTENTS

9.1   INTRODUCTION

Older adult falls are a common and costly public health problem that are increasing with the aging of the population and rising rates of chronic illness and associated medication use. People of all ages fall, yet older adult fallers have significantly more challenges in recovery from their fall-related injuries. Falls are the leading cause of injury and death from injury, among those 65 years of age and over.[1] More than one in four adults aged 65 and older fall each year with 20% of fallers experiencing injuries.[2] Falls can cause physical and psychosocial impairments, such as fractures of the hip, spine, and forearm, traumatic brain injuries, and the fear of falling presents with subsequent restriction in activities, possible nursing home admission, and death. Fall-related injuries can make it difficult for older persons to function normally and can significantly limit independent living. Problems with mobility, balance, sensory impairments, and loss of muscle mass and strength that occur with age contribute to the likelihood of falling. In addition, people are living longer with chronic conditions, such as cardiovascular disease, diabetes mellitus, and arthritis. These illnesses, as well as many of the medications used to treat them, can increase fall risk.

DOI: 10.1201/9781003197843-9

The effect of older adult falls on caregivers and families is tremendous, contributing to caregiver burden and impeding caregivers' ability to carry on with their regular activities, lost time from work, and negatively impacting quality of life. Fortunately, falls are also predictable with screening and assessment, and in most cases preventable with appropriate interventions that address identified risk factors.

Extensive research has been conducted on the scope and significance of the problem, factors that increase fall risk, and interventions to reduce falls and injury. Effective clinical interventions, evidence-based community programs, and clinical-community partnerships have been identified as strategies to reduce falls. These strategies must be supported and scaled to achieve a reduction in falls and related injuries among older adults.

The social-ecological model provides a public health framework for prevention that includes four impact levels for older adult falls reduction: individual, relationship, community, and societal.[3] As described in this chapter, each level comprises factors that interact with those at different levels. An approach that targets multiple levels or multiple risk factors with various stakeholders will result in sustained impact on reducing falls and their negative consequences among older adults.

9.2   NATIONAL STATISTICS ON FALLS AMONG OLDER ADULTS

Falls are common:

- 25–30% of adults aged 65+ falls each year
- 20% of older adult fallers experiences an injury

Falls are impactful:

- Falls are the leading cause of fatal injury and the most common cause of non-fatal trauma-related hospital admissions among older adults.
- Falls result in more than 3 million injuries treated in emergency departments annually, leading to over 800,000 hospitalizations.[4]
- Falls are the leading cause of hip fractures[5] and traumatic brain injuries.[6]
- Falls, with or without injury, also result in a serious quality-of-life impact. For example, fear falling has long been recognized as a significant negative impact of falls, resulting in older adults limiting their activities, mobility, and social engagements. This can result in physical decline, depression, social isolation, feelings of helplessness, and further increase in falls risk.[7]
- The older adult population is projected to increase by 55% by 2060, hence falls rates and associated negative consequences, including health care spending are projected to rise.

Falls are costly:

- Each year about $50 billion is spent on medical costs related to nonfatal fall injuries and $754 million is spent related to fatal falls.[8]

- Nonfatal fall injury costs have tremendous public and personal financial burdens: $29 billion paid for by Medicare; $9 billion paid for by Medicaid; and $12 billion paid for by private and out-of-pocket payors.
- The cost of treating falls is projected to increase to over $101 billion by 2030.

The Centers for Disease Control and Prevention (CDC) is the best source for up-to-date national data on falls among older adults;[9] state level data can also be obtained from the CDC[10] as well as state health departments, specifically the injury prevention units.

9.3 RISK FACTORS FOR FALLS

No one risk factor alone predisposes an older person to fall. Typically, the interplay or combination of multiple and diverse risk factors – both intrinsic and extrinsic – contribute to a fall event. Risk factors for falls are either nonmodifiable or modifiable. Nonmodifiable risk factors include age, especially for those 80 and over; gender with females being at higher risk for falls and men at greater risk of death associated with falls; and race/ethnicity with Native American elders experiencing the highest rate of falls and fall-related deaths. Those who have experienced falls in the past are at increased risk of experiencing subsequent falls; those who have fallen once are twice as likely to fall again.[11]

Modifiable factors include physical changes with aging such as impairment of balance and gait, muscle loss and weakness, and limitations of activities of daily living (ADL). Deconditioning is directly related to increased falls risk due to loss of muscle mass and strength that leads to weakness, instability, and associated decrease in balance, gait changes, and functional decline. Deconditioning has many causes with inactivity being a primary one; causes or correlates of inactivity include long recovery from acute illness, chronic illnesses (vision changes, low blood sugar, loss of sensation in feet), heart disease, and obesity; sarcopenia, or loss of muscle mass and functioning results in deconditioning and is also a major risk factor for falls. Fear of falling is another risk factor for deconditioning because the fear leads to a reduction in physical and social activities outside of the home. The COVID-19 pandemic with necessary social distancing and isolation led to deconditioning in many older adults.[12]

Other modifiable risk factors include:

- **Chronic conditions**, including those associated with deconditioning, have been shown to be associated with a higher risk of falls: all forms of arthritis, diabetes, heart disease, chronic pain, obesity, stroke, and Parkinson's disease. The presence of osteoporosis increases the likelihood of falling, fear of falling, and fractures.[13]
- **Postural hypotension**, a drop in blood pressure on standing or changing position, resulting from medication side effects, dehydration, or other causes
- **Sensory deficits**, such as vision and hearing impairments, as well as peripheral neuropathies leading to loss of sensation in the extremities.

- **Cognitive decline and memory loss** that occur with Alzheimer's disease and related dementias are associated with higher fall rates, no matter what setting including memory care in assisted living and nursing facilities; those with dementia are more than three times more likely to fracture their hip when they fall. The rate of death following a hip fracture for those with Alzheimer's is also high.
- **Mental health conditions**, such as depression, anxiety, and sleep disorders, found in about 20% of older adults have also been shown to lead to higher rates of falls.
- **Social isolation and loneliness** have been shown to be associated with higher rates of falls and hospitalizations.[14]
- **Medications** – both number and type of medications. The risk of falling has been shown to increase with the number of prescription and over-the-counter medications taken.[15] Older adults taking more than three or more medications are at increased risk for falls and recurrent falls.[15] Specific therapeutic categories of medications increase the risk that an older person will fall.[16] These therapeutic categories often cause side effects and adverse effects that predispose older persons to falls: postural hypotension causing dizziness, lightheadedness, and balance impairment; sedation, decreased alertness, confusion and delirium, blurred or impaired vision, compromised neuromuscular function, and anxiety. Psychoactive medications, especially benzodiazepines and antidepressants, cause these effects most frequently in older adults. Pharmacokinetic and pharmacodynamic changes with aging impact how our bodies handle medications and our sensitivity to medications. Pharmacokinetic changes refer to distribution, metabolism, and elimination of medications, that is, medications tend to be more extensively distributed and to remain longer in the body of an older person. Declining kidney function can result in accumulation of medications that are eliminated by the kidney, leading to possible toxic drug levels. Pharmacodynamic changes increase an older person's sensitivity to potential adverse effects. Medications that act on the central nervous system, such as benzodiazepines, antidepressants, analgesics, and medications with anticholinergic effects, result in an increased pharmacodynamic response and hence dosages must be decreased accordingly.
- **Environmental factors** – most people choose to remain in their homes as they age, yet homes that were once supportive may present problems over time that could lead to falls. Research shows that the optimal way to reduce fall risks for people at moderate to high risk of falls includes changes or modifications to the home to improve safety. Home modification entails changing the environment to reduce home hazards, such as removing clutter, eliminating or reducing slippery surfaces, improving lighting, making entryways more accessible, adding supports such as handrails and grab bars in the bathroom and stairs, and changing how or where activities occur in the home.

9.4 FALLS RISK SCREENING AND ASSESSMENT

Numerous measures of fall risk have been validated and are used in clinical and community practice. The most comprehensive tool for falls risk screening and assessment is the CDC's Stopping Elderly Accidents, Deaths, and Injuries (STEADI) Resources,[17] based on the American and British Geriatrics Societies' Clinical Practice Guidelines for prevention of falls in older persons published in 2011.[18] The STEADI algorithm begins with a recommendation of yearly screening of all older adults using a 12-question "Stay Independent" screen[19] or asking 3 simple questions:

- Do you feel unsteady when standing or walking?
- Do you worry about falling?
- Have you fallen in the past year?

Those who respond yes to one of these questions or 4 or more to the 12-question screen are considered at risk for falls and should be given a comprehensive assessment of modifiable risk factors that includes the following:

- Evaluate gait, strength, and balance, using common assessments such as the Timed Up and Go,[20] 30 Second Chair Stand Test,[21] and 4 Stage Balance Test[22]
- Identify medications that increase falls risk, using tools such as the Beers Criteria[23]
- Ask about potential home hazards, using tools such as those found in the NCOA-University of Southern California Home Assessment Tools for professionals and Providers Inventory[24]
- Measure postural or orthostatic blood pressure
- Assess feet/footwear
- Assess Vitamin D intake
- Identify comorbidities such as depression and osteoporosis

Once modifiable risk factors are identified, then tailored, and coordinated falls prevention interventions should be implemented.

9.5 A TEAM APPROACH TO FALLS PREVENTION

Owing to the multifactorial nature of falls risk, a team approach to identify it and intervene is critical. A falls prevention team should be interdisciplinary as each member brings a unique mix of expertise and familiarity that can provide important protections for older adults and their caregivers. First and foremost, the older adult and their family and friends must be engaged in falls prevention. They must be part of the conversation regarding how falls risks can be controlled through often simple interventions, which can reduce fear of falling and stigma, so commonly associated with falling. Family and friends provide extra hands and another set of eyes to, for example, check and rid the home of falls hazards, identify medication side effects that could lead to falls, ensure that older adults receive annual vison and hearing exams to correct for sensory deficits. Family and friends can also attend doctor visits and help ask questions and gather information about preventing falls.

- **Primary Care Physicians (PCP) and Physician Assistants (PA):** PCPs and PAs are often best positioned to work with patients on falls prevention by getting the process started with screening and assessment, such as through the Medicare annual wellness visit falls screen or the CDC's STEADI algorithm. PCPs and PAs can take into consideration medical history and best manage chronic illness like diabetes and arthritis that may put an older patient at risk. Further PCPs can spot hidden injuries resulting from a fall, evaluate side effects and other problems with medications that may increase risk of falling, refer patients to therapists, and recommend evidence-based falls prevention programs (EBFPPs).
- **Physical Therapists (PT):** Physical therapists are an important member of the falls prevention team. PTs are highly trained health care professionals, with an expertise in movement and exercise. Their skills and knowledge base are essential when dealing with the complexities of aging, including falls prevention and management. Physical therapy training includes assessment, exercise prescription, and progression, all with appropriate monitoring. All these skills can be utilized for older adults before a fall-related injury happens and for prevention of functional decline and disability. They work with older adults to improve balance, strength, mobility, endurance, and posture through appropriate exercises regimen.
- **Nurses:** Depending on practice setting, the role of the nurse will vary and is pivotal in preventing falls among older adult patients, and generally includes:
  - Educating the patient and family on fall prevention
  - Completing and documenting fall risk screening and assessment
  - Documenting patient-specific fall prevention practices
  - Monitoring the patient's medical condition for any changes
  - Reporting falls to the physician
  - Obtaining medical orders from the physician as needed
  - Supervising nursing aides for falls prevention interventions
  - Obtaining the supplies (e.g., cane, walker, bed alarm) needed to prevent patient falls
- **Occupational Therapists (OT):** OTs focus on ensuring optimal function of their patients and are vital part of any falls prevention team. They can conduct home assessments to ensure livability of the home and make recommendations for appropriate home modifications to maximum safety to reduce falls risk. OTs can also ensure that assistive devices, such as canes and walkers, fit needs of the individual and are used properly. OTs often collaborate with handy persons, home remodelers, and contractors to make home modifications. In addition, many community-based organizations have programs that engage OTs in conducting home assessments and making home modifications, including Rebuilding Together and Habitat for Humanity as well as smaller local programs.[25]

- **Pharmacists:** As described earlier. the impact of medications on falls risk is significant. Pharmacists, no matter what practice setting, play a key role in identifying medications that pose risk for older adults and recommending changes in medication regimens, such as dose reduction, switching to an alternative medication, or eliminating a medication. Pharmacists also play a critical role in educating older adults and caregivers about all aspects of medication use.
- **Fire Departments/Emergency Medical Technicians:** Local fire department play a key role in falls prevention and do more than help when a fall event occurs. Fire departments can help prevent falls by offering home safety checks to spot falls hazards. They also can assist in testing and replacing smoke alarms. Firefighters can also connect older adults with resources in the community to prevent falls.
- Other members of the falls prevention team may include:
  - Podiatrists whose role includes assessing and treating foot pain, identifying and correcting underlying biomechanical and gait abnormalities, prescribing exercise programs, and issuing foot health and footwear advice.[26]
  - Ophthalmologists and/or optometrists for identifying and treating of eye diseases and prescribing corrective lens.
  - Audiologists identify hearing loss and prescribe appropriate hearing aids; in addition, they identify, diagnose, and provide treatment options for persons with vestibular disorders that lead to dizziness and imbalance. Some audiologists specialize in vestibular disorders and may work on falls prevention teams along with physicians and physical therapists.
  - Community-based organizations, such as area agencies on aging, senior centers, and Parks and Recreation sites as well as faith-based organizations are community leaders in falls prevention and often conduct screening and assessment events, lead and/or actively participate in falls prevention coalitions, and offer EBFPPs described in the section that follows.
  - Handy persons, home remodelers, and contractors are important members of the falls prevention team making a range home modifications to maximize function and safety and reduce falls risk.

9.6 FALLS PREVENTION COALITIONS

Approximately 40 states have active falls prevention coalitions made up of individuals, government agencies, organizations, and health care professionals with a common goal of reducing older adult falls, fall-related injuries, and deaths in their communities or states. Since no single organization is responsible for addressing all aspects of falls prevention, these coalitions bring together organizations and providers that need to collaborate to reduce falls, identify state or community needs, recommend policy changes, and build capacity. Although every state or community falls prevention coalition is different, some common goals and example activities include the following:

- Increasing the availability and accessibility of community fall prevention programs and services.

- Increasing awareness of the issue and of the effective prevention strategies among stakeholders.
- Building and leveraging an integrated, sustainable fall prevention network.
- Increasing provider participation in fall prevention practices.
- Enhancing data surveillance collection, analysis, and data systems linkages.
- Improving fall prevention activities in places where older adults reside.
- Instituting ongoing evaluation of state efforts and outcomes.
- Increasing funding opportunities and investments for fall prevention.

Falls Prevention Awareness Week is an annual national health campaign observed during the week of the first day of fall to increase awareness around falls, falls risk, and injury prevention, and interventions to prevent falls. Falls prevention coalitions are key to organizing and conducting events and activities in observance of Falls Prevention Awareness Week. The National Council on Aging facilitates Falls Prevention Awareness Week, provides support to state and local falls prevention coalitions, and creates opportunities for networking and sharing of information among coalitions. A list of state falls prevention coalitions can be found at: www. ncoa.org/article/how-to-contact-your-states-falls-prevention-coalition.

9.7 COMMUNITY-BASED FALLS PREVENTION PROGRAMS

EBFPPs offer proven ways to decrease falls, fall-related injuries, and fall-related risks among older adults; by reducing falls and falls risk, the programs also help reduce falls risks, fear of falling, falls, and fractures. These programs have been designated to meet the following criteria established by the Administration for Community Living (ACL): (1) have demonstrated effectiveness in reducing falls or fall-related risk factors; (2) have been evaluated using a randomized controlled trial or quasi-experimental study with a control group; (3) have research findings published in a peer-reviewed journal; (4) be fully translated into the community setting; and (5) have dissemination material available for facilities to use. By helping to prevent falls, EBFPPs can help increase older adults' health status, confidence, independence, and quality of life.

EBFPPs have been translated into practical, effective community-based intervention and are packaged with a variety of supportive materials. As a result, a program's content and fidelity will be consistent in all settings and easy to deliver. Packages usually include implementation manuals and specialized training.

Since EBFPPs are based on rigorous study of interventions and tested with multiple populations in a variety of settings, such as senior centers, Parks and Recreation, YMCAs, libraries, senior housing, and churches, they produce positive changes or outcomes for people who participate. These programs are delivered in-person in a range of community settings as well as remotely through Zoom and other online platforms. Examples of the most commonly implemented EBFPPs are:

- **A Matter of Balance**, an eight-week structured group intervention that emphasizes practical strategies to reduce fear of falling and increase activity levels. Participants learn to view falls and fear of falling as controllable, set realistic goals to increase activity, change their environment to reduce fall risk factors, and exercise to increase strength and balance. This program is offered through MaineHealth in Portland, ME.

- **CAPABLE** or **Community Aging in Place-Advancing Better Living for Elders** addresses improvement of function, falls risk reduction, and cost. CAPABLE is a program offered through the Johns Hopkins School of Nursing for low-income seniors to safely age in place. The approach teams a nurse, an occupational therapist, and a handy worker to address the home environment and uses the strengths of the older adults themselves to improve safety and independence.
- **EnhanceFitness**, a long-term group exercise and falls prevention program that helps older adults at all levels of fitness become more active, energized, and empowered to sustain independent lives. This program is based at Project Enhance at Sound Generations in Seattle, WA.
- **Fit&Strong!**, an eight-week program targeting older adults with osteoarthritis that has demonstrated significant functional and physical activity improvements and falls risk reduction. This program is based at University of Illinois at Chicago, Institute for Health Research and Policy.
- **Otago Exercise Program (OEP)** was developed at the University of Otago in New Zealand and consists of 17 strength and balance exercises and a walking program, performed three times a week by the older adult in the home, out-patient, or community setting. The program is most effective for frail older adults; it is recommended that a PT assess and prescribe the initial exercises. The older adult does the exercises independently with visits from the PT or PT assistant and then transitions to a self-management phase for 4–10 months. OEP training is provided by the Carolina Geriatric Workforce Enhancement Program at the University of North Carolina, Chapel Hill.
- **Stepping On**, a seven-week program that offers older adults living in the community proven strategies, including exercise, medication, home, and community safety, to reduce falls and increase self-confidence. This program is based at the Wisconsin Institute for Healthy Aging.
- **Tai Chi for Arthritis** and Falls Prevention **and Tai Ji Quan: Moving for Better Balance** are programs that help to improve muscular strength, flexibility, balance, and stamina, using slow graceful movement that progressively challenges strength and balance over time.

9.8 PROGRAM BENEFITS TO OLDER ADULTS

- Reduced falls, fall-related injuries, or falls risks.
- Improved safety and quality of life.
- Increased self-efficacy in managing falls risks.
- Increased activity/exercise to increase strength and balance.
- Improved environments to reduce fall risk factors.
- Increased or sustained independence, positive health behaviors, or mobility.
- Reduced pain.

9.9   PROGRAM BENEFITS TO COMMUNITY-BASED AND HEALTH CARE ORGANIZATIONS

- Reduced health care costs associated with an emergency room, physician, hospital, and rehab visits.
- More efficient use of available resources.
- Facilitation of partnership development and community/clinical linkages.
- Better health outcomes and more positive health care experience.
- Ease of replicating and spreading programs.
- Greater opportunity for varied funding sources, as EBFPPs get proven results.

Community providers, clinicians, and older adults should think about falls risk as a continuum. While all EBFPPs have demonstrated a reduction in falls risk factors, not all falls prevention programs target the same outcomes. For example, some programs focus primarily on exercise to improve balance and strength, while others include educational and discussion-based components that help participants reduce their risk. Offering multiple programs to address different levels of risk should be considered when identifying and proposing EBFPP in the community. This reflects a more inclusive and comprehensive approach to falls prevention. Decisions on program adoption should be based on the following:

- The needs of the community and target population(s).
- A thorough review of the program costs and operation, as well as implementation site readiness.
- Current programs being implemented and ability to build on what is being offered.
- Number and type of partners that can be leveraged.
- The likelihood of the program being sustained after the grant award period has ended through new funding sources and/or being embedded into routine operations and budget at host and implementation sites.

9.10   FUNDING FOR FALLS PREVENTION INTERVENTIONS

Both the Welcome to Medicare Visit – available to new Part B enrollees – and Medicare's Annual Wellness Visit provide a health risk assessment. This reimbursable assessment includes an examination in which a physician or other qualified health care provider will perform safety screenings for fall risk and ability to perform ADL at home. Medicare will also pay for the services of physical therapists, occupational therapists for a home assessment, and certain other health care providers that doctors may refer their patients to, based on the risk assessment. However, Medicare will not pay for modifications to the home to increase accessibility or reduce risk of falling.

ACL is an operating division within the US Department of Health and Human Services whose mission is to maximize the independence, well-being, and health of older adults, people with disabilities across the lifespan and their families and caregivers. ACL has a long history of supporting health promotion and disease prevention efforts, including EBFPPs through funding from the Older Americans Act Title III-D.

The Affordable Care Act (ACA) established the Prevention and Public Health Fund (PPHF) to provide expanded and sustained national investments in prevention public health, to improve health outcomes, and to enhance health care quality. The PPHF is intended to ensure a coordinated, comprehensive, sustainable, and accountable approach to improving our country's health outcomes through the most effective prevention and public health programs. The PPHF has supported falls prevention efforts at ACL and the CDC, including grants to states to build capacity for and implement and sustain EBFPPs. The PPHF has also supported the National Falls Prevention Resource Center at the National Council on Aging.[27]

The CDC provides support for falls and injury prevention via development and support for surveillance and provider education programs, including the CDC STEADI Initiative, described previously. CDC and ACL complement each other in falls prevention efforts, prevention programs, and services. Health care entities including trauma centers and state and local public health agencies are key players in supporting falls prevention interventions and are important partners in falls prevention coalitions.

Over the past several years, health care entities have recognized the value of EBFPPs and have provided support for these programs. These health care entities include hospitals and trauma centers, health systems, managed care companies, and Medicare Advantage Plans. Some state Medicaid programs also reimburse community organizations that deliver EBFPPs to Medicaid enrollees.

9.11 NATIONAL EFFORTS TO REDUCE FALLS AMONG OLDER ADULTS

The National Council on Aging's 2015 National Falls Prevention Action Plan has provided the framework for action for falls prevention across the nation.[28] The Plan envisions older adults experiencing fewer falls and fall-related injuries, maximizing their independence and quality of life. The purpose of the Action Plan is to implement specific strategies and action steps to effect sustained initiatives that reduce falls among older adults. Further, the Action Plan was intended to help accomplish the falls-prevention-specific Healthy People 2020 objective to reduce the rate of emergency department visits due to falls among older adults by 10%. The Action Plan includes 12 broad goals, 40 strategies and over 240 action steps focusing on increasing physical mobility, improving medication management, enhancing home and environmental safety, increasing public awareness and education, and funding and expansion of falls risk screening, assessment, and interventions to prevent falls. The plan leverages the past many years of advancement in falls prevention, addresses gaps, and integrates new opportunities, such as those resulting from the Affordable Care Act.

The plan has been widely disseminated to an array of stakeholders from the public and private sectors with an interest in healthy aging and the capacity to implement action steps outlined in the plan. Implementation partners have included professionals in the health care and aging fields, federal and state agencies, professional associations, consumer and caregiver organizations, state and local falls prevention coalitions corporations, and foundations.

In 2019, the US Senate Special Committee on Aging issued a report and held a Congressional hearing on falls prevention, with a focus on policy strategies to reduce this growing public health problem, including ways in which federal agencies can collaborate to prevent falls.[29] The Senate Special Committee on Aging identified four key areas of work for policy makers, academics, and stakeholders to focus on in an effort to prevent falls and better address falls-related injuries:

- Raising awareness about falls-related risks, prevention and recovery at the national, state, and local levels.
- Improving screening and referrals for those at risk of falling so that individuals receive the preventive care necessary to avoid a fall or recover after one.
- Targeting modifiable risk factors, including increasing the availability of resources for home safety evaluations and modifications, so that older adults can remain in their homes and communities.
- Reducing polypharmacy so that health care providers and patients are aware of any potential side effects that could contribute to a fall.

9.12   INTERNATIONAL EFFORTS TO REDUCE FALLS

The World Health Organization's *Step Safely* is a multipronged strategy for preventing and managing falls across the life course.[30] On a global scale, falls take the lives of 684,000 people each year. Beyond the death toll, 172 million more people experience disabilities arising from a fall each year across all countries. Falls are not only a significant global public health problem but also are a rapidly growing one. Fall-related deaths have risen far faster than any other type of injury over the last two decades. WHO indicates that many factors account for this increase with aging populations, with patterns of urbanization chief among them. The *Step Safely* technical package provides concrete recommendations for evidence-based strategies to prevent and manage falls for children and adolescents, workers, and older people. Among others, effective prevention measures for older adults include strength and balance training for older people. Preventing falls can help achieve the Sustainable Development Goals linked to health and well-being, decent work, and safe, inclusive cities. The WHO advocates for the global community to actively look for opportunities to reduce the growing harm and suffering and loss that result from falls. A previous report published by the WHO focused on falls prevention among older adults that examined the global impact, determinants of older adult falls, challenges, and examples of effective policies and interventions.[31]

9.13   NOVEL TECHNOLOGIES FOR FALLS PREVENTION

Over the past two decades, in addition to advancements in personal emergency response systems, several new technologies have emerged that are used for falls detection, address specific falls risks, and prevent fall-related injuries. Examples of new technologies include artificial intelligence-enabled videos of residents with dementia in assisted living/memory care and nursing facilities that help detect and prevent falls. The video alerts help staff to respond to fall incidents in minutes, leading to quicker responses in which residents get the care they need when they need it. Other technologies use a scale to measure physical balance to correctly identify people who are at risk for falling, so that steps can be taken to reduce risk; ongoing balance tracking can help older adults stay motivated to make choices to maintain or improve their balance, like attending balance exercise classes more regularly. Another innovative technology uses a motor-cognitive approach to gait rehabilitation through the use of a virtual reality experience while training on a treadmill.

A complementary approach to preventing falls is to prevent serious injuries if someone falls. One strategy is the use of padded hip protectors, usually worn in pockets in specially designed underwear, to cushion the hip. However, the padding is not often sufficient to mitigate the forces generated during a fall, and hip protectors can move, and no longer protect the hip, or could increase the risk of other injuries. They can also be uncomfortable and bulky, leading to lack of adherence among older adults. Hip protectors have been studied in different settings including long-term care.[32,33] Alternatives have been developed such as hip-protecting airbags, designed to prevent hip fractures. Another variation is designed to detect a fall in progress, then deploys airbags from the head to hip in response. This garment monitors the user's center of gravity with tiny sensors. When the sensors detect an unrecoverable fall, airbags inflate in less than half a second to absorb the impact and reduce the risk of injury. The bags rapidly deflate to prevent a bounce-back effect, which can cause concussion. The rapid deflation gently lowers the user to the floor, like a deflating air mattress.

NOTES

1   Centers for Disease Control and Prevention, National Center for Injury Prevention and Control. Web-Based Injury Statistics Query and Reporting System (WISQARS) [online].
2   Bergen G, Stevens MR, Burns ER. Falls and Fall Injuries Among Adults Aged ≥ 65 Years – United States, 2014. MMWR Morbidity Mortality Weekly Report. 2016;65(37):993–998. doi: http://dx.doi.org/10.15585/mmwr.mm6537a2external icon.
3   Sallis JF, Owen N, Fisher EB. Ecological models of health behavior. In Glanz K, Rimer BK, Viswanath K (Eds.). Health behavior and health education: theory, research, and practice (pp. 465–486). San Francisco, CA: Jossey-Bass; 2008.
4   Centers for Disease Control and Prevention, National Center for Injury Prevention and Control. Web–Based Injury Statistics Query and Reporting System (WISQARS) [online]. Accessed August 5, 2016.
5   Parkkari J, Kannus P, Palvanen M, Natri A, Vainio J, Aho H, Vuori I, Järvinen M. Majority of hip fractures occur as a result of a fall and impact on the greater trochanter of the femur: a prospective controlled hip fracture study with 206 consecutive patients. Calcif Tissue Int. 1999;65:183–187. doi: 10.1007/s002239900679
6   Coronado VG, Thomas KE, Sattin RW, Johnson RL. The CDC traumatic brain injury surveillance system: characteristics of persons aged 65 years and older hospitalized with a TBI. J Head Trauma Rehab. 2005;20(3):215–228. doi: 10.1097/00001199-200505000-00005
7   Howland J, Peterson EW, Levin WC. Fear of falling among community-dwelling elderly. J Aging Health. 1993;5(2):229-243. doi: 10.1177/089826439300500205
8   Florence CS, Bergen G, Atherly A, Burns ER, Stevens JA, Drake C. Medical costs of fatal and nonfatal falls in older adults. J Am Geriatr Soc. 2018;66(4):693–698. doi: 10.1111/jgs.15304
9   www.cdc.gov/homeandrecreationalsafety/falls/adultfalls.html
10 www.cdc.gov/falls/data/falls-by-state.html
11 O'Loughlin J, et al. Incidence of and risk factors for falls and injurious falls among the community-dwelling elderly. Am J Epidemiol. 1993;137(3):342–354. doi: 10.1093/oxfordjournals.aje.a116681

12   Ferguson B, et al. Wider impacts of COVID-19 on physical activity, deconditioning and falls in older adults. A Report of Public Health England. August 2021.

13   Resnick B, Nahm ES, Zhu S, Brown C, An M, Park B, et al. The impact of osteoporosis, falls, fear of falling and efficacy expectations on exercise among community dwelling older adults. *Orthop Nurs.* 2014;33(5):277–286. doi: 10.1097/NOR.0000000000000084

14   Bu F, Abell J, Zaninotto, P. et al. A longitudinal analysis of loneliness, social isolation and falls amongst older people in England. *Sci Rep.* 2020;10:20064. doi: 10.1038/s41598-020-77104-z

15   Tinetti ME, Speechley M, Ginter SF. Risk factors for falls among elderly persons living in the community. *N Engl J Med.* 1988;319(26):1701–1707. doi: 10.1056/NEJM198812293192604

16   DeJong MR. VanDerElst M, Hartholt KA. Drug-related falls in older patients: implicated drugs, consequences, and possible prevention strategies. *Ther Ad Drug Saf.* 2013;4(4):147–154. doi: 10.1177/2042098613486829

17   www.cdc.gov/steadi/index.html

18   Panel on Prevention of Falls in Older Persons, American Geriatric Society and British Geriatrics Society. Summary of updated American Geriatrics Society/British Geriatrics Society clinical practice guideline for prevention of falls in older persons. *J Am Geriatr Soc.* 2011;59(1):148–157. doi: 10.1111/j.1532-5415.2010.03234.x

19   Rubenstein LZ, et al. Validating an evidence-based, self-rated fall risk questionnaire (FRQ) for older adults. *J Safety Res.* 2011;42(6):493–499. doi: 10.1016/j.jsr.2011.08.006

20   Beauchet O, Fantino B, Allali G, Muir SW, Montero-Odasso M, Annweiler C. Timed up and go test and risk of falls in older adults: a systematic review. *J Nutr Health Aging.* 2011;15(10):933–938. doi: 10.1007/s12603-011-0062-0

21   Bohannon RW. Five-repetition sit-to-stand test: usefulness for older patients in a home-care setting. *Percept Mot Skills.* 2011;112(3):803–806. doi: 10.2466/15.26.PMS.112.3.803–806

22   Southerland LT, et al. Accuracy of the 4-stage balance test and sensor-based trunk sway as fall risk assessment tools in the emergency department. *J Acute Phys Ther.* 2021;12(2):79–87. doi: 10.1097/JAT.0000000000000150

23   American Geriatrics Society. 2019 Updated Beers Criteria for potentially inappropriate medication use for older adults. *J Am Geriatr Soc.* 2019;67(4):674–694. doi: 10.1111/jgs.15767

24   https://homemods.org/materials/a-review-of-assessment-tools/

25   https://homemods.org/materials/evidence-based-programs-and-best-practices-in-home-assessment-and-modification-2/

26   Frankowski O. The role of the podiatrist in falls prevention. *J Foot Ankle Res.* 2010;3(Suppl 1):P10. doi: 10.1186/1757-1146-3-S1-P10

27   www.ncoa.org/professionals/health/center-for-healthy-aging/national-falls-prevention-resource-center

28   Cameron K, Schneider E, Childress D, Gilchrist C. National Council on Aging Falls Free® National Action Plan. 2015. www.ncoa.org/article/2015-falls-free-national-falls-prevention-action-plan

29   Falls Prevention: National, State and Local Solutions to Better Support Seniors. Special Committee on Aging, U.S. Senate. October 2019. www.aging.senate.gov/imo/media/doc/SCA_Falls_Report_2019.pdf

30   *Step safely: strategies for preventing and managing falls across the life-course.* Geneva: World Health Organization; 2021.

31   Global Report on Falls Prevention in Older Age. Geneva: World Health Organization; 2008.

32   https://pubmed.ncbi.nlm.nih.gov/31477556/#:~:text=Conclusions%20and%20implications%3A%20Hip%20protectors,at%20the%20time%20of%20falling.

33   www.cochranelibrary.com/cdsr/doi/10.1002/14651858.CD001255.pub5/full

BIBLIOGRAPHY

Agency for Health Care Research and Quality, Safety Program for Nursing Homes: On-Time Falls Prevention: www.ahrq.gov/patient-safety/settings/long-term-care/resource/ontime/fallspx/index.html

American Geriatrics Society. 2019 updated Beers Criteria for potentially inappropriate medication use older adults. *J Am Geriatr Soc.* 2019;67(4):674–694. doi: 10.1111/jgs.15767

Cameron K, Schneider E, Childress D, Gilchrist C. *National Council on Aging Falls Free® National Action Plan;* 2015. www.ncoa.org/article/2015-falls-free-national-falls-prevention-action-plan

Centers for Disease Control and Prevention. Stopping Elderly Deaths, Accidents, and Injuries Initiative of Resources for Inpatient and Outpatient Care, Pharmacists, Patients and Caregivers. www.cdc.gov/steadi/index.html

Falls Prevention: National, State and Local Solutions to Better Support Seniors. Special Committee on Aging, U.S. Senate. October 2019. www.aging.senate.gov/imo/media/doc/SCA_Falls_Report_2019.pdf

Joint Commission Center for Transforming Health Care. Targeted Solutions for Preventing Falls Quality Improvement Program. www.centerfortransforminghealthcare.org/products-and-services/targeted-solutions-tool/preventing-falls-tst/?_ga=2.65984942.1814601029.1639436943-1232345419.1620763069

Lord S, Sherrington C, Naganathan V (Eds.). *Falls in older people: risk factors, strategies for prevention and implications for practice.* 3rd Ed. Cambridge University Press; 2021. doi:10.1017/9781108594455

Montero-Odasso M, Camicioli R. *Falls and cognition in older persons: fundamentals, assessment and therapeutic options.* Springer; 2020.

Morse JM. *Preventing patient falls, establishing a fall intervention program,* 2nd Ed. Springer; 2008.

National Center on Assisted Living/American Health Care Association, A Guide to Managing Fall Risk: A Competency-Based Approach to Reducing Falls. 2013. www.ahcancal.org/Quality/Clinical-Practice/Documents/A%20Guide%20to%20Managing%20Fall%20Risks_2013.pdf#search=falls%20prevention

Panel on Prevention of Falls in Older Persons, American Geriatric Society and British Geriatrics Society. Summary of updated American Geriatrics Society/British Geriatrics Society clinical practice guideline for prevention of falls in older persons. *J Am Geriatr Soc.* 2011;59(1):148–157. doi: 10.1111/j.1532-5415.2010.03234.x

Tideiksaar R. *Falls in older people: prevention and management (essential falls management).* 4th Ed. Health Professions Press; 2010.

World Health Organization. *Global report on falls prevention in older age.* Geneva; 2008.

World Health Organization. *Step safely: strategies for preventing and managing falls across the life-course.* Geneva; 2021.

10 Home Modifications

Brianna Brim

CONTENTS

10.1   PLANNING TO AGE IN PLACE

Eighty-seven percent of adults aged 65 and older want to stay in their current home and community as they age.[1] While the desire to remain in the home and age in place is nearly universal, the ability to follow through on this goal may not be. Often, the ability to remain in the home is preceded by a fair amount of planning; home modification is one area that can pay dividends towards an independent future. Home modification is the process of changing or designing spaces that increase user functionality by increasing safety, accessibility, usage, and/or independence.[2] While this may sound daunting, evidence shows small-scale changes can have a large impact when it comes to safety, independence, and preventing injuries at home for older adults.[3] Home modifications can range from large-scale renovations including removing walls to increase hallway widths or installing elevators and lifts to small-scale tasks such as changing light bulbs and securing an area rug. Home modifications may be very costly, or cost nothing, except the time to reconfigure spaces to increase safety and functionality. Home modification is a necessary part of aging in place for many and has been shown to decrease injury and falls risk, contribute to well-being, decrease caregiver burden, and enable individuals to remain in their home longer[1-5]

Best practice in home modification occurs with an interdisciplinary focus and uses multiple professionals from healthcare, design, and construction. It factors in progression of client-known illnesses, preparation for potential genetic outcomes, likely impacts of the aging process, and the current and future usage of the home in question. The people who plan to live there, the home itself, and the use of the home, including daily routines, desires, and community access should be factored in determining what, if any, home modifications can support longevity in the home.[3]However, a highly individualized plan may be expensive and financially out of the reach of many individuals under the, often limited, current insurance reimbursement standards. In these cases, some general considerations related to home modification can support independence and will be discussed in this chapter.

10.2   INDIVIDUAL AND HOME USAGE

While many aspects of disability and disease can be managed or avoided by a healthy lifestyle, many genetic conditions and typical physical changes that accompany the aging process may be unavoidable. Common conditions – such as age-related macular degeneration, Parkinson's Disease, or elective joint replacement surgery – can limit independence, temporarily or permanently. Planning for known periods of disability and disease progression when considering a home purchase, rental, or modification is helpful for maintaining independence. While disease acquisition and progression are often variable and unpredictable on the individual level, understanding the common comorbidities and disease outcomes of known health conditions as well as family history and genetic predisposition can empower individuals to make the best choices for themselves. Anyone could become disabled at any time, though we rarely calculate that into our long-term plans, and keeping this in mind will help us to make better choices for the future. Social usage of the home is another important consideration that may encourage various modifications and changes. Framing home modification in terms of visitors (i.e., grandchildren, children, friends, and colleagues) may encourage environmental changes that support a variety of ability levels. These changes can help prepare older adults for situations that they may not foresee, which could limit or influence their ability to remain in the home.

Understanding how a person uses or plans to use their home is essential before investing in home modifications, which may be costly and divert finances away from other aspects of independence. Before investing in a home modification, ensure that attention is being focused on areas of the home that are most utilized by the individual. Consider the persons' daily routines, values, roles, and responsibilities. Costly kitchen renovations like adjustable height countertops increase ease of use for wheelchair users or those who need the ability to sit when cooking. However, if the older adult has never been responsible for cooking, then this modification would be a poor use of resources. For an adult who only takes showers, modifying a bathtub would not be a wise investment. A large living room may increase resale value, but if the space could be partially allocated to a new first-floor bathroom, the space becomes more valuable and useful to the older adult who may have mobility or urgency needs and prefers the convenience of a first-floor bathroom. Thought and resources should be prioritized to the most frequently used and items that are more easily and inexpensively modified. In most cases, universal accessibility is not achievable for the average older adult due to existing home features and finances. A home does not need to be modified for every possible circumstance or be fully compliant with national accessibility standards; instead, emphasis should be placed on changes that maximize accessibility and safety for the older adult.

10.3   HOME ENVIRONMENT

During any renovation, repair, or replacement of existing home structures, it is important to consider features that support aging in place. Consider the features being upgraded and apply the principles of Universal Design whenever possible.[6] Universal Design principles will translate into more manageable rooms within the home that could support a variety of physical, sensory, or cognitive changes that may accompany the aging process. These principles should also be considered when purchasing or renting a new home. Universal design ensures that home modifications are simple, intuitive, allow room for error, and can be used by people with a variety of abilities. These features are imperative with the potential for functional changes due to disease, disability, and decreased muscle mass often secondary to aging. The American Association of Retired People (AARP) has a free and easy-to-follow guide called HomeFit® which can help individuals apply concepts of universal design and evaluate their own home.[1]

Before embarking on any large-scale home modification, ensure a contractor has mapped load-bearing walls to avoid unexpected costs or threats to home stability. In addition, features that would increase costs (i.e., the need to move vents, electrical conduits, or relocating plumbing features) should be discussed before any work is initiated. Pay attention to the features of the home or home setup that have been helpful, useful, or enabled for the older adult. Work to recreate or capitalize on systems already in place to increase independence. Before renovations or purchase of new appliances, consider why the older versions were helpful or ineffective. For instance, if an older side-by-side refrigerator and freezer combination has allowed easy access to frozen foods, make sure the replacement does not force a person with decreased balance to bend over to retrieve items from a lower chest-style freezer. Be mindful of period-specific or personal design features that are meaningful to the client. Reusing these when possible will help ensure that the home modifications are utilized by the client. In addition, the similarity of features throughout the house, like faucets and light switches, decreases thought required for use and may prove valuable from a safety perspective, especially in relation to how hot water is accessed in each sink. Aesthetics may discourage some home modifications if the user views them as unsightly or a sign of their decreased abilities. To avoid this, consider the aesthetics

Principle 1: Equitable Use; The design is useful and marketable to people with diverse abilities.

1a. Provide the same means of use for all users: identical whenever possible; equivalent when not.

1b. Avoid segregating or stigmatizing any users.

1c. Provisions for privacy, security, and safety should be equally available to all users.

1d. Make the design appealing to all users.

Principle 2: Flexibility in Use; The design accommodates a wide range of individual preferences and abilities.

2a. Provide choice in methods of use.

2b. Accommodate right- or left-handed access and use.

2c. Facilitate the user's accuracy and precision.

2d. Provide adaptability to the user's pace.

Principle 3: Simple and Intuitive Use; Use of the design is easy to understand, regardless of the user's experience, knowledge, language skills, or current concentration level.

3a. Eliminate unnecessary complexity.

3b. Be consistent with user expectations and intuition.

3c. Accommodate a wide range of literacy and language skills.

3d. Arrange information consistent with its importance.

3e. Provide effective prompting and feedback during and after task completion.

Principle 4: Perceptible Information; The design communicates necessary information effectively to the user, regardless of ambient conditions or the user's sensory abilities.

4a. Use different modes (pictorial, verbal, tactile) for redundant presentation of essential information.

4b. Provide adequate contrast between essential information and its surroundings.

4c. Maximize "legibility" of essential information.

4d. Differentiate elements in ways that can be described (i.e., make it easy to give instructions or directions).

4e. Provide compatibility with a variety of techniques or devices used by people with sensory limitations.

Principle 5: Tolerance for Error; The design minimizes hazards and the adverse consequences of accidental or unintended actions.

5a. Arrange elements to minimize hazards and errors: most used elements, most accessible; hazardous elements eliminated, isolated, or shielded.

5b. Provide warnings of hazards and errors.

5c. Provide fail-safe features.

5d. Discourage unconscious action in tasks that require vigilance.

Principle 6: Low Physical Effort; The design can be used efficiently and comfortably and with a minimum of fatigue.

6a. Allow the user to maintain a neutral body position.

6b. Use reasonable operating forces.

6c. Minimize repetitive actions.

6d. Minimize sustained physical effort.

FIGURE 10.1   Principles of Universal Design.

Source: Adapted from the Center for Excellence in Universal Design.

Principle 7: Size and Space for Approach and Use; Appropriate size and space are provided
   for approach, reach, manipulation, and use regardless of user's body size, posture, or
   mobility.
   7a. Provide a clear line of sight to important elements for any seated or standing user.
   7b. Make reach to all components comfortable for any seated or standing user.
   7c. Accommodate variations in hand and grip size.
   7d. Provide adequate space for the use of assistive devices or personal assistance.

FIGURE 10.1   (Continued)

and/or cross-generational appeal when completing home modifications; in most cases
aesthetically pleasing or universally helpful alternatives are available at comparable
prices and should be considered in features that may remain in the home for the next
occupant (see Figure 10.1).

10.4   FUNDING HOME MODIFICATIONS

Home modifications can be as expensive as desired. However, there is evidence
that significant functional improvement and independence can be achieved with an
average of $636 (USD) for many older adults.[3] One of the most impactful home
modifications lies in decluttering spaces, where the cost is associated with labor to
move or store items. Commercial retail websites and big-box stores have products
readily available and competitively priced for home modification. If completing a
renovation project, or simply replacing appliances and home features, choosing items
with an aging-friendly or accessible version often adds no cost or is marginally more
expensive. Individuals may save money when converting a standard height toilet to
a comfort height toilet, depending on where it is purchased. Foresight and planning
during renovations will save money and ensure higher resale or rental value on the
property while simultaneously making it safer and more supportive of aging needs.

   While some modifications can be easily financed, other major home repairs can be
costly. For individuals with conditions that may rapidly decrease functional abilities,
there are governmental and nonprofit resources available. If a person carries private
insurance, speaking with a representative for Durable Medical Equipment (DME)
may give insight on how to finance items through insurance. Evidence of medical
necessity will be required, such as a prescription from a physician and/or a letter of
medical necessity. Items funded by insurance are often limited to commode chairs,
hospital beds, patient lifts (not including stair-chair lifts), and pressure-reducing
surfaces.[7] Note that this list does not include any permanent or physical changes to
the home and DME funding requires current significant functional limitations.

   In the United States, federally and state-funded regional Area Agencies on Aging
(AAA) offer regionally appointed nonprofit or government office grants and funding
to support large-scale home modifications and individualized assessments at no cost.
These organizations all have their own processes for the evaluation and completion
of home modification. In Philadelphia, Pennsylvania, for example, Philadelphia
Corporation for the Aging (PCA) operates a program called Senior Housing Assistance

Repair Program (SHARP) that allows for significant home modifications; however, waitlists for these programs may be long.[8] In addition, government-supported assistive technology programs may be an option for exploring technology or smart home supports, which are also housed at the state level.[9] Another free resource is the National Rehabilitation Information Center (NARIC), which has a host of funding options and can connect individuals with local, federal, and nonprofit means to fund home modifications.[10] It is important to distinguish the needs of the individuals; preventive or proactive home modifications are generally not funded by any of these organizations, due to limited resources. Assistance through these channels will often require current, permanent, or progressive loss of function to qualify.

In other countries, coverage for home modification in the event of progressive or permanent decreases in function may be easier to finance, as end-user funding is more common. This means that funds go directly to consumers, rather than third-party contracts or organizations. In Canada, there are a multitude of grants, loans, and other government-sponsored programs to fund home access, mobility, and functionality requirements for individuals who are disabled.[11] In the United Kingdom (UK), direct grant and loan funding is also common. Grants and loans are available to both consumers and landlords to promote accessibility in rental properties, grants specifically for individuals recently hospitalized with new disability, and general accessibility grants. These programs have varying scope and coverage, with some being offered to 300 UK citizens a year.[4]

10.5   HOME FEATURES BY ROOM

As previously stated, it is important to consider the individual and usage of the home when planning or executing home modification. Therefore, the following recommendations and considerations may not appeal to all older adults but serve as examples of small- and large-scale modifications. In addition, there are a number of smart home technologies covered elsewhere in this book that could significantly increase the safety and functionality of spaces.

10.6   GENERAL HOUSE FEATURES

General home features should be considered as a top priority in a home modification plan. However, many of these safety features are not considered by consumers for daily use, despite being instrumental in an emergency situation.[12] This list is a composite of features recommended by multiple organizations including AARP,[1] AOTA,[2] the Living in Place Institute,[13] the Health and Place Initiative (HAPI),[114] and Center for Universal Design,[15] as well as insights from occupational therapy practice in home modification.[12]

- Illuminate outdoor house numbers to be visible from the street
- Increase outdoor lighting for a safe approach and to deter burglary
- Ensure that smoke and carbon monoxide detectors are working
  - Consider a ten-year sealed combination smoke/carbon monoxide unit that requires no maintenance or battery changes

- Ask the local fire department whether they provide free smoke detectors to residents
- Ensure fire extinguishers are accessible and have not expired
  - Dispose of used or expired fire extinguishers which can leak slowly after initial use
  - Buy smaller fire extinguishers and store them on every level
- Secure or remove area rugs
  - Remove rugs from open spaces like walkways and in front of stairs
  - Keep and secure area rugs where water may accumulate (i.e., in front of a sink, tub, entryway)
  - Firmly secure rugs by using adhesive carpet tape, available at most home improvement stores
- Lower maximum water heater temperature to decrease burn risk, especially for individuals with decreased sensation
- Declutter to reduce tripping hazards
- Increase indoor lighting in every room of the house
  - Install night lights throughout the home (motion-activated night lights are helpful for individuals with sensitivity to light)
  - Look for [LED] lightbulbs with a higher lumen count; compared to traditional incandescent light bulbs, LED replacement bulbs rated "60-Watt Replacement" use fewer watts (how much energy is being used) than older incandescent light bulbs use, but often have less lumen (how bright a bulb appears); read the package to find the most lumen – 1200–1400 is best for most activities – and ensure that the watt usage is still below the watt rating for a particular fixture
  - Look for motion-activated or tap-style lights for high traffic, low light areas like stairwells, basements, closets, or cabinets
  - Look at lighting color
    - Daylight bulbs have a blue hue which has been shown to increase focus but may alter circadian rhythms; these lights are good for task lighting in a reading area
    - Soft white or warm white bulbs have a yellow hue and can help to support good sleep patterns; this color should be used in areas frequented before bed
    - Neutral or cool light bulbs can increase energy levels during the day and are good for the kitchen or workspace[16]
    - Some light bulbs have a three-way component not only for brightness (lumen) but also for light color, alternating daylight, soft white, and a neutral tone; this can give one fixture or space increased versatility
- Consider reinforcing walls during any renovation so that additional cross beams and studs can be utilized if grab bars or additional features are needed, particularly in front of doors with a step-up, in stairwells, and in bathrooms
- Hire an electrician to relocate floor-level electrical outlets to 18 inches if local building code allows, to limit bending and allow for easier view of loose electrical plugs

- Check for fire hazards; the American Red Cross has a comprehensive checklist that will help decrease fire risk[17]
- Install or look for first-floor full bathrooms for convenience, or install a first-floor powder room
  - If this is not feasible, purchase a stand-alone commode chair to use on the first-floor
  - There are also stand-alone tub/shower combination units that can be connected to a sink for use in homes without first-floor bathrooms
- Consider flexible use furniture for your home
  - Sofa beds can create a first-floor bedroom in the event of limited ability to climb stairs
  - Reclining chairs can be particularly useful for preferred post-surgical positioning-limited mobility
- Choose flooring wisely. Flooring should have low glare with matte finishes instead of gloss finishes wherever possible
  - Flooring that has "give" (compliance) to it with heavier padding is useful in bedrooms and living rooms as the shock absorbency of these surfaces may lessen impact during a fall[18]

10.7   LAYOUT, APPROACHES, ENTRYWAYS, AND STAIRWELLS

If in the market for a new residence, it is preferable to look for ranch style, elevator access, or other single-level living environments. However, that is not always feasible depending on the area desired, or the current housing situation. In general, look for spaces with minimal steps to enter.

- Decrease threshold height when possible to include level surfaces during any renovation; level entry surfaces between rooms and in the home
- Consider a ramp installation, elevator, outside wheelchair lift, or stair-chair lift in the event of progressive diagnoses; these are costly renovations but necessary in the event of needed wheelchair or scooter use
  - Plan ahead for these devices by looking for homes with
    - extra closet space that could be converted to an elevator
    - a larger front entry approach or front yard to support a ramp
    - straight staircases instead of curved or L-shaped staircases for easy installation of a stair-chair lift
- Ensure trash cans are stored with an easy route to the curb for trash pickup

10.8   BATHROOMS

Particular attention and funds should be considered regarding bathroom modification. One of the leading causes of older adult institutionalization is an inability to self-toilet. Self-toileting includes the ability to physically get on and off the toilet, and complete toilet hygiene. In addition, falls often occur in the bathroom, or en route to the bathroom due to urgency.[19, 20] Small changes can ensure the bathroom remains accessible or that there are alternatives to increase independence with bathroom activities.

- Install raised or comfort height toilets
  - If this is not an option or a quick alternative is needed, toilet seat risers with handles can be placed over existing toilets to increase height and are available in most drug stores
- Install grab bars in your bathroom. The most functional locations that could serve a variety of needs are:
  - One vertical bar at the entrance of the tub/shower
  - One horizontal bar, waist height on the faucet wall
  - One horizontal bar, waist height on the wall adjacent to the faucet in the tub/ shower
  - One vertical bar adjacent to the toilet
- Consider a bidet. Bidets have been shown to decrease hemorrhoids and urinary tract infections, but may not be a good option if the older adult has a history of bacterial vaginosis (VB).[21] There are a number of surgical, orthopedic, or neurological conditions that can impact bathroom hygiene and a bidet can lessen difficulty with perineal hygiene.
  - Several companies have easy-install model bidets that do not require electricity and can be retrofitted to any toilet
  - Some bidets also allow for a drying feature but do require electricity; install a Ground-Fault Circuit Interrupter outlet (GFCI) near the toilet, 18 inches from the floor for these devices
- Fix slow drains in tubs and showers immediately
  - These will allow soap to accumulate and create a slippery surface; prevent slowing of drains by using a hair-catching device in the drain
- Remove nonslip bath mats, where soap can accumulate and become slippery; instead, replace them with textured bath stickers to decrease slipping risk in the tub
- Consider a tub/shower bench/chair with a back and nonskid feet, or have a seat feature installed in a shower stall
- Add an additional hand-held shower head attachment, or install a hand-held head as the primary shower head to increase ease of cleaning and offer easier use

10.9 KITCHEN

The kitchen is another high-risk fall area. However, many other hazards such as fire, burns, and food-borne illness also should be considered when assessing any changes to the kitchen. In general, the kitchen should focus on functionality, cleanliness, and ease of use.[22]

- Use front panel control dishwashers for better visibility and adaptability potential
  - Hidden controls make it difficult to determine if the appliance is on or hot
  - Bump dots or other raised indicators can adhere over controls for decreased vision, touch, or cognitive deficits

- Determine how the refrigerator and freezer will be used
  - Purchase side-by-side refrigerator options if the freezer is frequently used to decrease bending requirements and ease of access with canes, walkers, or wheelchairs
  - Purchase top fridge models if the freezer is not frequently used
  - Look for single-door options for a wider entry, rather than French or side-by-side door options if a cane, walker, or wheelchair is likely to be needed
- Opt for a stove with front burner controls, instead of back controls to limit extended reaching which reduces the risk of falls and/or burns
- Look for stoves and ovens with hot surface light indicators
- Look for stoves and ovens with lock features or removable control knobs which provide a secure way to turn off the devices in the event an older adult becomes unsafe to use the device
- Ensure ample counterspace for meal preparation near the stove, oven, and/or microwave to decrease physical stress during meal preparation
- Look for solid, nonporous surface countertops instead of tile or butcher block to increase ease of cleaning and decrease the risk of food-borne illness
  - Ensure these surfaces have low glare; matte or satin finishes are preferable to gloss finishes
- Move frequently used items to accessible places like top drawers or cabinets
  - Frequently used heavy items, like slow cookers, should be placed in access-ible cabinets or on the counter
- Use removable faucet heads in kitchen sinks to increase ease when washing dishes
  - These provide the potential for a person or caregiver to wash hair on the first floor
- Consider installing trim from the top of cabinets to the ceiling to decrease the need to clean above refrigerators or cabinets

10.10 SPECIFIC RECOMMENDATIONS FOR FUNCTIONAL LIMITATIONS

While the previous recommendations capitalize on universal design and increase ease of use globally, it is important to consider physical and cognitive limitations and address those directly with home modification. The suggestions should be considered in addition to those already discussed.

10.10.1 DECREASED VISION (PRESBYOPIA)

- Increase lumen count in lightbulbs; refer to the general home modification list for safe ways to do this
- Add more task lighting in the kitchen, and high trip areas like basement steps
  - Use battery-powered tap lights or motion-activated lights, which can be wall-mounted with removable adhesives
- Work to increase contrast in common tripping areas
  - Add white tape or paint to stair risers of stairs and between transitions, or steps

- On basement steps use contrasting paint or tape on the edge to make the end of the step visible as you step down
- Add tactile cues to surfaces to help delineate features
  - Bump dots, puffy paint, or different textured stickers can be used on the control buttons for microwaves, washer/dryers, faucets, dishwashers, or other devices to increase ease of use

10.10.2 Decreased Balance, Frequent Falls, or Difficulty Bending Over

- Use wall-mounted surge protector strips mounted 18 inches from the floor to decrease falls risk when plugging/unplugging items
- Add grab bars in showers, tubs, and at entryways with steps
  - Ensure these are properly installed into studs, or in older homes where stud integrity may not be known use products like Wing-its ® to ensure the grab bar can hold the weight
  - Towel racks should not be used as grab bars
  - Avoid suction cup grab bars on vertical surfaces as they can easily detach with force
- Utilize bathroom equipment like a raised toilet or commode chair with handles over the toilet or shower chair to decrease bathroom falls risk
- Utilize motion-activated night lights along routes to the bathroom
- Purchase risers for underneath washers and dryers to decrease balance challenges

10.10.3 Decreased Memory

- Place needed items in visible spots to encourage use. Out of sight, out of mind is especially true for individuals experiencing memory changes
- Utilize smart home technology to automate processes
  - Smart locks can automatically be set to lock the front door at night
  - Timers can be set to turn on/off lights around sundown
  - Refer to the smart-home chapter in this book for more ideas
- Ensure potentially hazardous appliances like the stove, oven, or microwave have shut off features, removable knobs, lock features, etc. in the event the older adult becomes unable to use these safely

10.10.4 Decreased Grip Strength or Limited Shoulder Range of Motion

- Install a jar opener underneath upper cabinetry in the kitchen
- Replace key access locks with electronic locks that can use an entry code or be operated by a cell phone
- Replace door knobs with lever-operated door handles to decrease stress on hands
- Install touch-activated faucets in kitchen and bathrooms

10.11 CONCLUSION

Home modification can be a valuable tool and process to decrease falls, increase functionality, and help maintain independence for older adults. This process can require significant renovation, or be incorporated into home projects and repairs throughout the lifespan. Planning and foresight surrounding the older adult, their current health, current usage of the home, and potential for changing abilities due to aging or disease are imperative to allow older adults to remain in their home. Small changes can have a large impact but funding home modifications may pose significant challenges depending on an older adult's finances, current abilities, and country of residence. While individualized home assessment improves outcomes, using Universal Design principles can encourage changes that increase use for a variety of functional abilities and needs.

REFERENCES

1. AARP HomeFit guide. AARP. www.aarp.org/livable-communities /housing/info-2020/homefit-guide.html. September 2020. Accessed October 5, 2021.
2. Fagan LA, Sabata D. Home modifications and occupational therapy. American Occupational Therapy Association. www.aota.org/about-occupational-therapy/professionals/pa/facts/home-modifications.aspx. 2011. Accessed October 5, 2021.
3. Stark S. Removing environmental barriers in the homes of older adults with disabilities improves occupational performance. *OTJR (Thorofare N J)*. 2004;24(1):32–40. doi:10.1177/153944920402400105.
4. Zhang B, Zhou P. An economic evaluation framework for government-funded home adaptation schemes: a quantitative approach. *Healthcare (Basel)*. 2020;8(3):345. doi:10.3390/healthcare8030345
5. Carnemolla P, Bridge C. Housing design and community care: how home modifications reduce care needs of older people and people with disability. *Int J Environ Res Public Health*. 2019;16(11):1951. doi:10.3390/ijerph16111951
6. The 7 principles. National Disability Authority. https://universaldesign.ie/What-is-Unive rsal-Design/The-7-Principles/. 2020. Accessed October 5, 2021.
7. Durable medical equipment (DME) coverage. Medicare.gov. www.medicare.gov/coverage/durable-medical-equipment-dme-coverage. Accessed October 5, 2021.
8. Home repairs & modifications. Philadelphia Corporation for the Aging. www.pcaca res.org/services/help-in-the-home/home-repairs-modifications/#. 2021. Accessed October 17, 2021.
9. Home. Pennsylvania Assistive Technology Foundation. https://patf.us/. 2021. Accessed October 17, 2021.
10. Independent living and community participation. National Rehabilitation Information Center. www.naric.com/?q=en/node/50. Accessed October 17, 2021.
11. Canadian government grants for disabled homeowners: nearly 100 grants for home renovations. ShowMeTheGreen.ca. https://showmethegreen.ca/home/home- improvement/canadian-government-grants-for-disabled-homeowners-2018/. 2021. Accessed October 17, 2021.
12. Brim B, Fromhold S, Blaney S. Older adults' self-reported barriers to aging in place. *J Appl Gerontol*. 2021;40(12): 1678–1686.: doi:10.1177/0733464820988800
13. Home. Living in lace Institute. https://livinginplace.institute/. 2021. Accessed October 17, 2021.

14. Health and Places Initiative. Mobility, universal design, health, and place: a research brief. Harvard.edu. http://research.gsd.harvard.edu/hapi/files/2014/10/HAPI_Univer sal Design_ResearchBrief-102814.pdf. 2014. Accessed October 17, 2021.
15. Young, LC. Residential rehabilitation, remodeling and universal design. Raleigh, NC: Center for Universal Design, College of Design, NC State University; 2006. https://projects.ncsu.edu/ncsu/design/cud/pubs_p/docs/residential_remodelinl.pdf. Accessed October 17, 2021.
16. The color of light affects the circadian rhythms. Centers for Disease Control and Prevention. www.cdc.gov/niosh/emres/longhourstraining/color.html. April 1, 2020. Accessed October 24, 2021.
17. Is your home a fire hazard? American Red Cross. www.redcross.org/get-help /how-to-prepare-for-emergencies/types-of-emergencies/fire/is-your-home-a-fire-hazard.html. 2021. Accessed October 17, 2021.
18. Lachance CC, Jurkowski MP, Dymarz AC, et al. Compliant flooring to prevent fall-related injuries in older adults: A scoping review of biomechanical efficacy, clinical effectiveness, cost-effectiveness, and workplace safety. *PLoS One.* 2017;12(2): e0171652. doi: 10.1371/journal.pone.0171652
19. Foley AL, Loharuka S, Barrett JA, et al. Association between the geriatric giants of urinary incontinence and falls in older people using data from the Leicestershire MRC incontinence study. *Age Ageing.* 2012;41(1):35–40. doi:10.1093/ageing/afr125
20. Matsumoto, M, Inoue K. Predictors of institutionalization in elderly people living at home: the impact of incontinence and commode use in rural Japan. *J Cross Cult Gerontol.* 2007;22(4):421–432. doi:10.1007/s10823-007-9046-2
21. Kiuchi T, Asakura K, Nakano M, Omae K. Bidet toilet use and incidence of hemorrhoids or urogenital infections: A one-year follow-up web survey. *Prev Med Rep.* 2017;6:121–125. doi:10.1016/j.pmedr.2017.02.008
22. Hrovatin J, Sirok K, Jevsnik S, Oblak L, Berginc J. Adaptability of kitchen furniture for elderly people in terms of safety. *Hem Ind.* 2012;63(2):113–120. doi:10.5552/drind.2012.1128

11 Advancements in Technology to Promote Safety and Support Aging in Place

George Mois and Wendy A. Rogers

CONTENTS

DOI: 10.1201/9781003197843-11

11.1   INTRODUCTION

Enabling autonomy and independence throughout the aging process is essential to promoting and supporting a high quality of life for the older adult population. However, many of the homes and communities that older adults live in have not been developed with the aging process in mind. Communities often lack the housing resources to meet the needs of a shifting population demographic.[1,2] Lack of resources and adequate housing presents many challenges that are becoming threats to older adults' safety. However, recent advancements in emerging technologies are pushing the boundaries of access to resources capable of adapting home environments to minimize risks, support security, and enable independence.[3,4]

11.2   AGING IN PLACE

Aging in place is defined as "One's journey to maintain independence in one's place of residence as well as to participate in one's community."[5] The process of aging in place involves three key components: the individual, the home environment, and the community that surrounds them.[6] A common misconception in the discussion surrounding the concept of aging in place is that the individual must remain in their current home. Although many older adults are choosing to remain in their home, individuals may choose to move to a different home or choose alternative forms of housing.[5] For example, in the process of preparing for aging in place, older adults may move from a two-story home into a single-story home.

Furthermore, the concept of aging in place is not limited to physical spaces. The most important goals are to maintain the individual's autonomy, independence, and support for social relationships developed across the lifespan.[6,7] Over a lifetime, individuals develop symbiotic relationships and connections with the communities they inhabit, thereby creating an ecosystem. The ecosystem serves as the structure that sustains and enables the process of aging in place.[8] Alteration to the ecosystem can have a direct effect on not only older adults' aging in place but also on other members of their community. Therefore, protecting and enabling individuals to remain part of this ecosystem is essential in stimulating and supporting both the safety and well-being of older adults and that of the community.

11.3   THE HOME ENVIRONMENT

The home environment is an important part of aging in place. The places older adults call home are diverse and include a wide array of living arrangements (e.g., house, apartment, condo, townhome, shared housing, retirement community).[5] Whatever home environment older adults choose to call home comes with its own set of unique challenges and concerns related to safety and the ability to age in place. For example, multi-level homes, stairwells in apartment complexes, small doorways, carpeted floors, and low-access bathrooms are just a few of the safety concerns experienced by individuals wanting to age in place. The lack of housing that meets the needs of individuals can create frustration and limit their ability to age in place successfully.

11.4   IMPLICATIONS FOR AGING IN PLACE

To support the process of aging in place, professionals must consider how the home environment affects older adults' mobility and accessibility. Professionals working with older adults need to be attentive and sensitive in implementing strategies that promote a home environment which is adaptive and supportive of aging in place. For example, educating and assisting older adults to plan and implement preventative strategies to promote safety are important steps towards improving their ability to age in place successfully.

11.5   NEEDS OF OLDER ADULTS

Inadequate resources and limited access to services present a wide assortment of challenges for older adults.[9-12] These challenges include isolation, mobility limitations, impairments (e.g., vison, hearing, locomotion), poor nutrition, limited healthcare access, and lack of access to transportation.[13-15] Moreover, older adults may experience physical, perceptual, and cognitive declines.[11,16] The aging process is unique for everyone.

The experience of aging is often the result of a wide array of biological, psychological, social, and environmental factors that impact the individual's well-being.[17] Changes in one's functional abilities can span over long periods of time whereas other individuals may experience sudden changes in health and functional ability.[11,13] For example, strokes, falls, and other acute injuries can alter and create sudden change in an individual's functional ability in addition to any chronic conditions. These sudden changes require adaptation to the environment and the methods through which individuals interact with their community.[17,18] Furthermore, these changes can affect the three categories of everyday activities. These categories include activities of daily living (ADL), instrumental activities of daily living (IADL), and enhanced activities of daily living (EADL).

Activities of daily living (ADLs) refer to the basic self-care tasks that an individual engages in to carry on with day-to-day life. These tasks include feeding, walking, dressing, grooming, bathing, and transferring.[5,8,19,20] Maintaining one's ability to carry out ADLs is an important part of promoting safety and supporting the individual's ability to age in place. Identifying challenges to an individual's functional ability to complete ADLs can help identify areas in the home to maintain and enhance safety. Across the literature, the most common challenges reported by older adults related to ADLs included walking, getting in and out of bed and chairs, bathing, and showering.

Instrumental activities of daily living (IADLs) refer to the tasks the individual engages in and carries out to maintain independence and age in place. IADLs differs from ADLs as they require more advanced capabilities such as critical thinking and organization. IADLs include managing finances, transportation, shopping, meal preparation, managing medications, and cleaning.[19,20] Determining challenges related to IADLs and identifying tools that can help support older adults is essential in promoting independence and autonomy. The most common challenges reported by older adults related to IADLs included housekeeping, meal preparation, and home maintenance.

Enhanced activities of daily living (EADLs) refer to the ability to engage and participate in social and enriching activities. EADLs include activities such as learning new skills (e.g., play the violin, photography); volunteering and mentoring; or engaging in hobbies (e.g., gardening, music, drawing).[21,22] Supporting older adults' ability to engage in enhanced activities of daily living can help promote safety and maintain meaningful participation in activities.

11.6 UNDERSTANDING SAFETY ISSUES IN THE HOME

As individuals enter older adulthood, changes in lifestyle and individual factors such as health and functional abilities can alter the suitability of their current home vis-à-vis a safe and functional living environment. The changes individuals experience across the aging process vary, and every individual experiences age-related changes in a unique manner. However, changes in one's functional ability can be progressive or sudden. For example, a fall that results in a broken hip may completely alter how an individual is able to complete ADLs, IADLs, and EADLs. The home environment needs to be adaptable to the needs of the individual throughout the aging in place process, regardless of the changes they may experience.

As professionals working with older adults, we must be able to identify and help support the individuals we serve to ensure safety in the home. Educating the individuals we serve, and their families, about safety issues in the home is an important step towards developing preventative strategies and enhancing one's ability to age in place successfully. Here we outline five general areas for consideration: minimizing risks, maximizing safety, security, comfort, and health.

Minimizing Risks. Minimizing risks in a home environment includes:

1. Falls – reducing tripping hazards, enhancing bathroom safety and walking surfaces
2. Environment – preventing fires and carbon monoxide poisoning
3. Health – enhancing medication adherence and exercise
4. Security – preventing robberies, abuse, fraud
5. Products – providing information about recalls, preventing equipment failure, and misapplication

Maximizing Safety. Maximizing home safety includes:

1. Planning – developing a personalized plan to meet individual wants and needs
2. Resources – connecting with support agencies, identifying potential sources for financial support, and identifying tools to support aging in place

3. Education – informing about home risks (e.g., falls, emergencies) and translating safety strategies to enhance accessibility
4. Support – personalizing support depending on the individual's plan, assisting in accessing tools and resources

Security. Enhancing security includes:

1. Personal security – identifying and addressing concerns of individuals living alone, preventing crime and abuse
2. Home security – developing strategies for the prevention of home break-ins, burglaries, and robberies that place both the individual and their valuable possessions at great risk

Comfort. Supporting home comfort includes:

1. Temperature control – enhancing control over home temperature to address safety concerns associated with age-related changes, chronic disease (e.g., thyroid disease, diabetes), and medications that can impact individuals' ability to maintain a healthy body temperature
2. Air quality control – supporting air quality control in the home to create a comfortable home environment, removing allergens and pollutants, and reducing irritation, asthma attacks, and respiratory illnesses

Health. Supporting individual health includes:

1. Support – identifying health resources, providing support for underlying conditions and potential sudden changes in one's health, and supporting recovery
2. Planning – developing plans to prevent or manage health condition, identifying resources to support for changes in one's health, planning home environment to enable healthy aging

11.6.1 POTENTIAL FOR TECHNOLOGY

Technology advancements are changing the way individuals participate in day-to-day life and carry out daily routines. As a result, new opportunities to access resources and tools have emerged that can help promote safety and enable individuals to age in place. Research indicates that technologies have been particularly useful in adapting and maintaining an individual's support with ADLs, IADLs, and EADLs.[6,21,23] For example, technologies such as digital personal assistant, wearables, robots, and telepresence may provide individuals with support in remaining active, completing home-related chores, participating in hobbies, and staying socially engaged with family and their community.

In recent years, adoption rates of technologies among the older adult population have grown exponentially. For example, in 2000, 12% of older adults aged 65 and over were using the internet, whereas in 2016 approximately 67% of older adults reported using the internet.[24] However, the adoption rate of technology by older adults remains significantly lower than that of the general adult population.[24,25] This gap in adoption is often attributed to a lack of access, described as the digital divide, which refers to the growing gap in access experienced by vulnerable and underprivileged members of society. For example, older adults who live in rural communities or have a lower socioeconomic status have less access to technological resources.

The goal of this chapter is to provide an overview of technologies and how they relate to maintaining safety and supporting the home activities needs of older adults. We identify various classes of technologies and their role in fostering a safe home environment. Furthermore, we provide considerations for professionals working with older adults on the adoption and implementation of technologies to promote safety and support the process of aging in place.

11.6.2   CRAFTING A SMART HOME ENVIRONMENT

The multitude of technologies embedded in the home environment can be thought of as an ecosystem to meet the diverse set of needs of the individual. The ecosystem is made possible through the integration of devices, which is also known as the internet of things (IoT).[23,26] The IoT is a network of connected devices that communicate via the internet.[23,26] The communication across devices includes home, personal, and healthcare devices and can enable improved autonomy and safety. For example, using home sensors that can detect falls and personal devices that can detect sudden changes in heart rate could help reduce the time required for individuals to get support and assistance. The IoT can also provide opportunities to enable individuals to remain connected within the community they are choosing to age in place.[27] For example, the use of video conferencing systems available in their home may allow older adults to participate in community events, catch up with friends, or have dinner with their family. The selection and implementation of technologies within the IoT should be aimed to directly meet the needs of the individual. Furthermore, the technologies implemented in the home should be adaptive to meet the needs of the individual across various functionality levels.[23,27] This is particularly important because introducing new technologies to meet changes in ones' ability may create new challenges as the individual may have to learn how to use a new technology.

The home ecosystem may include technologies such as home security, fall detection devices, health technologies, home comfort technologies, smart appliances, personal emergency response systems (PERS), and robots.[28] We review each category with a focus on their potential to support safety and aging in place for older adults.

11.7   HOME SECURITY

Home security entails a set of technologies that are aimed at enhancing the living environment by promoting safety and security.[29] Older adults often express concerns regarding medical emergencies, environmental disasters (e.g., burst pipes, fires), and predators (e.g., burglars, home intruders). To help address these concerns, developers and designers have created a diverse set of technologies such as video doorbells, sensors (e.g., smoke, carbon monoxide, motion detectors), cameras (e.g., home monitors), and lights (e.g., wireless lights) to improve resources and service access. Access to home security technologies can provide the users with peace of mind regarding both the individual's safety and the protection of their home. Technologies such as video doorbells, smart door locks, surveillance cameras, and motion sensors can enable the older adult to see who is at their front door without having to open the door. An older adult with limited mobility can use these technologies to allow a visiting nurse to enter their home without having to go to the front door.

Other applications for home security technologies such as motion sensors and wireless lighting are used to improve in-home safety of the user. For example, wireless lights that are activated through either a voice command or motion can increase visibility when someone is going up or down a flight of stairs. Unlike traditional home security systems that can be costly and require a recurring fee, home security technologies function through internet connectivity. These devices are often controlled through a hub (e.g., Google Home, Amazon Echo), which enables the user to access them though a centralized control point.

11.8   FALL DETECTION

Fall detection devices are designed to help maximize safety and support older adults' ability to age in place. These devices are typically described as medical alert systems and often take the shape of wearables (e.g., necklaces, bracelet, watches).[30] Wearables are typically equipped with a button or set of buttons that allow the user to get immediate assistance through a medical alert operator. These devices provide connectivity to medical operators who have access to the user's emergency contacts and are able to get the right assistance for the user in case of a fall. The user can push the button located on the wearable device and get in touch with a medical operator. The operator can dispatch emergency professionals such as paramedics, firefighters, or police officers to the residence of the older adult. Some medical alert systems can also advance features that can detect falls automatically. These devices can initiate the call with the medical operator, without any input from the user. However, features such as automatic fall detection can require additional monthly subscriptions.

11.9   HEALTH TECHNOLOGIES

Health technologies are an expanding field that includes medical devices, services, procedures, and intervention delivery systems designed to support and deliver healthcare. Some functionalities include assistance with health monitoring, health tracking, reminders, and remote service delivery (e.g., telehealth). This class of technologies promotes safety and security by leveraging other technologies to support planning and delivery of health interventions.[31] For example, a rural dwelling older adult who is experiencing heart disease can get consultations through telepresence from a cardiologist located in a different city, state, or country. Furthermore, health technologies located in the home such as health monitoring devices (e.g., smart watches, wireless blood pressure machine) can give the healthcare provider crucial information that would be collected during an in-person consultation. Health technologies can also be used to remind individuals to take their medication, engage in exercise, and track routine changes (e.g., walking, nutrition) that could be impacting health outcomes.

11.10   COMFORT TECHNOLOGIES

Comfort technologies include any devices or systems that support temperature control or air quality in the home. Comfort technologies include devices such as smart thermostats, sensors (e.g., air quality sensors), and smart air purifiers.[32,33] These technologies can promote safety by improving health outcomes and control over the home environment. For example, installing a smart thermostat can detect when someone is home, learn patterns in temperature preferences, make temperature adjustment based on real time weather patterns, and reduce energy consumption. Use of comfort technologies such as smart air purifiers can monitor air quality, enhance air quality, reduce allergens and odors, and adjust operation based on the user preferences.

11.11   SMART APPLIANCES

Smart appliances are devices that incorporate connectivity to a network to provide enhanced functionalities that promote safety, autonomy, and the ability to age in place.[34] Smart appliances include devices such as refrigerators, washers, dryers, dishwashers, ovens, ranges, microwave ovens, and toasters. Unlike traditional appliances, smart appliances can be controlled remotely through a centralized hub (e.g., Google Home, Amazon Echo) and enable to user to control the functionalities of the device through voice. For example, the user may be able to preset the oven via a device app where the set temperature is indicated without ever having to touch the physical device. Other smart appliances, such as smart refrigerators, can track the groceries currently available, groceries to be ordered, have built-in displays, and track expiration dates of the foods located inside. Integration and accessibility to smart appliances can support safety and enhance control over a home environment.

11.12   PERSONAL DEVICES

Personal devices are technologies that can be worn by older adults to enhance safety by providing fast access to assistance in case of an emergency and assistance with day-to-day activities (e.g., calls, health tracking).[30] These devices can be carried by the person or take on a more familiar shapes such as watches, bracelets, necklaces, and rings. The functionalities of personal devices can range widely. For example, a smart watch may provide the individual with functionalities such as texting, calling, notification, heart rate tracking, and fitness tracking. However, devices such as PERS are primarily designed to provide immediate support in case of emergency. These devices enable the user to have immediate access to emergency support by pressing a button located on the device.

11.13   ROBOTS

Robots are a diverse and evolving class of technologies that are being designed to promote safety, support ADLs, IADLs, and EADLs, and enable aging in place.[35]

Robots are broadly defined as "an autonomous system that exists in the physical world, can sense its environment, and can act on it to achieve some goals."[36] The most common types include social robots, assistive robots, socially assistive robots, and telepresence robots.[8] Robots can provide social engagement (e.g., conversation, reminders), assist with household chores (e.g., vacuuming), connect (e.g., video calls, remote visits through telepresence), and enable mobility and assistance (e.g., feeding, locomotion). Using a robotic vacuum can improve safety by reducing the need to pick up and move around heavy and unwieldy hand-held vacuums. Furthermore, socially assistive robots can engage individuals in social activities, provide reminders about medication and doctors' appointments, and serve as companions.

11.13.1 TECHNOLOGY CONSIDERATIONS

Across the classes of technologies identified, there are themes that should be considered when reviewing specific products for inclusion in a particular older adult's home. They include:

11.13.1.1 Accessibility
- Does the device take into consideration accessibility challenges (e.g., cost, maintenance, individual needs)?
- What are the accessibility challenges and benefits of the device for short-term and long-term ownership?

11.13.1.2 Accommodations
- What are the accommodations needed to support adoption of the device (e.g., internet access, home modifications)?
- How do these accommodations impact the individual ability to use the technology?

11.13.1.3 Compatibility
- What are the compatibility requirements of the technology of interest (e.g., adaptors, software updates, network)?
- Is the technology of interest compatible with other technologies previously adopted (e.g., smart appliances, home hub)?

11.13.1.4 Features and Functionalities
- What are the features (e.g., weight, range, battery life) and functionalities (e.g., ADLs, IADLs, EADLs) provided by the device?
- Is the device able to expand its features and functionalities (e.g., modularity, connectivity, updates for hardware and software)?

11.13.1.5 Hazards
- Are there any potential hazards associated with adopting or using the technology (e.g., privacy, safety, tripping)?
- Does the technology have features that help prevent or mitigate the hazards (e.g., sensors, motion detection, mobile alerts, self-docking, password protections)?

11.13.1.6 Interface

- How will the user interact with the technology (e.g., phone application, device display, voice activation)?
- How does the technology interface support diversity of the user preferences and needs?

11.13.1.7 Unintended Consequences

- Will the technology discourage older adults from remaining physically active (i.e., become overreliant on technology tools)?
- How will the technology influence social connection (e.g., reduction in person visits)?

11.13.2 ENABLING OLDER ADULTS TO TAKE ADVANTAGE OF THESE TECHNOLOGY INNOVATIONS

Functional abilities of the older adults can shift due to age-related challenges such as perception, motor, and cognitive decline. To enable older adults to take advantage of technology, professionals serving older adults should take into account the following personal considerations.

11.13.2.1 Customizability

As noted earlier in this chapter, the experiences and needs of individual older adults vary. Therefore, the technologies that individuals adopt need to be customizable to meet their diverse set of needs. For example, interfaces that are customizable can enable individuals to meet their functional abilities and individual preferences. Furthermore, customizability can also entail the ability to select and disable functionalities that are of interest to the individual.[37,38] For example, the ability to disable an idle camera or turn off location tracking are among just a few of the customization features that can enhance user control and autonomy.

11.13.2.2 Adaptability

When considering the adoption of technology, it is important to consider the adaptability of the devices to meet the diverse set of user needs and functional abilities across the aging process.[39,40] For example, the experience of a stroke could affect an individual's functional ability and require that technology adapt to meet a new set of user needs to maintain safety and well-being. The capability to adapt is particularly important as individuals develop comfort and familiarity with the technology. For example, devices that provide a diverse set of user interfaces (e.g., voice user interface, menu driven user interface, touch user interface) can help reduce the need to adopt new technologies. Having to introduce new technologies during pivotal moments of need can influence adoption and utilization of the technology.

11.13.2.3 Maintenance

Taking into consideration the maintenance requirements of technologies can play an important role in how older adults adopt and use them long term. High maintenance requirements (e.g., poor customer support, manual updates) can result in decreased use over time and frustration with the device.[41,42] As professionals serving older adults, we must disclose maintenance requirement of technologies and ensure that older adults are informed when choosing the technologies they want to adopt.

11.13.2.4 Privacy and Security

Privacy and security of the individual is an essential part of being able to support one's safety and autonomy. When thinking about the implementation of technologies within a home environment, older adults should have full control and autonomy over the devices they are adopting.[43] Implementing external controls for turning off sensors or cameras and control over who can access personal data (e.g., heart rate, location in the home) are essential components in promoting the user's security, privacy, and independence.[44-46] As professionals, we must understand the challenges experienced by the older adults and think critically about the technologies we can leverage to support their needs.

11.13.2.5 Trust

Developing trust in a device is an important part in helping individual to adopt and use a device long term. Helping individuals develop trust in technologies they adopt requires transparency regarding the functionalities of the technologies, clear outline of their benefits and concerns, and formation of realistic expectations of technology capabilities.[30,47] Our role as individuals serving older adults is to help educate and provide them with the resources and knowledge so they may make an informed decision based on what they think would be best for their circumstances.

11.13.2.6 Technology Acceptance Models

Now that we have identified the personal consideration in enabling older adults to take advantage of technologies to support safety and enable aging in place, we will explore the factors contributing to the adoption of technologies. Adoption of technologies can look quite different across individuals. However, identifying commonalities enables us to understand better in how we should present resources and become their advocates.

Technology acceptance models have been developed to identify and organize the factors contributing to their adoption. The two most used models include the Technology Acceptance Model (TAM) and the Unified Theory of Acceptance and Use of Technology (UTAUT).[42,48,49] These frameworks indicate technology adoption can be based on perceived ease of use, perceived usefulness, perceived enjoyment, intent to adopt, performance expectancy, effort expectancy, social influences, and facilitating conditions.[50,51] However, across theoretical frameworks, prominent factors contributing to the adoption of technology by older adults are perceived ease of use, perceived usefulness, perceived enjoyment, and facilitating conditions.

1. **Perceived ease of use.** This factor is defined as the degree to which a piece of technology is free of effort to be used by an individual.[50] Consequently, professionals working with older adults would want to consider if the

technology would be easy to use in the day-to-day life of that individual. In that context, they might think about what makes the technology easy to use (or not) and how they might support the person's use.

2. **Perceived usefulness.** This factor refers to the degree to which a piece of technology could help improve an individual ability to carry out or complete a task.[50] Professionals working with older adults should identify if the technology being considered for adoption could be useful in helping or enhancing the ability to complete tasks such as ADLs, IADLs, or EADLs. In this context, professionals may consider what makes the technology useful (or not) in helping the individual complete a task.

3. **Perceived enjoyment.** This factor is defined as the degree to which the process of using a piece of technology is enjoyable in its own right, separately from the perceived benefits of the technology.[52] Professionals may want to consider if older adults feel like using the technology is fun, pleasant, and enjoyable. In this frame of reference, they might reflect on what makes the technology enjoyable (or not) to use.

4. **Facilitating conditions.** An important aspect of the technology adoption process is consideration of the facilitating conditions. Facilitating conditions refer to degree to which older adults feel like they meet conditions such as skills, resources, and knowledge to be able to successfully use the technology that they intent to adopt.[42,53] This presents professionals serving older adults with a tremendous opportunity to help bridge the gap and meet the needs for adoption. For example, providing older adults with instructional support (e.g., developing educational materials) and social support (e.g., listening to their needs, supporting them in the decision to adopt a technology) can play an important part in successful adoption of technology. As professionals, we have the obligation to inform and provide individuals with available options and resources and support the decision they make regarding the adoption of technologies.

The uptake and implementation of technologies by older adults is an important step in supporting the process of aging in place.[54,55] Awareness of the factors contributing to adoption can enable professionals to leveraging available technology resources to support their needs.

11.13.3 RECOMMENDATIONS FOR PROFESSIONALS

As helping professionals (e.g., gerontologists, nurses, social workers) working with older adults, it is essential to understand the role of technology to promote safety and support aging place. Technology is constantly evolving. This means that as professionals, we must engage lifelong learning and seek knowledge and understanding of existing and emerging technologies. Connecting and educating older adults on the use and application of technologies is an important step toward leveraging it to promote safety and support aging in place. Furthermore, we must serve as advocates on the behalf of the older adults we serve. For example, facilitating communication with designers, developers, and researchers to ensure that the needs of older adults are addressed in the design and development of technologies. Increasing the involvement of older adults in the design and development process can help enhance designers' and developers' understanding of their needs.

When thinking about technology and its role in supporting aging in place, there are five key areas of consideration, namely, selection, installation, training, maintenance, and support for continued use. To help professionals plan and implement technologies to promote safety and support aging in place, we provide a structured tool to guide the process of implementing technologies (see Table 11.1). This tool outlines areas for consideration and helpful questions and tips to guide discussion and consideration of technology application with older adults.

TABLE 11.1

Tool to Guide Technology Implementation to Promote Safety and Support Aging in Place

Area	Considerations

Selection	• **Involve older adults in the technology selection process.** Including the client in the selection of the technology can help them feel empowered and enable the professional to better understand the individual needs.
	• **Important to consider**
	  • What are the main functions of the technology?
	  • Do the functions of the technology meet the needs of the individual who is going to use it?
	• This should be an open discussion regarding how the technology fits the current needs of the individual and what is its role in the plan to age in place.
	  • Benefits, concerns, and challenges of the technology.
	• Create a list that helps identify client's perceptions and thoughts of the technology.

Installation	• **Requirements.** Once a technology is selected, it is important to outline how the technology is going to be implemented and used in the home environment of the older adult.
	• **Important to consider**
	  • Who is going to be installing the technology?
	  • Where in the home will the technology be installed?
	• It is important the client has full autonomy in making the decision where the technology will be located. Consider how the location of the technology may impact their safety (e.g., tripping hazard).
	  • What settings and features should the technology have enabled?
	• This should be an open discussion with your client. The client should be made fully aware of how the functions and features are going to impact their safety, security, and privacy.

(continued)

TABLE 11.1 (Continued)
Tool to Guide Technology Implementation to Promote Safety and Support Aging in Place

Area	Considerations

Training	• **Developing the training guide.** Once the installation of the selected technology takes place, develop a training guide that provides step-by-step directions on how the technology can be used. To support this process, available online training can be used as a foundation. However, materials should be adapted to the needs and capabilities of the individuals being served.
	• **Important to consider**
	  • Is the training guide easy to follow and understand?
	  • Does the guide provide the older adults with easy steps in troubleshooting common challenges?
	• Reconnecting to the internet, enabling and disabling features.
	• Does the guide provide steps for access to additional assistance?
	• Customer service phone number, technology support

Maintenance	• **Maintaining the technology.** Many of the market-ready technologies available require consistent updates to maintain full functionality. The client should be made aware of how these updates are carried and potential risk of not completing an update (e.g., security breach, loss in functionality).
	• **Important to consider**
	  • What are the maintenance requirements of the technology?
	  • Are the technology updates conducted manually or automated?
	  • Who can assist with the maintenance needs of the technology?

Supporting continued use	• **Supporting continued use.** Successful implementation and application of the technology is dependent on continued use to serve the needs of the individual. The technology selected should adapt to the needs of the user and the home environment.
	• **Important to consider**
	  • How does the technology fit in the plan of aging in place?
	  • Will the technology still serve a purpose if there's a change in the individual's functional ability?
	  • How does the technology integrate with both home and personal devices?

11.14 SUMMARY

Promoting the autonomy and independence of the older adults is an important step in helping support quality of life and healthy aging. However, challenges (e.g., individual, environmental) and resource gaps present threats to individuals' safety and ability to age in place successfully. As professionals, we must understand challenges experienced in the home (e.g., falls, health, security, comfort, health) and identify helpful resources to enhance safety and meet the needs of the older adults we serve.

Advancements in technology have presented us with a wide spectrum of resources capable of enhancing safety and one's ability to age in place successfully.

In this chapter we identified classes of technologies that can help create a smart home environment to promote safety in the home. We organized these technologies in the following categories: home security, fall detection, health technologies, comfort technologies, smart appliances, personal devices, robots, and IoT. Clearly there is much potential for advanced technologies to support older adults in their homes. However, as professionals we must consider the personal considerations of the individuals we serve and adoptions frameworks informing the adoption of technologies. We identified five specific areas for technology adoption considerations (customizability, adaptability, maintenance, privacy/security, trust) and outlined the leading frameworks for technology acceptance (TAM, UTAUT). Finally, we provided a tool to help identify areas for consideration along with helpful questions and tips to guide discussion and consideration of technology application with older adults.

ACKNOWLEDGMENTS

The authors were supported in part by the National Institute on Aging (National Institutes of Health) Grant R44 AG059450. We appreciate the insights of Bob Wolf during the development of the chapter.

REFERENCES

1. Blanchard J. Aging in community: communitarian alternative to aging in place, alone. *Generations*. 2013;37(4):6–13.
2. Ewen HH, Lewis DC, Carswell AT, Emerson KG, Washington TR, Smith ML. A Model for Aging in Place in Apartment Communities. *J Hous Elderly*. 2017;31(1):1–13.
3. Arthanat S, Vroman KG, Lysack C. A home-based individualized information communication technology training program for older adults: a demonstration of effectiveness and value. *Disability and Rehabilitation-Assistive Technology*. 2016;11(4):316–324. doi:10.3109/17483107.2014.974219
4. Campos Antunes TP, Bulle de Oliveira AS, Hudec R, et al. Assistive technology for communication of older adults: a systematic review. *Aging Ment Health*. 2019;23(4):417–427. doi:10.1080/13607863.2018.1426718
5. Rogers WA, Ramadhani WA, Harris MT. Defining aging in place: the intersectionality of space, person, and time. *Innov Aging*. 2020;4(4):igaa036. doi:10.1093/geroni/igaa036
6. Kim K, Gollamudi SS, Steinhubl S. Digital technology to enable aging in place. *Exp Gerontol*. 2017;88:25–31. doi:10.1016/j.exger.2016.11.013
7. Satariano WA, Scharlach AE, Lindeman D. Aging, place, and technology: toward improving access and wellness in older populations. *J Aging Health*. 2014;26(8):1373–1389. doi:10.1177/0898264314543470
8. Mois G, Beer JM. Robotics to support aging in place. In: *Living with robots: emerging issues on the psychological and social implications of robotics*. Academic Press; 2019.
9. Ahn M, Kwon HJ, Kang J. Supporting aging-in-place well: findings from a cluster analysis of the reasons for aging-in-place and perceptions of well-being. *J Appl Gerontol*. 2020;39(1):3–15. doi:10.1177/0733464817748779

10. Black K. Health and aging-in-place: implications for community practice. *J Comm Pract*. 2008;16(1):79–95. doi:10.1080/10705420801978013

11. Fausset CB, Kelly AJ, Rogers WA, Fisk AD. Challenges to aging in place: understanding home maintenance difficulties. *J Hous Elderly*. 2011;25(2):125–141. doi:10.1080/02763893.2011.571105

12. Iecovich E. Aging in place: from theory to practice. Anthropological Notebooks. 2014;20(1):21–32..

13. Fernández-Mayoralas G, Rojo-Pérez F, Martínez-Martín P, et al. Active ageing and quality of life: factors associated with participation in leisure activities among institutionalized older adults, with and without dementia. *Aging Ment Health*. 2015;19(11):1031–1041. doi:10.1080/13607863.2014.996734

14. Kelly ME, Duff H, Kelly S, et al. The impact of social activities, social networks, social support and social relationships on the cognitive functioning of healthy older adults: a systematic review. *Systematic Reviews*. 2017;6(1):259. doi:10.1186/s13643-017-0632-2

15. Seplaki CL, Agree EM, Weiss CO, Szanton SL, Bandeen-Roche K, Fried LP. Assistive devices in context: cross-sectional association between challenges in the home environment and use of assistive devices for mobility. *Gerontologist*. 2014;54(4):651–660. doi:10.1093/geront/gnt030

16. Crose R, Leventhal EA, Haug MR, Burns EA. The challenges of aging. In: *Health care for women: psychological, social, and behavioral influences*. American Psychological Association; 1997:221–234. doi:10.1037/10235-014

17. De L d'Alessandro E, Bonacci S, Giraldi G. Aging populations: the health and quality of life of the elderly. *Clin Ter*. 2011;162(1):e13–8.

18. Kerschner H, Pegues JAM. Productive aging: a quality of life agenda. *J Am Diet Assoc*. 1998;98(12):1445–1448. doi:10.1016/S0002-8223(98)00327-7

19. Lawton MP, Brody EM. Assessment of older people: self-maintaining and instrumental activities of daily living. *Gerontologist*. 1969;9(3):179–186.

20. Lawton MP. Aging and performance of home tasks. *Hum Factors*. 1990;32(5):527–536. doi:10.1177/001872089003200503

21. Rogers WA, Mitzner TL, Bixter MT. Understanding the potential of technology to support enhanced activities of daily living (EADLs). *Gerontechnology*. 2020;19(2):125–137. doi:10.4017/gt.2020.19.2.005.00

22. Rogers WA, Meyer B, Walker N, Fisk AD. Functional limitations to daily living tasks in the aged: a focus group analysis. *Hum Factors*. 1998;40(1):111–125. doi:10.1518/001872098779480613

23. Choi YK, Thompson HJ, Demiris G. Internet-of-things smart home technology to support aging-in-place: older adults' perceptions and attitudes. *J Gerontol Nurs*. 2021;47(4):15–21. doi:10.3928/00989134-20210310-03

24. Anderson M, Perrin A, Jiang J, Kumar M. 10% of Americans don't use the internet. Who are they? Pew Research Center. Published 2019. Accessed April 6, 2020. www.pewresearch.org/fact-tank/2019/04/22/some-americans-dont-use-the-internet-who-are-they/.

25. Heo J, Kim J, Won YS. Exploring the relationship between internet use and leisure satisfaction among older adults. *Act, Adapt Aging*. 2011;35(1):43–54.

26. Carnemolla P. Ageing in place and the internet of things – how smart home technologies, the built environment and caregiving intersect. *Vis Eng*. 2018;6(1):7. doi:10.1186/s40327-018-0066-5

27. Tural E, Lu D, Austin Cole D. Safely and Actively Aging in Place: Older Adults' Attitudes and Intentions Toward Smart Home Technologies. *Gerontol Geriat Med.* 2021;7:23337214211017340. doi:10.1177/23337214211017340

28. Fattah SMM, Sung NM, Ahn IY, Ryu M, Yun J. Building IoT services for aging in place using standard-based IoT platforms and heterogeneous IoT products. *Sensors (Basel).* 2017;17(10):2311. doi:10.3390/s17102311

29. Rantz MJ, Porter RT, Cheshier D, et al. TigerPlace, a state-academic-private project to revolutionize traditional long-term care. *J Hous Elderly.* 2008;22(1–2):66–85. doi:10.1080/02763890802097045

30. Rupp MA, Michaelis JR, McConnell DS, Smither JA. The role of individual differences on perceptions of wearable fitness device trust, usability, and motivational impact. *Appl Ergo.* 2018;70:77–87. doi:10.1016/j.apergo.2018.02.005

31. Mois G, Fortuna KL. Visioning the future of gerontological digital social work. *J Gerontol Soc Work.* 2020;63(5):1–16. doi:10.1080/01634372.2020.1772436

32. Ghahramani A, Galicia P, Lehrer D, Varghese Z, Wang Z, Pandit Y. Artificial intelligence for efficient thermal comfort systems: requirements, current applications and future directions. *Front Built Environ.* 2020;6:1–16. Accessed January 14, 2022. www.frontiersin.org/article/10.3389/fbuil.2020.00049

33. Cheng CC, Lee D. Enabling smart air conditioning by sensor development: a review. *Sensors (Basel).* 2016;16(12):2028. doi:10.3390/s16122028

34. Silverio-Fernández M, Renukappa S, Suresh S. What is a smart device? – a conceptualisation within the paradigm of the internet of things. *Vis Eng.* 2018;6(1):3. doi:10.1186/s40327-018-0063-8

35. Beer JM, Mitzner TL, Stuck RE, Rogers WA. Design considerations for technology interventions to support social and physical wellness for older adults with disability. 2015;5:249–264. doi:10.5875/ausmt.v5i4.959

36. Mataric MJ. *The Robotics Primer.* MIT Press; 2007.

37. Gardan J. Definition of users' requirements in the customized product design through a user-centered translation method. *IJIDeM.* 2017;11(4):813–821. doi:10.1007/s12008-015-0275-2

38. Herrero RP, Fentanes JP, Hanheide M. Getting to know your robot customers: automated analysis of user identity and demographics for robots in the wild. *IEEE Robot Autom Lett.* 2018;3(4):3733–3740. doi:10.1109/LRA.2018.2856264

39. Shore L, Power V, De Eyto A, O'Sullivan LW. Technology acceptance and user-centred design of assistive exoskeletons for older adults: a commentary. *Robot.* 2018;7(1):3. doi:10.3390/robotics7010003

40. Wu YH, Damnée S, Kerhervé H, Ware C, Rigaud AS. Bridging the digital divide in older adults: a study from an initiative to inform older adults about new technologies. *Clin Interv Aging.* 2015;10:193–201. doi:10.2147/CIA.S72399

41. Czaja SJ, Lee CC, Nair SN, Sharit J. Older adults and technology adoption. *Proceedings of the Human Factors and Ergonomics Society Annual Meeting.* 2008;52(2):139–143. doi:10.1177/154193120805200201

42. Magsamen-Conrad K, Upadhyaya S, Joa CY, Dowd J. Bridging the divide: using UTAUT to predict multigenerational tablet adoption practices. *Comput Human Behav.* 2015;50:186–196. doi:10.1016/j.chb.2015.03.032

43. Harrefors C, Axelsson K, Sävenstedt S. Using assistive technology services at differing levels of care: healthy older couples' perceptions. *J Adv Nurs.* 2010;66(7):1523–1532. doi:10.1111/j.1365-2648.2010.05335.x

44. Ting TC. Privacy and confidentiality in healthcare delivery information system. In: *Proceedings 12th IEEE Symposium on Computer-Based Medical Systems (Cat. No.99CB36365)*; 1999:2–4. doi:10.1109/CBMS.1999.781239

45. Niemela M, van Aerschot L, Tammela A, Aaltonen I. A telepresence robot in residential care: family increasingly present, personnel worried about privacy. In: Kheddar A, Yoshida E, Ge SS, et al., eds. *Social Robotics, Icsr 2017*. Vol 10652; 2017:85–94. doi:10.1007/978-3-319-70022-9_9

46. Davis BH, Shehab M, Shenk D, Nies M. E-mobile pilot for community-based dementia caregivers identifies desire for security. *Gerontechnology*. 2015;13(3):332–336. doi:10.4017/gt.2015.13.3.003.00

47. Pitardi V, Marriott HR. Alexa, she's not human but … unveiling the drivers of consumers' trust in voice-based artificial intelligence. *Psychology & Marketing*. 2021;38(4):626–642.

48. Chao CM. Factors determining the behavioral intention to use mobile learning: an application and extension of the UTAUT model. *Front Psychol*. 2019;10. Accessed January 13, 2022. www.frontiersin.org/article/10.3389/fpsyg.2019.01652

49. Nistor N, Lerche T, Weinberger A, Ceobanu C, Heymann O. Towards the integration of culture into the unified theory of acceptance and use of technology. *Br J Educ Technol*. 2014;45(1):36–55. doi:10.1111/j.1467-8535.2012.01383.x

50. Davis FD. Perceived usefulness, perceived ease of use, and user acceptance of information technology. *MIS Q*. 1989;13(3):319–340. doi:10.2307/249008

51. Venkatesh V, Morris MG, Davis GB, Davis FD. User acceptance of information technology: toward a unified view. *MIS Q*. 2003;27(3):425–478. doi:10.2307/30036540

52. Davis FD, Bagozzi RP, Warshaw PR. Extrinsic and intrinsic motivation to use computers in the workplace. *J Appl Soc Psychol*. 1992;22(14):1111–1132. doi:10.1111/j.1559-1816.1992.tb00945.x

53. Dai B, Larnyo E, Tetteh EA, Aboagye AK, Musah AAI. Factors affecting caregivers' acceptance of the use of wearable devices by patients with dementia: an extension of the unified theory of acceptance and use of technology model. *Am J Alzheimers Dis Other Demen*. 2020;35:1533317519883493. doi:10.1177/1533317519883493

54. Peek STM, Wouters EJM, Luijkx KG, Vrijhoef HJM. What it takes to successfully implement technology for aging in place: focus groups with stakeholders. *J Med Internet Res*. 2016;18(5):e98. doi:10.2196/jmir.5253

55. Rantz M, Skubic M, Miller S, et al. Sensor technology to support aging in place. *J Am Med Dir Assoc*. 2013;14. doi:10.1016/j.jamda.2013.02.018

12 Supporting Community Engagement with Assistive Technology as a Means of Health Promotion for Older Adults

Anna Y. Grasso, OTD, MS, OTR/L, CAPS, ECHM and Beth E. Davidoff

CONTENTS

12.1 INTRODUCTION

As individuals age, changes in motor, cognitive, and sensory capabilities occur that put older adults at risk of functional impairment in all aspects of daily living. Routine activities such as remembering to take medications, bathing, dressing, preparing and eating meals, walking, driving, making phone calls, reading, writing, and engaging in social and leisure activities may be affected. These limitations can be few or many, yet any such impairment can potentially reduce the person's quality of life.

Age-related declines in functioning can negatively impact community engagement and lead to safety concerns. Adding a medical condition or disability may create obstacles to leaving home that seem insurmountable. Older adults may have one or more chronic health conditions that can affect participation in daily life activities. Chronic illnesses are defined as "conditions that last a year or more and require ongoing medical attention and/or limit activities of daily living."[1,2] These can include medical conditions such as arthritis, asthma, hypertension, diabetes, as well as aphasia (a language impairment due to stroke), dementia, or other cognitive impairments. Multimorbidity (i.e., having multiple chronic health conditions) increases with age and more than 78% of adults 65–74 years of age and more than 87% of adults 75 years and older have ≥ 2 chronic conditions.[3] Impairments may include motor limitations, sensory limitations involving vision and/or hearing, communication limitations, and cognitive/memory limitations. These impairments can impact the older adult's ability to safely and independently perform activities of daily living (ADLs), such as bathing, toileting, walking, eating, and dressing.[4] They may also impact how frequently an older adult leaves their bedroom, home, neighborhood, and community, which is referred to as life space constriction.[4]

Owing to retirement from the workforce and/or the loss of a spouse or close friends, older adults may find that their social networks become smaller over time. When this is coupled with the presence of a disability or one or more chronic health conditions, social isolation can ensue. Getting out of the home and into the community is associated with greater social engagement and physical activity, both important determinants of health.[5,6] A 2017 study by AARP found an estimated $6.7 billion in additional Medicare spending for socially isolated individuals, as well as increased mortality rates compared to their socially well-connected counterparts.[7] Please refer to Chapter 19 in this book ("Social Isolation and Loneliness") for a comprehensive view on the impact of social isolation on older adults and the role of healthcare providers in addressing this challenge.

Awareness and utilization of healthcare professionals, assistive technologies, and public policies and services may help bridge the gap between an individual's current motor, cognitive, or sensory capacities and the demands of the community spaces where meaningful social engagement can take place.

12.2 BARRIERS TO COMMUNITY ENGAGEMENT FOR AGING ADULTS

12.2.1 Self-Care Challenges

For older adults with chronic health conditions and disabilities, each element of their self-care routine may require significant time, effort, or external support from a caregiver. An estimated 20–25% of adults over age 65 have difficulty with ADLs including bathing, dressing, toileting, and/or grooming.[8] If the perceived or actual effort it takes to perform ADLs is too great, or the financial and emotional costs of formal or informal caregiving support are too high, staying home may be the default solution.

12.2.2 COGNITIVE CHALLENGES

As a result of normal aging, older adults may experience subtle changes in certain aspects of cognition, while some aspects remain relatively constant or may increase. *Crystalized intelligence* refers to overlearned and rehearsed information, such as vocabulary and general knowledge that an individual accumulates over their lifetime. These abilities do not decline with age. Conversely, *fluid intelligence*, which refers to an individual's ability to process information, attend selectively to stimuli, problem solve, and acquire new information, does decline with age.[9] Research has indicated that "successful cognitive aging" may be related to older adults engaging in certain activities that are intellectually stimulating and involve physical exercise and social engagement. [10,11] Other adults may have chronic medical conditions such as dementia (including Alzheimer's disease and Parkinson's disease) as well as traumatic brain injury and aphasia, which result in more substantial changes in cognitive function.

12.2.3 VISION CHALLENGES

Of adults aged 65–74, 12.2% report some degree of vision loss, which increases to 15.2% for adults age 75 and older.[12] Vision loss can impact a variety of community mobility elements including driving, shopping, reading, and recognizing environmental obstacles. One study showed that even with relatively mild visual acuity issues, older adults had statistically significant decreases in community participation.[13]

12.2.4 COMMUNICATION CHALLENGES WITH HEARING AND SPEAKING

As adults age, changes can occur in how they hear and understand others as well as how others hear and understand them. Twenty-nine million Americans experience hearing loss and it increases with age, with 77% of adults 60–69 years of age reporting hearing impairment.[14] In the normal course of aging, healthy older adults undergo a normal aging process that affects the pitch, loudness, quality, and rate of speech (presbyphonia), yet researchers have found that older adults are able to produce speech that is understood by most conversation partners.[15] These normal changes can be magnified if the older adult has a hearing loss because the individual may have difficulty listening to or monitoring their own voice. Medication taken for medical conditions can also affect voice characteristics.[16] Furthermore, chronic acquired medical conditions such as chronic obstructive pulmonary disease (COPD), head and neck cancer, aphasia, Parkinson's disease, Alzheimer's disease, and amyotrophic lateral sclerosis (ALS) present significant challenges to communication which can impact social participation and community engagement.

Older adults may report communication difficulties involving conversations with a range of communication partners such as familiar adults, children, unfamiliar adults or strangers; and in different environments, such as in quiet, in noisy places, over the phone, in a car, in a group, close by, or at a distance.[17] Reduced communicative effectiveness can occur from normal aging processes or from hearing, speech, or language disorders, and may result in substantial reductions in the quality of life for older adults. In a recent study of 240 community-dwelling adults 65–94 years of age, Palmer et al. found that "communication impairment was a significant independent predictor for fewer friends in the social network, a reduction in certain components of social support, and reduced social participation ... and predicted higher levels of loneliness and depression and reduced self-efficacy."[17] Research also suggests that social participation of older adults with others in varied activities can positively impact cognitive decline.[18,19]

12.2.5 MOBILITY CHALLENGES

Mobility is a foundational factor that influences if and how an older adult will engage in their community. The Centers for Disease Control recognize that mobility is critical for health and well-being.[20] Mobility includes ambulation, use of a mobility device, driving, and use of public transportation. One mobility factor that significantly impacts community engagement is falls self-efficacy.[21] If an older adult lacks confidence in their ability to navigate the community without falling, they may avoid leaving home. Chapter 4 of this book covers fall prevention in depth.

The physical environment in a community can provide significant barriers and safety hazards, limiting engagement and access. A single step up onto a curb or into a home or business can be an insurmountable obstacle for someone with mobility limitations. Steep slopes, uneven or slippery terrain, lack of safe pedestrian walkways and crosswalks, and poor lighting can prove treacherous. Commercially available mobility devices such as canes, walkers, rollators, and wheelchairs may be purchased by an older adult with the intention of compensating for poor balance or other mobility limitations; however, without proper device selection, fitting, and training by a professional, these devices may become additional dangers.

Driving a personal vehicle and accessing public modes of transit are other key elements of community mobility. As adults age, their ability to safely drive may be compromised by a variety of physical, sensory, and cognitive factors. Driving cessation may be a personal choice, one prompted by family or medical providers, or perhaps by an accident or close call. Driving rules for older adults vary with every state, with some requiring more frequent license renewal and testing restrictions such as vision screening.[22] Safely maintaining the ability to drive or shifting to alternative modes of transportation are critical for community engagement and well-being.[22] Availability of convenient public transportation options like buses, taxis, trains, and subways varies by geographic region. Accessibility options on public transportation including clear signage and ability to accommodate those in wheelchairs or persons requiring physical assistance to board may also vary. Ride-share services such as Uber and Lyft are available in many communities and known to most older adults; however, only 29% reported ever using the services, and 68% reported no plans to use the services, possibly as a result of privacy or safety concerns.[23]

12.3 SUPPORTS FOR COMMUNITY ENGAGEMENT

12.3.1 THE ROLE OF ASSISTIVE TECHNOLOGY

The Technology-Related Assistance for Individuals with Disabilities Act of 1988 was reauthorized by Congress in 1998[24,25] and in 2004.[26] It defined assistive technology (AT) device as "any item, piece of equipment, or product system, whether acquired commercially off the shelf, modified, or customized, that is used to increase, maintain, or improve functional capabilities of individuals with disabilities." This legislation also identified the value of AT:

> For some individuals with disabilities, assistive technology is a necessity that enables them to engage in or perform many tasks. The provision of assistive technology devices and assistive technology services enables some individuals with disabilities to – (A) have greater control over their lives; (B) participate in and contribute more fully to activities in their home, school, and work environments, and in their communities; (C) interact to a greater extent with nondisabled individuals; and (D) otherwise benefit from opportunities that are taken for granted by individuals who do not have disabilities.

According to the World Health Organization,[27] it is estimated that across the world more than 1 billion people need at least one AT device, yet only one in ten people have access to the AT they need. By the year 2030, the WHO estimates that this will increase to more than 2 billion. Furthermore, the WHO indicates that many older adults may benefit from more than two AT devices to improve their functional abilities. Categories of AT devices include the following: mobility products, seeing/vision products, hearing products, communication, cognition, self-care, and environment.[28]

AT devices range from inexpensive, simple tools to expensive, sophisticated computer software and hardware. Many items are ready-made and can be purchased commercially through online and retail sources; other ready-made items need to be customized or modified to fit the individual; and other AT devices will require fabrication to meet the individual's unique capabilities.

AT, then, helps people perform or participate in activities that are otherwise out of reach. Czaja stated that, "access to technology is not sufficient. The technology must be useful to and usable by older people."[29] When AT is not identified based on the individual needs of the user, the risk of continued functional limitations and device abandonment increases.

12.3.2 TYPES OF AT DEVICES

AT devices will be discussed along the continuum from no- or low-tech devices, mid-tech devices, and high-tech devices. Information regarding the specific types of technology, when provided, is based on current information. The reader should be aware of the fact that technology changes rapidly, including commercial availability and cost, and as a result, the focus of the remaining discussion will be primarily on categories and types of AT rather than on specific products. When the latter are discussed, it may be possible that the model or version may no longer be available, but a newer version may be. The World Health Organization developed the rapid Assistive Technology Tool (rATA) to screen an individual's need for and satisfaction with AT.[30] However, it is extremely important that older adults with functional limitations in any of the areas mentioned (i.e., self-care, cognition, vision, hearing, communication, and/or mobility) consult with knowledgeable AT specialists who are familiar with current AT and can conduct a comprehensive assessment of the person's specific abilities. These AT specialists will then make recommendations, provide training, and perform follow-up evaluation to ensure that the AT meets the individualized needs of the older adult. Without these procedures, AT may not be appropriate and may be abandoned.[31]

12.3.3   AT SELF-CARE SUPPORTS

A variety of AT may improve the ability of older adults to perform basic ADLs with greater safety, independence, and confidence. Performance of self-care tasks is necessary in preparation for community engagement, and in many cases, self-care will be performed while in the community (i.e., toileting, eating). Here are some common AT for self-care listed from low-tech to mid-tech. It is recommended that older adults consult with an occupational therapist (OT) for evaluation of their ADL performance to best determine customized recommendations for strategies and assistive technologies to meet their needs.

- **Low-tech**
  - **Long-handled sponge** – allows the user to wash body parts such as their feet and back with minimal effort.
  - **Sock aid** – this tool assists in donning socks when bending or reaching is difficult.
  - **Shoe horn** – assists the user in sliding their foot into a shoe without collapsing the shoe collar. This is supportive for those with difficulty bending.
  - **Reacher** – multipurpose tool helpful for picking up lightweight items from the floor or elevated surfaces. This is useful for donning clothing over the feet, or for pushing off clothing when bending and reaching are challenging.
  - **Universal cuff** – supports grasp of objects such as a toothbrush, comb, or silverware for those with impaired grip.
  - **Built-up handle utensils** – supports grasp of silverware for those who cannot make a tight grip around standard tools. This can be made at home according to the needs of the individual or purchased and added to utensils; prefabricated utensils with wide grips are also available.
- **Mid-tech**
  - **Bidet/washlet** – allows for greater independence in toilet hygiene. A variety of styles exist at various price points, depending on features (i.e., heated water, dryer, remote control).
  - **Handheld shower** – allows the user to direct the flow of water when bathing, requiring less mobility. This pairs well with shower chairs.
  - **Electric toothbrush** – may decrease effort required to complete effective toothbrushing.

- **Magnifying mirror** – supports grooming for those with low vision.
- **Hands-free blow dryer** – allows a user with limited upper extremity strength or function to perform hair styling.
- **Adapted clothing fasteners** – for those limited by arthritis or limited dexterity, solutions include magnetic or velcro buttons and elastic shoelaces.
- **Bed shaker alarm** – for those with hearing loss, this can augment traditional alarm clocks to ensure timely wake up for an appointment or outing.

12.3.4   AT COGNITIVE SUPPORTS

There are numerous AT supports for cognition that can help older adults compensate for declines in memory and orientation to time, person, and place. In addition to those listed here, communication supports that will be discussed later help the more complex cognitive domain of language that may be impaired in older adults with more severe chronic medical conditions.

- **Memory supports** can be used to provide reminders for daily activities. These include no- or low-tech supports such as compartmentalized medication boxes or prefilled medication blister packs labeled with the days of the week and/or times of day (i.e., morning, noon, evening).
- **Orientation to Time, Person, and Place:** Clocks with large displays with the hour, minute, and date and/or announcements about time. The WatchMinder 3 has preprogrammed reminders that can be selected (https://watchminder. com/shop/watchminder3) as needed. The MEMO Timer provides a more concrete way to visualize time that uses buttons with numerals (e.g., 5, 10, 15, 20) corresponding to the number of minutes for the timer and small lights on the device to provide a countdown for elapsed time. (www.abilia.com/en/our-products/cognition/time-and-clocks/memo-timer). Other clock configurations for older adults with sensory impairments are available from MaxiAids (www. maxiaids.com/clocks).

12.3.5   AT VISION AND HEARING SUPPORTS

For older adults with vision loss, interventions through medical providers, optometrists, and low-vision specialists may mitigate many common disorders. Visual acuity issues may be resolved with corrective lenses, and increased lighting and contrast in the environment within and outside the home can promote safety and access.[31] Low-vision specialists can prescribe customized solutions such as magnification to improve participation. When an older adult experiences difficulty hearing the speech of others or environmental sounds such as music, television, and other entertainment media, a complete evaluation by an audiologist will be needed to determine whether the individual has a hearing loss, and if so, the type and severity of the hearing loss and whether the individual would benefit from amplification in the form of hearing aids or assistive listening devices. Assistive listening devices use a receiver that is worn by a communication partner; the listener (the older adult) has a speaker that amplifies the partner's voice. It functions to reduce the signal-to-noise ratio so that the background noise is reduced and the partner's voice is amplified in order to improve the understandability of the partner's speech. Chapter 23 provides detailed information on these AT supports for hearing. Listed here are some common assistive technologies for landline and mobile telephones to support users with vision and hearing impairment in social and community engagement.

- **High-Tech**
  - **Telephone accessibility features** – These are supports for those with vision and hearing impairments to allow greater use for scheduling outings/appointments, arranging for transportation, and communicating with others.
    - **Low vision:** Commercially available landline telephones can be purchased with large, high contrast font, speed dial buttons with photographs, and voice command options to support those with vision impairment. Smartphones have customizable, embedded accessibility features that allow the user to adjust font and color contrast and use voice dialing, along with screen reader features.
    - **Impaired hearing:** Specialty landline telephones can support those with hearing loss through features such as sound amplification, flashing lights associated with the ringer, and captioning. Smartphones can pair with hearing aids and may also have options for live captioning of phone calls.
  - **Smartphone software applications (apps)** – many apps offer compensatory strategies for low vision or hearing loss.
    - **Low vision:** Live narration apps such as "Be My Eyes" can support older adults with blindness or low vision to engage in the community by utilizing a network of volunteers who can provide live sighted guidance for tasks like reading signage/labels, navigating a detour, or distinguishing colors (www.bemyeyes.com/).
    - **Impaired hearing:** Captioning apps such as "Ava" provide real time captioning of conversations for individuals who are deaf or hard of hearing (www.ava.me/).

12.3.6 AT Communication Supports – Speaking

If an older adult indicates that they have difficulty speaking or communicating with others in person, over the phone, in noisy environments, in the car or other vehicle, AT for speaking may be helpful. These communication supports are called augmentative and alternative communication (AAC). As with all areas of AT, it is imperative that a team of experienced AT professionals conduct a comprehensive assessment of the communication needs and challenges of the older adult, determine the barriers and supports to communication, trial different types of AAC systems, recommend specific products, provide AT training to help the older adult and their communication partners implement the technology in daily activities, and provide ongoing evaluation of the effectiveness of the device and strategies.[32]

The AAC devices that follow are listed from no- or low-tech to high-tech solutions.

- **No- or Low-Tech AAC Supports**
  - **Gestures, facial expressions, body language:** Most communicators across the world use these very natural ways to supplement speech. Gestures such as pointing, showing, and reaching as well as conventional gestures like nodding one's head YES or shaking one's head NO, and placing an index finger by the lips to indicate BE QUIET are all natural ways of augmenting speech. Facial expressions, body language, and tone of voice can also be used to supplement spoken language. These no-tech strategies are always available, have no cost, and can be used anywhere.
  - **Piece of paper or notebook and pen or pencil:** An effective no-tech strategy to augment or supplement spoken communication is for an older adult to write words on a piece of paper while speaking. This is particularly useful if their handwriting is legible. In some situations, it may be helpful to use a notebook to write letters, words, and short phrases, and these can be organized by topics (e.g., *medical, banking, restaurant, family, holidays,* etc.). This method is highly portable, as well.
  - **White board with dry erase markers:** Using a small white board with dry erase markers is as effective as using a piece of paper and pen or pencil, and the written information can be erased when no longer needed. These can be purchased online or at office supply retailers. A product called a Boogie Board® is a "reusable LCD writing tablet" that is available in different sizes (https://myboogieboard.com/pages/products).
  - **Communication boards** with concepts represented by symbols (e.g., letters and/or words and phrases, pictures, photographs) and organized in different ways (alphabetically; categorically by people, actions, places, things, etc.; grammatically by parts of speech like pronouns, nouns, verbs, prepositions, etc.; functionally by topics such as meals, activities, medical/dental appointments, playing cards, etc.). Communication boards can be created with a computer using a variety of software applications and then laminated for durability. Ready-made communication boards can be downloaded from different sources or purchased commercially. Many of these communication boards have a *traditional grid display* with symbols arranged in rows and columns. Another type of communication board can utilize a *visual scene display* where a personal photograph important to the older adult communicator has concepts written on it to which the older adult can point to communicate important information.
  - **Patient-Provider Communication** is critical for older adults to discuss medical needs with their providers, including physicians, personal care attendants, home healthcare nurses, and other medical personnel. The Patient-Provider Network (www.patientprovidercommunication.org) is a valuable resource for communication relating to medical care. This website has numerous ready-made communication tools available for free in English (www.patientprovidercommunication.org/covid-19-tools/free-engl ish-tools/) as well as in many languages paired with English (www.patientpr ovidercommunication.org/covid-19-tools/free-bilingual-tools/).

- **Communication books** consist of similar features as communication boards. Commercially available communication books such as the Large Communication Book or Portable Communication Book are available from Attainment Company (www.attainmentcompany.com/).
- **Mid-Tech (Battery-Operated) AAC Supports**
  - **Voice amplifiers** use a microphone and amplifier/speaker to make the person's voice louder. Depending on the size, the speaker can be worn by the older adult so that it projects their amplified voice toward the communication partner, or the speaker can be placed on a table or floor. Several different microphone styles are available, ranging from lavalier-style that clips to a shirt collar to hand-held and headset microphones. Commercially available voice amplifiers include the ChatterVox (www.chattervox.net/) and Winbridge (www.winbridgestore.com/). A recent study compared two voice amplifiers and a personal communication FM system used as a voice amplification system with no device for older adults with hypophonia (reduced vocal intensity) due to Parkinson's disease.[33] Compared to no device, all three devices were found to improve intelligibility of the older adult's speech as judged by their communication partners and naive listeners. The personal communication system differs from voice amplifiers in that the older adult with the soft voice speaks into a microphone and the communication partner wears a headset and hears the speaker's amplified voice.
  - **Communication devices with recorded speech output** have a limited number of squares or cells with different symbols (i.e., letters, words, phrases, pictures, and/or photos) on a static display. The number of symbols on the display range from 1 to 32. Some devices have one display only while others have multiple displays, which would need to be physically changed by the user or communication partner. Messages corresponding to the symbols are digitally recorded by a communication partner. When the symbol is selected, a message is spoken out loud by the device. The recorded voice should be appropriate for the age, gender, and cultural and linguistic background of the adult who requires the AAC system. As with communication books, symbols are organized in different ways depending on the needs of the older adult. A few manufacturers of mid-tech AAC devices include AbleNet, Inc. (www.ablenetinc.com/speech-generating-devices/all-speech-generating-devices/), Adaptivation (www.adaptivation.com/communication-aids), Attainment Company (www.attainmentcompany.com/technology/communication-aids), and Enabling Devices (https://enablingdevices.com/product-category/communication-devices/).
- **High-Tech (Electronic) AAC Supports**
  - **Display options:**
    Like static display devices, high-tech (electronic) AAC supports have a range of squares or cells with different symbols representing concepts (i.e., letters, words, phrases, pictures, and/or photos). In a traditional grid display, the number of symbols on the display can be arranged in grids of rows and columns ranging from 1 to as many as 140 cells per display. This category of AAC devices has communication displays that change dynamically and

automatically when a square or cell is activated that has been programmed to link to another page. One advantage of high-tech AAC devices with dynamic displays is that the user does not have to physically change the display to navigate to additional vocabulary and messages. This is helpful for older adults who have fine motor or visual impairments. Another advantage is that the user would be able to combine sequences of symbols to produce generative and novel utterances. High-tech devices have communication software apps that often utilize computer-generated text-to-speech.

Another type of display on high-tech AAC devices is a visual scene display (VSD), which incorporates motivating and meaningful photographs that can be programmed with messages embedded in "hot spots" on key areas of the photo. Printed text can be incorporated as well. VSDs may be beneficial for individuals with cognitive impairments or aphasia.[34]

- **Mainstream technology:** There are a wide variety of communication software apps that can be used with mainstream technology like smartphones, tablets, and personal computers, which are beneficial because these devices can service multiple purposes, not just for communication. Some communication apps are specifically designed for older adults with complex communication needs; however, some funding sources will not pay for mainstream technology to be used for AAC.

- **Dedicated speech-generating devices** are used only for communication purposes and meet the Medicare criteria of *durable medical equipment*.[35] Communication apps are also installed on dedicated AAC devices. There are several AAC device manufacturers that market both types of mainstream or open devices and dedicated or closed devices that can qualify as durable medical equipment. Because of the wide variety of AAC devices and communication apps available, comprehensive AT services from AAC specialists will ensure that older adults obtain AAC devices that meet their current and future needs.

- **Alternative Access** – This is an area that requires the expertise of the AAC team that consists of speech-language pathologists (SLPs) occupational therapists (OTs), and physical therapists (PTs) who assess, conduct trials with devices, recommend a specific AAC system, provide ongoing training to implement the AAC system, and evaluate its effectiveness for the older adult who has complex communication needs. Alternative ways to access AAC systems may need to be explored. Additional information on AAC can be found at www.asha.org/public/speech/disorders/aac/.

12.3.7   AT MOBILITY SUPPORTS

Those who are confident that they can avoid a fall are more likely to participate in meaningful activities.[21] Consulting a mobility specialist such as a PT may yield positive outcomes for those concerned about falling. Balance strengthening and endurance exercises may be enough to improve mobility and instill confidence, or a specialist may provide training in the use of an appropriate assistive device such as a cane, walker, or wheelchair to allow for safe community engagement and mobility.

For those who may benefit from long-term use of a wheelchair, consulting with an Assistive Technology Professional (ATP) is key for ensuring proper features, fit, and funding of the device. A directory of ATPs can be found at www.resna.org/Certified-Professionals-Directory.

The Americans with Disabilities Act of 1990 (ADA) outlines standards for public accommodations such as restaurants, movie theaters, and doctors' offices which are designed to prevent discrimination on the basis of disability.[36] Newly constructed or altered public accommodations are required to comply with ADA principles as long as the modifications do not fundamentally alter the nature of the offerings of the business, or pose undue hardship to the business. Accessibility features such as ramps, elevators, grab bars near toilets, braille signage, flashing alarm lights, and accessible parking spaces are examples of accommodations that comply with ADA. Awareness of this law and advocacy for its implementation by community members may improve accessibility of public spaces for older adults and all those with disabilities.

Physical spaces designed with Universal/Inclusive Design principles can also promote safe community engagement for people with wide-ranging abilities.[37] Features such as automatic doors, gently sloping ramps, and signage that includes high contrast words and symbols support access for those with varying physical, cognitive, and sensory abilities. Although overlap exists, Universal/Inclusive Design differs from the ADA as it is not enforced by law and does not specifically seek to avoid discrimination of those with disabilities, but rather it seeks to be inclusive of people of all ages and abilities. A caregiver pushing a baby stroller, a traveler towing luggage, and a person using a mobility scooter all benefit equally from Universally Designed environments. Architects, city planners, and green space designers who incorporate Universal/Inclusive Design principles can support community engagement for all. Apps such as Access Earth provide information about the accessibility of public spaces around the world through crowdsourcing of information from users (www.access.earth/).

AARP offers several driver safety courses designed for older adults, and experts such as Driving Rehabilitation Specialists are a resource for individuals who may be at risk of unsafe driving.[38,39] These specialists can assess skills and competencies necessary for safe driving and provide remediation, vehicle adaptations, and non-driving alternatives as appropriate. The American Occupational Therapy Association maintains a searchable database of Driving Rehabilitation Specialists that can be found at https://myaota.aota.org/driver_search/.

Alternative modes of public or private transportation that are affordable and easily accessible can promote social engagement for older adults who no longer drive.[6] Public transportation services are required to accommodate individuals with disabilities under Title II of the ADA.[40] Riders should have access to facilities and services featuring information in various forms (i.e., braille, large print, electronic, announcements), lifts, ramps, and devices to secure wheelchairs, support from trained vehicle operators, and ample time for boarding/exiting vehicles. Under the ADA, paratransit services must exist where fixed route services are offered.[40] These services can be scheduled for those who cannot use the traditional fixed route services, with individual scheduling available for trips near traditional fixed routes, at a fare of no more than double traditional rates. Some ride-share services also offer an option for wheelchair accessible vehicles such as vans with ramps or lifts.

As emphasized throughout the chapter, health professionals such as optometrists, audiologists, PTs, OTs, SLPs, driver rehabilitation specialists, and AT specialists should be consulted to provide in-depth support and recommendations for older adults experiencing any challenges impacting daily living skills and community engagement. Additional resources for older adults include:

12.4   ADDITIONAL RESOURCES FOR OLDER ADULTS

- Area Agencies on Aging (AAA)
- State Departments of Aging
- National Council on Aging (NCOA): www.ncoa.org/
- American Association for Retired Persons (AARP): www.aarp.org/
- National Assistive Technology Act Technical Assistance and Training (AT3) Center: www.at3center.net/stateprogram

12.5   CONCLUSION

The risk of functional impairment, chronic illness, and mortality is higher for older adults who engage in limited physical and social activity in their community. However, utilization of appropriate assistive technologies, accessibility laws and principles, resources designed for older adults, and the expertise of healthcare professionals can promote safe community engagement and social participation.

REFERENCES

1. U.S. Department of Health and Human Services. *Multiple chronic conditions – a strategic framework: optimum health and quality of life for individuals with multiple chronic conditions.* Washington, DC. www.hhs.gov/sites/default/files/ash/initiatives/mcc/mcc_framework.pdf. Published Dec 2010. Accessed October 31, 2021.
2. Warshaw G. Introduction: advances and challenges in care of older people with chronic illness. *Generations.* 2006;30(3):5–10.
3. Ornstein SM, Nietert PJ, Jenkins RG, Litvin CB. The prevalence of chronic diseases and multimorbidity in primary care practice: a PPRNet report. *J Am Board Fam Med.* 2013 Sep–Oct;26(5):518–524. doi: 10.3122/jabfm.2013.05.130012. PMID: 24004703.
4. Ankuda CK, Freedman VA, Covinsky KE, Kelley AS. Population-based screening for functional disability in older adults. *Innov Aging.* 2020;5(1):igaa065. Published 2020 Dec 22. doi:10.1093/geroni/igaa065
5. Gough C, Lewis LK, Barr C, Maeder A, George S. Community participation of community dwelling older adults: a cross-sectional study. *BMC Public Health.* 2021;21(1):612. Published 2021 Mar 29. doi:10.1186/s12889-021-10592-4
6. Levasseur M, Généreux M, Bruneau JF, et al. Importance of proximity to resources, social support, transportation and neighborhood security for mobility and social participation in older adults: results from a scoping study. *BMC Public Health.* 2015 Dec;15(1):1–9.

7. Flowers L, Houser A, Noel-Miller, C et al. Insight on the issues 125: Medicare spends more on socially isolated older adults. AARP Public Policy Institute. www.aarp.org/content/dam/aarp/ppi/2017/10/medicare-spends-more-on-socially-isolated-older-adults.pdf. November 2017. Accessed October 29, 2021.

8. Freedman VA, Spillman BC, Andreski PM, et al. Trends in late-life activity limitations in the United States: an update from five national surveys. *Demography.* 2013;50(2):661–671.

9. Harada CN, Natelson Love MC, Triebel KL. Normal cognitive aging. *Clin Geriatr Med.* 2013;29(4):737–752. doi:10.1016/j.cger.2013.07.002

10. Bourassa KJ, Memel M, Woolverton C, Sbarra DA. Social participation predicts cognitive functioning in aging adults over time: comparisons with physical health, depression, and physical activity. *Aging Ment Health.* 2017;21(2):133–146. doi:10.1080/13607863.2015.1081152

11. American Foundation for the Blind. Special report on aging and vision loss. www.afb.org/research-and-initiatives/aging/special-report-aging-vision-loss. Published 2013. Accessed October 26, 2021.

12. Perlmutter MS, Bhorade A, Gordon M, Hollingsworth HH, Baum MC. Cognitive, visual, auditory, and emotional factors that affect participation in older adults. *Am J Occup Ther.* 2010;64(4):570–579. doi:10.5014/ajot.2010.09089

13. Ding D, Lawson KD, Kolbe-Alexander TL, et al. Lancet physical activity series 2 executive committee. The economic burden of physical inactivity: a global analysis of major non-communicable diseases. *The Lancet.* 2016 Sep 24;388(10051):1311–1124.

14. Mahboubi H, Lin HW, Bhattacharyya N. Prevalence, characteristics, and treatment patterns of hearing difficulty in the United States. *JAMA Otolaryngol Head Neck Surg.* 2018;144(1):65–70. doi:10.1001/jamaoto.2017.2223

15. Hooper CR, Cralidis A. Normal changes in the speech of older adults: you've still got what it takes; it just takes a little longer! *Perspect Gerontol.* 2009;14(2):47–56. https://doi.org/10.1044/gero14.2.47

16. Rojas S, Kefalianos E, Vogel A. How does our voice change as we age? A systematic review and meta-analysis of acoustic and perceptual voice data from healthy adults over 50 years of age. *J Speech Lang Hear Res.* 2020; 63(2):533–551. doi.org/10.1044/2019_JSLHR-19-00099

17. Palmer AD, Carder PC, White DL, et al. The impact of communication impairments on the social relationships of older adults: pathways to psychological well-being. *J Speech Lang Hear Res.* 2019;62(1):1–21. doi:10.1044/2018_JSLHR-S-17-0495

18. Bassuk SS, Glass TA, Berkman LF. Social disengagement and incident cognitive decline in community-dwelling elderly persons. *Ann Intern Med.* 1999;131(3):165–173. doi:10.7326/0003-4819-131-3-199908030-00002

19. Hikichi H, Kondo K, Takeda T, Kawachi I. Social interaction and cognitive decline: results of a 7-year community intervention. *Alzheimers Dement (N Y).* 2016;3(1):23–32. doi:10.1016/j.trci.2016.11.003

20. Centers for Disease Control and Prevention. *The State of Aging and Health in America 2013.* Atlanta, GA: Centers for Disease Control and Prevention, US Dept of Health and Human Services; 2013. www.cdc.gov/aging/pdf/state-aging-health-in-america-2013.pdf. Accessed October 31, 2021.

21. Schepens S, Sen A, Painter JA, Murphy SL. Relationship between fall-related efficacy and activity engagement in community-dwelling older adults: a meta-analytic review. *Am J Occup Ther.* 2012;66(2):137–148. doi:10.5014/ajot.2012.001156

22. Stav WB. Updated systematic review on older adult community mobility and driver licensing policies. *Am J Occup Ther.* 2014;68(6):681–689. doi:10.5014/ajot.2014.011510

23. Binette J, Vasold K. *Home and community preferences: a national survey of adults ages 18-plus.* Washington, DC: AARP Research, August 2018. https://doi.org/10.26419/res.00231.001. Accessed October 29, 2021.

24. Technology-Related Assistance for Individuals with Disabilities Act of 1988, Pub. L. 100–407, 102 STAT. 1044 (1988). www.congress.gov/100/statute/STATUTE-102/STATUTE-102-Pg1044.pdf

25. Assistive Technology Act of 1998, Pub. L. No. 105–394, 112 Stat. 3627 (1998). www.congress.gov/105/plaws/publ394/PLAW-105publ394.pdf

26. Assistive Technology Act of 2004. Pub. L. No. 108–364, 118 Stat. 1707 (2004). www.congress.gov/108/plaws/publ364/PLAW-108publ364.pdf

27. World Health Organization. Fact sheet: Assistive technology. www.who.int/newsroom/fact-sheets/detail/assistive-technology. Published 2018 May 18. Accessed November 1, 2021.

28. World Health Organization. Assistive product specifications and how to use them. Published 2021. Accessed October 31, 2021.

29. Czaja SJ, Sharit J, Boot WR, Charness NH, Rogers WA. The role of technology in supporting social engagement and social support among older adults. *Innov Aging.* 2017;1(Suppl 1):1026–1027. Published 2017 Jun 30. doi:10.1093/geroni/igx004.3737

30. World Health Organization. Rapid assistive technology assessment tool (rATA). https://apps.who.int/iris/handle/10665/341939. Published 2021. Accessed October 31, 2021.

31. Cook A, Polgar JM, Encarnação P. *Assistive technologies: principles & practice.* 5th ed. St. Louis, Missouri: Elsevier; 2020.

32. Beukelman DR, Light JC. *Augmentative & alternative communication: supporting children and adults with complex communication needs.* 5th ed. Baltimore, MD: Paul H. Brookes; 2020.

33. Knowles T, Adams SG, Page A, Cushnie-Sparrow D, Jog M. A comparison of speech amplification and personal communication devices for hypophonia. *J Speech Lang Hear Res.* 2020;63(8):2695–2712. https://doi.org/10.1044/2020_JSLHR-20-00085

34. Beukelman DR, Hux K, Dietz A, McKelvey M, Weissling K. Using visual scene displays as communication support options for people with chronic, severe aphasia: a summary of AAC research and future research directions. *Augment Altern Commun.* 2015;31(3):234–245. doi:10.3109/07434618.2015.1052152

35. U.S. Centers for Medicare and Medicaid Services. *Durable medical equipment.* www.medicare.gov/coverage/durable-medical-equipment-dme-coverage. Accessed November 1, 2021.

36. Department of Justice. *Americans with Disabilities Act Title III regulations.* www.ada.gov/regs2010/titleIII_2010/titleIII_2010_regulations.pdf. September 15, 2010. Accessed October 28, 2021.

37. Institute for Human Centered Design. History. www.humancentereddesign.org/index.php/inclusive-design/history. Accessed October 28, 2021.

38. Freed S. Older drivers can be safer drivers- with support. AARP. www.aarp.org/auto/driver-safety/info-2021/older-drivers-fewer-accidents.html. October 21, 2021. Accessed October 28, 2021.

39. American Occupational Therapy Association. Driving and community mobility. *Am J Occup Ther.* 2016;70(Suppl. 2), 701241150:1–19.

40. ADA National Network. ADA & accessible ground transportation. https://adata.org/sites/adata.org/files/files/Accessible%20Ground%20Transportation%20final2019.pdf. Published 2018. Accessed October 30, 2021.

13 Older Adult Driver Safety

Casey Rust MD and Alice Pomidor, MD, MPH, AGSF

CONTENTS

13.1 INTRODUCTION

Older adults aged 65 years and higher make up a growing proportion of the population, with 80% licensed to drive. This age group is more susceptible to physical and cognitive changes which could lead to impaired driving safety, including vision loss, arthritis, neuropathy, seizures, syncope, dementia, and other medical conditions. Older adults may also have medications which alter mentation and impair driving safety. The goals of this chapter are to learn to recognize at-risk drivers, provide resources for older adults to help maintain driving skills for as long as possible, and when driving is no longer a viable option, provide tools for talking with older adults about retiring from driving.

Assessing driving safety in older adults is the responsibility of all older adult service professionals, not just those in healthcare. Anyone who comes into contact with older adults, including friends, families, and community members, may identify signs that an older adult is at risk for unsafe driving. When an older adult mentions problems with seeing the road at night or family members express concerns that the older driver is unsafe on the road, these can be red flags that driving skills should be evaluated. When available, evaluating and managing driving safety in older adults is best done by an interprofessional team which may consist of a clinician, occupational therapist, physical therapist, speech therapist, nurse, social worker, driving rehabilitation specialist, government officials, and other community members.

Rather than wait until older adults are no longer able to drive to initiate conversations about driving safety, older adult service professionals and caregivers

DOI: 10.1201/9781003197843-13

should strive to identify risk factors and intervene early to prevent driving disability. This is considered primary prevention. Secondary prevention involves addressing issues already affecting driving skills and trying to restore those skills through healthcare interventions or programs like driving rehabilitation. Tertiary prevention involves identifying losses of driving skills which are irreversible and recommending alternatives to minimize potential harm to older adults and others on the road. Driving cessation can lead to social isolation, depression, lack of access to healthcare and reduced nutrition resources, which can adversely affect the well-being of older adults. By focusing on primary and secondary prevention, we can help maximize the time older adults can safely drive.

13.2 RISK FACTORS FOR UNSAFE DRIVING

The first step in addressing driving safety in older adults is recognizing when an older adult is at risk for unsafe driving. The risk factors include impaired physical function, cognitive impairment, certain medical conditions, and use of multiple medications. When interacting with older adults one should be alert to signs and symptoms which should prompt further investigation. For example, a history of traffic incidents or observer concerns about an older adult's driving may indicate at-risk driving. Age alone is not an indicator to stop driving.

Sensory impairment such as decreased vision and loss of sensation in the extremities is an important risk factor for unsafe driving. Older adults may have loss of vision related to glaucoma, macular degeneration, or retinopathy. Visual acuity, or sharpness of vision, is important in order to read road signs and see traffic signals from a safe distance. Cataracts and glare can cause reduced contrast sensitivity which affect the ability to differentiate objects on the road. Drivers with impaired contrast sensitivity may mention difficulty driving in fog or at night. Peripheral neuropathy, or decreased sensation in the extremities, impairs driving safety as it reduces the ability to sense the pedals. Peripheral neuropathy is often associated with diabetes but also can be considered with other conditions such as neurologic impairment or tobacco/alcohol abuse.

Physical changes in older adults which may impair driving safety include difficulty walking, abnormal reflexes, or restricted neck range of motion such as caused by arthritis. Any physical limitations which may affect the ability to use the pedals, turn the steering wheel, or visualize other drivers and road hazards should warrant further investigation. An abnormal Get Up and Go Test, defined as inability to rise from a seated position and walk 10 feet and back in under 11 seconds, is associated with impaired driving performance.

Cognitive changes which affect driving safety include impairment in memory, executive function, visuospatial skills, attention, and processing speed. A diagnosis of dementia should not automatically lead to a discussion of immediate driving retirement, although most drivers with the diagnosis for over two years will have serious enough progression of disease to need to stop driving. Drivers who have gotten lost while driving in familiar areas are at risk even if they have not had another incident. Executive function involves decision-making and planning which are important parts of on-the-road driving skills. Visuospatial skills and processing speed influence the ability to react to other drivers on the road and take appropriate action in a timely manner.

Medications can affect driving skills by causing sedation, confusion, or changing reaction time. Clinical providers and staff should carefully review medications in older adults as part of evaluating driving safety. Medications that may impair driving safety include opioids, muscle relaxers, anticonvulsants, antidepressants, antiemetics, antipsychotics, benzodiazepines, stimulants, and antihistamines. Alcohol, marijuana,

TABLE 13.1
Medications Which May Impair Driving Safety

Medication Class	Examples	Effects
Anticholinergics	Bladder medications such as oxybutynin Sea sickness medications such as atropine or meclizine	Sedation, blurred vision, impaired cognition
Anticonvulsants	Seizure medications such as valproic acid, gabapentin, levetiracetam	Sedation, impaired cognition
Antidepressants	Paroxetine, amitriptyline, mirtazapine, duloxetine	Sedation, blurred vision, impaired cognition, lightheadedness
Antihistamines	Allergy medicines such as diphenhydramine or cetirizine	Sedation, blurred vision, impaired cognition
Antihypertensives	Amlodipine, hydrochlorothiazide, furosemide, metoprolol	Dizziness, syncope
Antiparkinsonian agents	Carbidopa-levodopa, pramipexole, ropinirole	Sedation
Antipsychotics	Haloperidol, risperidone, lurasidone, quetiapine	Sedation, blurred vision, impaired cognition, tremor
Benzodiazepines	Anxiety medications such as alprazolam, diazepam, lorazepam	Sedation, impaired vision, impaired cognition, dizziness
Hypoglycemics	Insulin, glipizide	Impaired concentration, shakiness, lightheadedness
Muscle relaxants	Cyclobenzaprine, baclofen	Sedation, blurred vision, impaired cognition
Opioid analgesics	Hydrocodone, oxycodone, morphine, hydromorphone	Sedation, impaired vision, lightheadedness
Sedative hypnotics	Sleeping pills such as zolpidem, eszopiclone	Sedation, impaired cognition

Source: Modified from: "Marottoli R, Gray S. Table 9.13 – Potentially Driver-Impairing (PDI) Medications," in "Medical conditions, functional deficits, and medications that may affect driving safety." In: Pomidor A, ed. *Clinician's Guide to Assessing and Counseling Older Drivers, 4th Edition.* New York: American Geriatrics Society; 2019.

and other drugs can also impair driving safety. Adults taking medications for hypertension and diabetes may be at risk if they are experiencing low blood pressure or hypoglycemia. The American Geriatrics Society's Beers Criteria lists potentially hazardous medications for older adults. People taking three or more potentially driving-impairing medications are 1.87 times at risk of a crash (see Table 13.1).

Certain medical conditions can affect driving safety and as a result may require temporary or permanent driving cessation. Adults with a history of seizures will typically be required to have a seizure-free period of six or more months before being allowed to resume driving; these regulations vary by state. Similarly, adults with a history of syncope from any cause, including cardiac arrhythmias or orthostatic hypotension, should be carefully assessed for driving safety due to the potential for loss of consciousness behind the wheel. Diabetes can impair driving safety in a number of ways including impaired cognition related to hypoglycemia or severe hyperglycemia, vision loss from diabetic retinopathy, or diabetic neuropathy. Strokes may impair vision, mobility, or cognition which may interfere with driving. Adults with neurodegenerative diseases such as Parkinson's, multiple sclerosis, and dementia should all undergo periodic driving safety assessments as their disease progresses over time. Delirium, an acute change in cognition and level of consciousness which impacts attention, is incompatible with driving. Adults with mild or moderate dementia should be carefully assessed for driving safety, while adults with severe dementia should not be permitted to drive. Sleep disorders such as obstructive sleep apnea (OSA) and narcolepsy impair driving safety due to drowsiness or impaired wakefulness. Adults with sleep disorders should be under the care of a clinician and should not be permitted to drive unless the conditions are well-controlled (e.g., an apnea hypoxia index of 19 or less may indicate control of OSA). If there is any uncertainty as to whether an older adult's medical conditions may impact their ability to drive safely, recommend that they discuss these questions with their clinician.

13.3   RISK ASSESSMENT

Risk assessments for driving can be performed by the individual older adult, family, caregivers, or others who interact with the older adult. In the case of cognitive impairment or dementia, the older adult should not be responsible for assessing their own driving risk. A screening assessment for driving safety should include components of vision, hearing, cognition, and motor skills. Deficits in any of these areas should warrant further comprehensive evaluation.

Community health screenings which include vision, mobility, and memory assessments can help to identify at-risk drivers. Visual acuity, assessed with a Snellen eye chart, and visual field testing can identify older adults needing referral to optometry or ophthalmology. Mobility can be assessed with a Get Up and Go Test, either timed or untimed. Range of motion testing including neck rotation, shoulder and elbow flexion, finger curl, and ankle plantar/dorsiflexion can identify physical limitations. A simple screening test for cognitive impairment is the Mini-Cog which includes a three-item delayed word recall task and a clock-drawing test. An abnormal Mini-Cog, defined as an abnormal clock or forgetting more than one item on delayed recall, should lead to more thorough cognitive assessment.

Cognitive assessments can be performed by anyone with the knowledge of how to correctly conduct these tests including primary care providers, nurses, speech therapists, and social workers. Neuropsychological tests which include visuospatial ability and executive function are better predictors of driving safety than global measures of cognition. Clock Drawing, Trails Making Test-B (Trails-B), and the Snellgrove Maze test are useful tools for evaluating driving safety. Clock Drawing is included in both the Montreal Cognitive Assessment (MoCA) and St. Louis University Mental Status (SLUMS) exam. The Clock Drawing Test and Trails-B are sensitive tests for driving safety, identifying 92% of those who failed the behind-the-wheel-test. Copying complex drawings, as in the MoCA, can also give information on visuospatial ability. A score of 18 or less on the MoCA may predict impaired driving safety and warrants additional care. The Mini-Mental State Examination (MMSE) does not consistently predict motor vehicle crashes or traffic violations.

Risk assessments should also consider medical history, review of medications, and an assessment of activities of daily living (ADLs). Gathering a driving history from the older adult, including history of accidents or close calls, gives information regarding an adult's driving habits and safety. Generally speaking, if an adult is unable to perform complex ADLs such as managing finances and medications, they are frequently not fit to drive.

There are several screening tools which can be used by patients and families to bring awareness to driving safety issues and start a conversation about driving safety. The Heath in Aging Foundation has an interactive driver safety questionnaire at their online Driver Safety Center (www.healthinaging.org/driving-safety) and provides education plus suggestions based on the responses. The Fitness-to-Drive Screening (FTDS) evaluation is a research-validated questionnaire available online for caregivers or other family members who have driven with the older adult within the last three months. American Association of Retired Persons (AARP) has safe driving resources and tips which are accessible to older adults.

There are also a number of resources for clinicians for assessing driving safety. The Clinical Assessment of Driving Related Skills (CADReS) is a collection of assessments including visual, cognitive, motor, and sensory tools which may be used by the clinical team to assess an older driver and its use is described in the Plan for Older Driver Safety, both of which are freely available in the 4th edition of the *Clinician's Guide to Assessing and Counseling Older Drivers*.

The gold standard for evaluating driving safety includes a medical exam as described above as well as an on-road driving evaluation performed by a certified driving rehabilitation specialist (CDRS). A CDRS is someone who has advanced education in driving evaluation and safety. Although the road tests offered by the state Department of Motor Vehicles (DMV) can be screening tests, they are not the same as the comprehensive on-road driving evaluations conducted by driving rehabilitation specialists. Some rehabilitation centers offer driving simulators which can provide additional information when on-road assessment is not available. Though driving schools may be able to instruct drivers on driving skills, they are not able to assess medical conditions that may interfere with driving in order to provide targeted interventions.

13.4   INTERVENTIONS

When older drivers are identified as being at risk for loss of driving skills, initiate conversations about ways to address the underlying issues . Risk factor interventions are likely to include a combination of education and referrals. Older adults may benefit from physical therapy to improve strength and mobility, occupational therapy to improve skills and ADLs, ophthalmology referral to treat eye conditions, speech therapy or neuropsychological evaluation for cognitive training, or driving rehabilitation for comprehensive driving interventions.

The first step in addressing driving safety may be talking to older adults about avoiding risky situations such as driving at night or driving in unfamiliar settings. Older adults may lack awareness of how their medical conditions and medications can impact driving safety. Older adults should be encouraged to discuss these questions with their medical provider. The older adult's clinical team can also help address their current medical conditions in order to reduce the risk for complications which would impact driving. For example, adults with diabetes may need education on managing diabetes to reduce the risk of diabetic retinopathy and neuropathy.

For older adults with mild cognitive impairment or early dementia, referral to speech therapy or occupational therapy may be appropriate. Medical providers should evaluate for reversible causes of cognitive impairment such as thyroid disease, untreated OSA, or adverse effects from medications. Anyone can help older adults with mild dementia by initiating conversations about planning ahead for alternative means of transportation.

Secondary intervention involves attempting to restore driving ability after a medical condition has led to impaired driving safety. Driving rehab aims to maintain driving ability through training and adaptive equipment for older adults with disabilities and impairments. Adaptive equipment may include extended gear shift levers, foot pedal extenders, or modified steering wheels or mirrors. When driving rehabilitation is unavailable, occupational therapy can be used to improve some functional skills necessary for driving safety.

It is important to consider that some adults will need to stop driving temporarily while attempts are made to restore their driving skills over time. Repeating assessments to monitor for improvement over time is necessary in such situations.

13.5   ADVISING OLDER ADULTS ABOUT DRIVING RETIREMENT

If efforts to restore driving safety are unsuccessful, or older adults suffer from conditions which are not compatible with safe driving such as uncontrolled seizures or advanced dementia, it is the responsibility of all clinical team members to discuss retiring from driving. Conversations about retiring from driving are best done with at least one clinical team member, the older adult, and their primary caregiver(s) or support network member. The clinical team member should identify the conditions or loss of skills which prevents driving safely rather than state that the older adult is an "unsafe driver," since most older adults have been safe drivers for most of their long lives. Emphasize that it is the condition which is leading to the loss of driving privileges, not the caregiver or the clinical team member "taking away the keys." Keep in mind that only the state DMV can actually suspend driving privileges, although it may be strongly recommended by the clinical team member.

When an older adult retires from driving, it is important to discuss how they will continue to access the community. Alternative means for transportation should be discussed including public transportation, transportation services specifically designed for older adults, or ride hailing services through transportation network companies. The national Eldercare Locator at https://eldercare.acl.gov is a public service of the U.S. Administration on Aging that connects older adults and their caregivers to local services, including the local Area Agency on Aging, and transportation. A useful guide to creating a transportation plan for how to get around when unable to drive is available online at https://planfortheroadahead.com. Several states also have their own coalitions for the support of older adult drivers and safe mobility which can provide more information and resources, such as community transportation coordinators. If available, a social services referral can be very helpful. For some older adults, driving retirement may lead to seeking alternative housing that is within walking distance of stores or other needs, housing which is accessible to public transportation, or housing which includes transportation options. Options for grocery and medication delivery may also be appropriate to discuss.

After driving retirement is discussed with an older adult, close follow-up is needed to monitor for social isolation and depression. Identifying strategies for continuing social interactions may help combat isolation and loneliness. The Geriatric Depression Scale (GDS) or Patient Health Questionnaire (PHQ-9) can be used to screen for depression in older adults. If an older adult has a positive screen for depression, they should follow up with their primary care provider to discuss options for management.

13.6   LEGAL AND ETHICAL ISSUES

Unfortunately, in some situations, the older adult will continue to drive despite being advised to stop. In these situations, it is necessary to intervene in order to protect the older adult as well as the public. When repeated efforts at communicating the need to stop driving are futile, caregivers and family members may have to take action to prevent unfit drivers from getting behind the wheel. This can include making keys inaccessible, removing vehicle access by disabling or selling the car, grinding keys down, or removing the battery from a key fob.

Motor vehicle crashes are the second leading cause of injury-related death, after falls, in adults ages 65 and older. When an older adult does not follow a recommendation to retire from driving, and their caregiver or support person is unable to prevent them from driving, the unfit driver should be reported to the DMV. Requirements for reporting unfit drivers vary by state and local jurisdiction. Some states have mandatory reporting requirements. Information about state reporting requirements and reporting forms can be accessed online through the state's DMV website, or in person at the DMV.

Following a report, typically a medical advisory board with the DMV will request medical information to verify the validity of the report. Drivers are usually required to submit a report or their driving privilege may be revoked until they do so. If the concerns are validated, then an examination at the driver license office may be required. If the evaluator does not find the complaint valid, no further action is taken. Drivers can often provide updated information from their medical provider or a driving rehab specialist at any time.

In order to maintain relationships with the older driver, it is best to be open and honest regarding concerns about driving safety and plans to report someone to the DMV. It is important to keep in mind that if an older adult maintains decision-making capacity, a caregiver or support person should not be contacted without the older adult's permission. It is essential to document all conversations with the older adult about recommendations to stop driving.

13.7 CONCLUSION

Driving safety in older adults is an important topic requiring collaboration among various professionals and community members. Anyone who feels that an older adult may be unsafe driving can take advantage of resources for screening for driving safety and strive to connect older adults at risk with appropriate services for support and intervention. The role of the professional is to assist the older adult in maintaining driving skills for as long as possible. If the time comes that an older adult is no longer fit to drive, having honest conversations about their limitations and connecting them with alternative means for transportation will allow the older adult to stay connected to their community.

BIBLIOGRAPHY

American Association of Motor Vehicle Administrators (AAMVA) Driver Fitness Working Group. *Driver fitness medical guidelines. American Association of Motor Vehicle Administrators.* Washington, DC: National Highway Traffic Safety Administration; 2009.

American Geriatrics Society Beer Criteria Update Expert Panel. American Geriatrics Society 2019 updated AGS Beers Criteria® for potentially inappropriate medication use in older adults. *J Am Geriatr Soc.* 2019;67(4):674–694.

Aksan N, Anderson S, Dawson J, Johnson A, Uc E, Rizzo M. Cognitive functioning predicts driver safety on road tests 1 and 2 years later. *J Am Geriatr Soc.* 2011;60(1):99–105.

Anderson SW, Aksan N, Dawson JD, et al. Neuropsychological assessment of driving safety risk in older adults with and without neurologic disease. *J Clin Exp Neuropsyc.* 2012;34(9):895–905.

Center for Statistics and Analysis. *Older population: 2018 data (traffic safety facts. Report no. DOT HS 812 928).* Washington, DC: National Highway Traffic Safety Administration; 2020.

Crizzle AM, Classen S, Bedard M, et al. MMSE as a predictor of on-road driving performance in community dwelling older drivers. *Accident Anal Prev.* 2012;49:287–292. (p. 7)

Elgin J, Owsley C, Classen S. Vision and driving. In: Maguire MJ, Davis ES, eds. *Driving and community mobility: Occupational therapy strategies across the lifespan.* Bethesda, MD: AOTA; 2012:173–219.

Hollis AM, Duncanson H, Kapust LR, et al. Validity of the mini–mental state examination and the Montreal cognitive assessment in the prediction of driving test outcome. *J Am Geriatr Soc.* 2015;63(5):998–992.

Larner AJ. Screening utility of the Montreal Cognitive Assessment (MoCA): In place of – or as well as – the MMSE? *Int Psychogeriatr.* 2012;24(3):391–396.

Leroy A, Morse L. *Multiple medications and vehicle crashes: Analysis of databases.* Washington, DC: National Highway Traffic Safety Administration; 2008.

Markwick A, Zamboni G, de Jager CA. Profiles of cognitive subtest impairment in the Montreal Cognitive Assessment (MoCA) in a research cohort with normal Mini-Mental State Examination (MMSE) scores. *J Clin Exp Neuropsyc.* 2012;34(7):750–757.

McCarthy DP, Langford D. *Process and outcomes evaluation of older driver screening programs: The Assessment of Driving-Related Skills (ADReS) older-driver screening tool. Department of transportation.* Washington, DC: National Highway Traffic Safety Administration; 2009.

Murphy SL, Xu JQ, Kochanek KD, Arias E. *Mortality in the United States, 2017. NCHS data brief, No 328.* Hyattsville, MD: National Center for Health Statistics; 2018. www.cdc.gov/nchs/data/databriefs/db328-h.pdf. (p. 12)

Older drivers. IIHS-HLDI crash testing and highway safety. www.iihs.org/topics/older-drivers#by-the-numbers. Published 2020. Accessed November 1, 2021.

Owsley C, McGwin Jr G, Searcey K. A population-based examination of the visual and ophthalmological characteristics of licensed drivers aged 70 and older. *J Gerontol A Biol Sci Med Sci.* 2013;68(5):567–573.

ADDITIONAL RESOURCES

American Occupational Therapy Association: www.aota.org/Practice/Productive-Aging/Driving.aspx

Association for Driver Rehabilitation Specialists: http://aded.site-ym.com/?page=725

Clinician's Guide to Assessing and Counseling Older Drivers: https://geriatricscareonline.org/application/content/products/B047/pdf/AGS-Clinicians_Guide_4th_edition.pdf

Eldercare Locator: 1-800-677-1116: https://eldercare.acl.gov/Public/Index.aspx

Fitness to Drive Screening Measure: http://fitnesstodrive.phhp.ufl.edu/us/?intcmp=AE-ATO-ADS-ASSESS-ROW2-SPOT

Geriatrics Care Online: http://geriatricscareonline.org

Health in Aging: Driver Safety: www.healthinaging.org/driving-safety

National Highway Traffic Safety Administration: Older Driver information: www.nhtsa.gov/road-safety/older-drivers

14 Travel and Seasonal Safety for Older Adults

Maysoon Agarib, MD

CONTENTS

14.1 TRAVEL

A significant portion of the population who travel and explore the world are older adults, especially after retirement. According to US travel statistics, baby boomers are the most likely to travel, whether domestically or internationally. This is likely due to their financial stability, technological savvy, and relative personal freedom since the

DOI: 10.1201/9781003197843-14

youngest baby boomers no longer have children at home. Whether traveling by airplane, train, cruise, or car, preparation should include several essential elements. The topic of driving safety will be covered in Chapter 13.

14.2   ESSENTIAL ELEMENTS FOR TRAVEL PREPARATION

Planning travel ahead of time should be a robust process and ideally double-checked by another involved and responsible person. Decisions regarding destination choices (especially during pandemics), vaccines requirements, timing of travel, and accommodations should be carefully considered in light of an older adult's comorbidities. All travel should follow guidelines and recommendations from federal and local organizations to ensure safe travel for older adults.

14.3   INTERNATIONAL AND AIR TRAVEL

Air travel brings additional unique risks for older adults and requires an additional level of planning. There is no emergency support network in the air. Diversion or unplanned landings for emergency medical cases are uncommon, with a diversion rate estimated to be between 8% and 13% of all in-flight medical emergencies. Airline personnel are not permitted to assist passengers with impaired activities of daily living who have potential risk for injuries, such as falls. Restrooms within the airplane itself are small and the bathroom may not be equipped with grab bars.

14.3.1   HEALTH CARE PROVIDER VISIT

Older adult should consider scheduling a visit with their health care provider, at least one month prior to their departure, to discuss travel plans, review any modifications to their current chronic illness and/or medication management, and check their destination's vaccine requirements, most common illnesses, and health risks.

14.3.2   VACCINATIONS

Vaccination requirements for international travel should be reviewed well in advance of planned travel. Many areas have travel medicine clinics which specialize in vaccinations and information regarding destination-specific health recommendations.

14.3.3   MEDICATIONS

A complete medication list should be prepared which includes prescription, over-the-counter medications., and herbal medications. All medications should be carried in their original bottle and labeled appropriately. An adequate supply of medicines to last the entire trip, including extra for unanticipated delays, should be in carry-on luggage. Older adults should pay attention to medication timing according to the original or destination time.

14.3.4 CURRENT HEALTH CONDITIONS

A list of medical conditions and treatment plans should be carried by all older adults. They should consider laminating the information and attaching it to a travel carry-on bag. This ensures the information will be easily apparent to personnel if an emergency occurs during flight.

14.3.5 TRAVELLING WITH FUNCTIONAL IMPAIRMENTS

Functional impairments, including difficulty with walking or standing for a prolonged period, are very common among older adults. Airports will provide wheelchairs, with assigned personnel, for movement throughout the airport and at security check points for adults with functional impairments. Recommendations are to check with the airline and the departure and arrival airports, in advance of travel, to ensure you understand the services which are available and make reservations as needed.

14.3.6 TRAVELING WITH OXYGEN

Older adults with chronic medical conditions requiring oxygen should schedule an appointment with their physician to discuss management during travel. They need to review the requirements of airports regarding the use of supplemental oxygen prior to air travel, as well as contacting the airlines to arrange oxygen during the flight. Medical documentation from a physician is generally required.

14.3.7 PREVENTING DEEP VENOUS THROMBOSIS

Deep Venous Thrombosis (DVTs) are blood clots which develop in the deep veins, usually in the legs. The blood clots can break off and travel to the lungs; when this occurs, it is called a pulmonary embolus and is a life-threatening medical emergency. Older adults are at greater risk of having a DVT, especially during long-distance airplane flights, due to medical comorbidities. Some experts recommend the use of below-the-knee graduated compression stockings (GCS) during prolonged travel to help prevent DVTs. CGS can be bought at any medical supply house or large pharmacy. In addition, older adults should be encouraged to exercise and move their legs, take frequently walks in the aisle as appropriate, and if possible, sit in an aisle seat which allows easier movement.

14.3.8 MEDICAL INSURANCE

Older adult travelers should consider international travel insurance which may provide easier access to emergency medical resources and travel assistance. It is important to check your existing medical insurance policy to see if you have coverage for emergency medical and dental care in the case of a sudden illness or injury on your trip. Special attention should be paid to coverage for high medical costs that result from a recurring condition. Except in rare cases, Medicare will not cover health care while travelling outside of the United States.

14.3.9 TRAVEL-ACQUIRED INFECTIONS AND DEHYDRATION

Traveling via airplane, may constitute a risk for dehydration due to long wait times in airports, decreased access to water, and the dry air inside the airplane. Older adults are encouraged to take a water bottle with them, if permitted, and to request adequate water from flight attendants to meet their needs.

14.3.10 TRAVELLING WITH DEMENTIA

Travelling for older adults with dementia requires additional planning and coordination. Adults with dementia may have increased episodes of confusion and anxiety related to the stress of the travel and the new environment. Plans for travel should be discussed with health care providers so that the risks and benefits of travel can be explored. Recommendations for safe travel include attempting to keep the trip as simple as possible, with a focus on limiting additional activities and avoiding disruptions in the person's daily routines. For example, the time of the travelling should be planned around the person's sleep schedule. Generally, evening and nighttime travel should be avoided. Planning for travel delays is particularly important for adults with dementia, and backup supply of medications, clothing, snacks, as well as detailed medical information should be prepared. With advance notice, airports will provide wheelchairs with an assigned employee for persons with disabilities. This can help the person with dementia manage both security check-ins and navigation across the airport. The Alzheimer's Association has additional information on adults with early-stage Alzheimer's disease who wish to travel alone.

14.3.11 EMERGENCY TRAVEL KIT

Older adults are at significant risk for the most severe consequence of an emergency during travel, such as a motor vehicle accident, losing travel documents, getting lost, prolonged travel delays or cancellations. Preparing in advance for an emergency may decrease the risk of morbidity and mortality. Older adults should be encouraged to "pack smart" and place all critical items in their carry-on bag. This would include their medications, medical history, as well as destination-specific items such as insect repellent, sunscreen, alcohol-based sanitizer, water disinfectant tablets, and appropriate clothing (see Chapter 15, "Emergency Preparedness and Fire Safety").[1-3]

14.4 SEASONAL SAFETY FOR OLDER ADULTS

Older adults are vulnerable to seasonal variations in temperature, climate changes, and extreme weather because of increased prevalence of impaired cognitive ability, functional decline, and comorbid medical conditions. Weather-related factors, such as reduced daylight hours during winter, slippery surfaces caused by rain or snow, affect independence in mobility and may also lead to increased social isolation (see Chapter 19)

14.4.1 Hot Weather or Heat Extreme

Heat extreme is defined as summertime weather that is significantly hotter and more humid than average. Older adults are at increased risk of hospitalization due to dehydration, fluid and electrolyte disturbances, heatstroke, kidney failure, and death during periods of extreme heat.[1] Estimates are that nearly 80% of the approximately 12,000 people in the United States who die of heat-related causes annually are aged 60 years and older. As climate change increases and the earth continues to be hotter, predications are that the older adult population will suffer from more heat-related adverse events.[2,3] Older adults are unable to adjust like young adults to changes in temperature. Many older adults have normal age-related changes in organ function, chronic health conditions, and take medications which may contribute to heat intolerance. Older adults and their caregivers need to understand the precautions required to prevent excess heat exposure and know the danger signs of when medical assistance is required. They should be counselled to follow updates about hot weather and be prepared to take the required steps to remain safe. Most extreme heat injuries and hospitalization are preventable, and it is critical to stay informed, recognize warning signs, and adjust to the environment to prevent injuries.

Signs of dehydration due to excess heat include dizziness, weakness, muscle cramps, confusion, and nausea. This condition can progress to heat exhaustion and ultimately heat stroke which is a medical emergency. Signs of heat stroke include fainting or becoming unresponsive, change in behavior, very elevated temperature > 104 F°, dry flushed skin. The emergency response system should be activated via 911 for concerns about possible heat stroke.

14.5 EXPERT RECOMMENDATIONS TO PREVENT HEAT-RELATED ILLNESS IN OLDER ADULTS

- Stay in a cool area and indoors as much as possible
- Seek cooling centers or air-conditioned public places like malls or library
- Limit outdoor and exercise activities
- Drink water regularly
- Take cool showers or sponge baths
- Avoid using direct fans, especially when the temperature is in the high 90s
- Avoid using the stove and oven during heat waves
- Wear appropriate clothing including a hat when you are outdoors[4-7]

14.6 COLD WEATHER

During the winter months, older adults are susceptible to hypothermia, frostbite, and nonfreezing cold injuries. The ability to feel temperature change is diminished in older adults; so it is important to ensure that indoor thermostats are easy to access and read to facilitate checking changes in temperature. Adults age 65 years and older are at increased risk of death from hypothermia. Hypothermia is a medical emergency; it may lead to cardiovascular and respiratory-related death. Initial signs of hypothermia include cold hands or feet, continuous shivering, temperature < 95° F. These symptoms may progress to increased confusion and an unconscious state. Recommendations for older adults and caregivers to prevent hypothermia include wearing appropriate clothing including a warm hat, maintaining adequate heat, and ensuring supervision for adults with cognitive and functional decline. Maintaining older adults' safety during winter weather includes preparing for cold and winter storms ahead of time by checking supplies and resources, staying safe during the storm by having adequate heat and shelter and staying indoors, bundling up with multiple layers of clothing/bedding, avoiding excessive alcohol which can exacerbate cold-related illness, and preventing cold-related injuries.[3]

During extreme cold and winter weather, it's advisable to avoid travel. However, if travel is required, an emergency kit that matches the physical and cognitive status of older adults should be prepared (see Chapter 15).

14.7   EMERGENCY CAR KIT

- Mobile phone with a portable charger and extra battery supply
- Warming items such as extra hats, coats, mittens, and blankets
- Flashlight with extra batteries
- Water and snack food
- First aid kit
- Hearing aids and eyeglasses
- Hazard or other reflectors
- Bright colored flag or help signs
- Road maps

14.8   SNOW AND ICE CONDITIONS

Some of America's major cities have snow on the ground for more than three months each year. Studies have found that older adults have significantly more difficulty getting out of their homes during winter weather. Older adults are at increased risk of falling during winter. If they fall, they are at higher risk of developing severe consequences of falls (see Chapter 9). Recommendations to prevent falls in inclement weather include simply avoiding going outdoors. If this is not possible, additional recommendations include wearing boots with nonskid soles and adding an ice-tip attachment to canes and other mobility devices to increase traction on ice.[8,9]

14.9   FLOOD CONDITIONS

Flooding related to heavy rains, hurricanes, or coastal storms place older adults at increased risk of injuries and death. During recent flood emergencies older adults died at significantly higher rates. For example, over half of the mortality during Hurricane Katrina was in adults aged 75 and older, and in Superstorm Sandy, almost half of those that died were aged over 65 years. Tragically, there have been many nursing home deaths too. Following this, there is a new state and national focus on emergency preparedness for vulnerable older adults during flood disasters (see Chapter 15).[10,11]

14.10 WINDY CONDITIONS

The average daily wind speed in the United States is between 6 and 12 miles per hour. Older adults are at greater risk of complications from higher than usual wind speeds. Safe wind speed for walking outdoors for an older adult is generally lower, due to age-related changes in gait and balance. Walking on a windy day affects both mobility and balance and places older adults at risk of falling and fractures. Hence older adults should avoid walking outside during windy days.[12]

BIBLIOGRAPHY

[1] Shen S, Koech W, Feng J, Rice TM, Zhu M. A cross-sectional study of travel patterns of older adults in the USA during 2015: implications for mobility and traffic safety. *BMJ Open.* 2017 Aug 11;7(8). doi: 10.1136/bmjopen-2016-015780

[2] Alzheimer's Association. "Travel" www.alz.org/help-support/caregiving/safety/traveling. Accessed 02/02/2023

[3] wwwnc.cdc.gov/travel/page/senior-citizens

[4] Bobb JF, Obermeyer Z, Wang Y, Dominici F. Cause-Specific Risk of Hospital Admission Related to Extreme Heat in Older Adults. *JAMA.* 2014;312(24):2659–2667

[5] Harvard Medicine. The Magazine of Harvard Medical school. Aging. Autumn. 2021hms.harvard.edu/magazine/aging/effects-heat-older-adults

[6] United States Environmental Protection Agency (EPA) "Climate Chage and the Health of Older Adults." https://19january2017snapshot.epa.gov/sites/production/files/2016-10/documents/older-adults-health-climate-change-print-version_0.pdf. Accessed 1/23/2023

[7] Heat stress in older adults. http://emergency.cdc.gov/disasters/extremeheat/olderadults-heat.asp Accessed 1/23/2023

[8] Clarke PJ, Yan T, Keusch F, Gallagher NA. The Impact of Weather on Mobility and Participation in Older U.S. Adults. *Am J Public Health.* 2015 Jul;105(7):1489–94.

[9] Mayo Clinic. "Geriatric winter falls: when a simple bump on the ice can be serious" 01/02/2019. www.mayoclinic.org/medical-professionals/trauma/news/geriatric-winter-falls-when-a-simple-bump-on-the-ice-can-be-serious/mac-20451017. Accessed 02/02/2023

[10] Pairin Yodsuban, Khanitta Nuntaboot,Community-based flood disaster management for older adults in southern of Thailand: A qualitative study, International Journal of Nursing Sciences, 2021; 8(4):409–17.

[11] CDC's disaster planning goal: protect vulnerable older adults. www.cdc.gov/aging/pdf/disaster_planning_goal.pdf Accessed 1.23.2023.

[12] Lin WQ, Lin L, Yuan LX, Pan LL, Huang TY, Sun MY, Qin FJ, Wang C, Li YH, Zhou Q, Wu D, Liang BH, Lin GZ , Liu H. Association between meteorological factors and elderly falls in injury surveillance from 2014 to 2018 in Guangzhou, China. Heliyon. 2022 Oct 4;8.

15 Emergency Preparedness and Fire Safety

Robert Wolf, JD, MUP

CONTENTS

Older adults die at disproportionately higher rates in disasters such as earthquakes, fires, hurricanes, and power outages. The frequency of these events is increasing due to climate change and other factors. Older adults' ability to prepare for and respond to emergencies may be diminished due to impairments in function, cognition, and social conditions. This is likely to have contributed to the fact that during Hurricane Katrina in 2005, approximately 50% of fatalities were in people aged 75 years and older. Similarly, in the 2004 tsunami in Indonesia, the highest percentages of deaths occurred in adults aged 60–69 (23%) and those over 70 years old (28%).

Research has shown that the majority of older adults has no plans for emergencies and does not have a supply of food, water, medications, and medical supplies for emergency use (Regina et al. 2018). This is true even in "age-friendly" communities that have a broad range of policies and programs to support older adults.

DOI: 10.1201/9781003197843-15

The adverse effects on the elderly during these emergencies can often be mitigated through advance planning. Emergency preparedness is the responsibility of a complex web of local, state, and federal organizations. Importantly, the role of the health professionals in planning for emergencies affecting community-dwelling older adults is not clearly defined. Particularly in geographic areas prone to natural disasters, health professionals need to understand both the fundamentals of emergency planning and the vulnerabilities of older adults in the face of disasters.

15.1   BASIC ELEMENTS OF EMERGENCY PLANNING FOR OLDER ADULTS

Planning for emergencies should be an ongoing and evolving process that incorporates regular reviews and updates that account for changes in older adults' health, living situation, and threat level. Emergency preparedness plans must incorporate the basic elements that are listed in the upcoming pages.

15.2   RISK ASSESSMENT AND PLANNING

The ability to envision a range of emergency scenarios is critical for the development of effective emergency preparedness plans. A key consideration is whether to evacuate or shelter in place. Evacuation plans need to take into consideration any potential physical damage to the structure, as well as loss of power and telecommunications. The plan must also consider physical disabilities or other personal limitations that will impede rapid evacuation. Conditions that impair mobility, such as arthritis or Parkinson's disease, and the related reliance on assistive devices (canes, walkers, wheelchairs) can make evacuating more challenging. All plans should be shared with the person's support network in advance. The plan should also include a strategy for other household members, personal care workers, and pets.

15.3   RISK FACTORS FOR OLDER ADULTS IN AN EMERGENCY

Any emergency plan for older adults must focus on the factors and circumstances that place the older adults at greatest risk during a disaster.

- **Chronic health conditions**: Approximately 80% of older adults have at least one chronic health condition and 50% have two or more, leaving them especially vulnerable during a disaster. Conditions such as stress, lack of food or water, and extreme heat or cold can rapidly worsen the symptoms of chronic illnesses.
- **Durable Medical Equipment (DME):** Older adults may be dependent on DME that require electric power, such as oxygen concentrators and motorized wheelchairs. Most older adults do not have a source of back-up power for these devices and should contact utility companies and local officials in order to be placed on emergency lists of power-dependent customers.
- **Visual, auditory, or cognitive impairments:** Sensory and cognitive impairment can make it difficult for older adults to receive and understand critical communications during emergencies.

- **Physical disabilities and low mobility:** Many emergencies require rapid evacuation. Older adults who have low mobility due to frailty or physical disabilities may not be able to exit their home quickly without assistance.
- **Alzheimer's disease and related Dementia:** Patients with dementia, even if mild, often have worsening cognitive and behavioral symptoms when daily routines are disrupted.
- **Psychological:** Both depression and anxiety are common in older adults and symptoms may be heightened in disasters. Underlying mental health disorders may lead to a diminished or inappropriate response in the face of disasters.
- **Living alone:** Approximately 50% of adults who are 85 years or older live alone. In the event of an emergency requiring urgent evacuation, the condition of living alone may impede the older adult's ability to respond appropriately. Fear and anxiety may also be exacerbated when facing a crisis alone.
- **Reluctance to leave home**: Many older adults are reluctant or afraid to leave their homes when an emergency necessitates evacuation.
- **Poverty**: One in three older adults in the United States are financially insecure with incomes below 200% of federal poverty level. This may limit their ability to repair physical damage and access medicines and mental healthcare during and after a disaster.

15.4 COMMUNICATION

There are three elements to an emergency communication plan: contact information for everyone living in the residence, a defined meeting place in the event of evacuation, and medical and insurance information. The communication plan should be available in both a hard copy that can be carried in a wallet as well as an electronic copy that can be stored in a smart phone.

15.4.1 CONTACT INFORMATION

The phone numbers and email addresses of everyone in the household should be recorded. For older adults with hearing impairment or speech disability, information about any special communications services should be included. Contact lists should have at least one person who resides outside of the community, who could act as a central point of contact in the event of local communication disruption. Emergency contacts should be clearly identified in cellular devices as "In Case of Emergency (ICE)."

15.4.2 MEETING PLACE

It is important to choose a safe and familiar place where household members can meet to reunite in the case of evacuation. In addition to a local meeting place, a secondary location, outside of the neighborhood, should be identified if the community is no longer safe.

15.4.3 MEDICAL AND INSURANCE INFORMATION

An up-to-date list of all medications, allergies, and major medical conditions for each Older adult should be created. Contact information for healthcare providers and health insurance information should also be listed.

15.5 EDUCATION AND DRILLS

Public health recommendations are that all older adults, household members, and personal care workers receive ongoing education about what to do in an emergency. Since medical and functional needs and household and family composition can change, the following activities should take place once a year, and more often if risk levels are high:

- Emergency evacuation drills should be conducted. Practice drills are essential to help those responsible for executing the plan to become confident and be prepared.
- Communications plans should be updated.
- Disaster supplies should be checked and all perishable items such as food, water, batteries, and medications should be refreshed.
- Individualized plans for older adults should be developed. For example, older adults who are legally blind and use a probe cane should be instructed to keep an extra cane by their bed, with an attached whistle. They should also be educated to exercise caution when moving in an emergency, as items in the home may have shifted. For older adults who rely on home health aides and other in-home services, plans should take into account the potential interruption in these services.

15.6 FIRE SAFETY

Older adults are at a higher risk for injury and death from residential fires. According to the National Center for Injury Prevention and Control, adults aged 65 years and older are three times more likely to die in a residential fire than younger adults. Older adults who have physical, sensory, or cognitive impairments, abuse drugs or alcohol, live with smokers, and/or reside in substandard housing are even more vulnerable. Fire risks also increase during cold weather, reflecting the use of space heaters.

Correct installation of smoke and carbon monoxide alarms inside and outside every sleeping area has been demonstrated to be the most important risk reduction for morbidity and mortality from fire. The most effective systems, which are required for new home and multifamily dwellings, include both interconnected alarms and an interface with the fire department. Older homes and buildings should have alarms that are powered by long-life lithium batteries. Alarms should be tested regularly. Older adults with sensory impairment require specialized adaptive equipment. For example, for older adults with impaired hearing, options include strobe fire alarms and bed shakers for nighttime use.

Other fire safety fundamentals are these:

- Space heaters should be at least 3 feet away from flammable materials and plugged directly into a wall outlet, not into an extension cord or power strip.
- Portable generators should not be used indoors, and gas ovens should never be used as a source of heat. Both pose a serious risk for carbon monoxide poisoning to which older adults with chronic health conditions are highly vulnerable.
- Home Oxygen – patients and caregivers should be counseled regarding oxygen safety, including risks due to smoking, sedating medications, alcohol, and environmental risks.
- Fire evacuation plans should be practiced and fire escape routes need to be free from barriers and not blocked by furniture or clutter.

15.7 THE DISASTER SUPPLY KIT

Experts in disaster preparedness suggest storing enough items for three days to enable self-sufficiency, including food, water, batteries, clothes, and medicine. The kit should be modified to account for the individualized needs of older adults.

15.7.1 What to Include in a Disaster Supply Kit ("Go-Bag")

- Cell phone and a fully charged external battery and charging cable.
- Water: one gallon of water per person per day for at least three days, for drinking and sanitation.
- Food: at least a three-day supply of nonperishable food and a can opener.
- A warm blanket or sleeping bag.
- A complete change of clothing, including additional layers of clothing for cold weather climates.
- An up-to-date list of all medications, allergies, and major medical conditions. If there is advance notice, pharmacies should be contacted for emergency medication supplies.
- Contact information for medical and social service providers and health insurance details.
- Eyeglasses
- Hearing aids and hearing aid batteries.
- Battery-powered or hand-crank radio and extra batteries.
- Battery-powered or hand-crank flashlight and extra batteries.
- First aid kit.
- Whistle.
- Cash and credit or debit cards.
- Keys.
- Dust masks and eye protection.
- Personal care products such as sanitary supplies and incontinence supplies,
- Pet carriers and supplies. (In an emergency, it is better for older adults to take their pets with them if they evacuate. However, the pet's safety should never compromise their own safety).

- Identification cards, legal and insurance information, including copies of important identification and health documents such as Social Security cards, driver's licenses or other forms of ID, and healthcare and property insurance information.

15.8   DECIDING WHETHER TO EVACUATE OR SHELTER IN PLACE

The decision to evacuate or to shelter in place is central to surviving a disaster. Whenever possible, older adults should follow the instructions of local officials. Other factors to consider are the likelihood that dwellings will be damaged, making it difficult to shelter in place, and the presence of emergency back- up power. It is critically important that older adult sheltering- in- place should make themselves known to neighbors, houses of worship, community-based organizations (CBOs), and local authorities.

15.9   POST-DISASTER RISK AWARENESS

In the period immediately following a disaster, older adults are at particular risk of financial fraud. These should be reported to the U.S. Justice Department's National Center for Disaster Fraud (www.justice.gov/disaster-fraud) or the Federal Emergency Management Agency (www.fema.gov/about/offices/security/disaster-fraud).

15.10   DISASTER PLANNING FOR CBOS

A 2006 report from the American Association of Retired Persons (AARP) Policy Institute found that there was a troubling lack of clarity in the roles and responsibilities of the many different entities participating in emergency management, including federal, state, and local governments. The report recommended better engagement with aging and disability organizations and experts in planning for and responding to disasters.

The lack of emergency preparedness by older adults can be addressed, in part, through public education and emergency preparedness campaigns that encourage behavior change and measure and evaluate the efficacy of these communications. Studies show that the most common way older adults receive information during an emergency is through television, and that many older adults do not have access to online resources or smart phones. CBOs can play a key role in notifying vulnerable older adults about an impending disaster and assist them in responding appropriately.

When CBOs are unable to sustain operations during emergencies, individuals who rely on them for food and support are exponentially impacted. Organizations Preparing for Emergency Needs (OPEN) is a free training program, provided by Federal Emergency Management Agency (FEMA), that helps CBOs develop organizational disaster response plans. CBOs can also enhance their emergency capacities by working in tandem with other community service providers. Ideally, these important relationships should be established in advance of an emergency.

There are strict federal and state emergency preparedness requirements for Medicare and Medicaid providers and suppliers. This requirement focuses on four aspects of emergency preparedness: risk assessment and planning, policies and procedures, communication plans, and training and testing. Emergency plans must include procedures for evacuation and for sheltering in place, and in both circumstances must integrate a system for tracking patients. Nursing homes are required to provide for the subsistence needs of staff and residents, whether they evacuate or shelter in place, including food, water, medical and pharmaceutical supplies, backup power, and emergency lighting and alarm systems.

The special needs and circumstances of older adults and persons with disabilities need to be incorporated into community-wide disaster planning. Careful consideration should be given to the diversity of functional abilities among people with disabilities, rather than simply aggregating groups into categories such as "special needs" or "disabled." This allows more targeted emergency planning by identifying specific categories of functional needs that need to be addressed during an emergency.

15.10.1 SOCIAL DETERMINANTS OF HEALTH

In addition to functional abilities, disaster vulnerability is affected by social determinants of health. Knowledge of where vulnerable groups are concentrated within communities, and the general nature of their circumstances, is an important step towards effective emergency management. Emergency planners and policymakers should identify the most socially vulnerable older adults: those who are also poor, live in substandard housing, are immigrants, or have other social risk factors. It may be possible to have governmental or healthcare agencies or CBOs create "special needs" registries that identify, register, and track older persons who cannot evacuate on their own. However, registries can be problematic, because many people do not like to have their names on a government list. Los Angeles County suspended its registry operation because it only captured a small fraction of the county's 10 million people. There are also privacy and Health Insurance Portability and Accountability Act (HIPAA) concerns, the need for continuous updating, and the possibility of unrealistic expectation that responders will come to registrants' houses first. As an alternative, "Community Vulnerability Maps" integrating this information can be drawn up as part of this community planning process.

15.11 CONCLUSION

Older adults have multiple risk factors that contribute to their greater vulnerability during disasters. Emergency plans can help reduce both morbidity and mortality, but many older adults do not have emergency plans in place. An emphasis on enhancing the education and preparedness of older adults, caregivers, and professionals involved in their care is critical for reducing risks to older adults during a disaster.

BIBLIOGRAPHY

Acosta D, Shih RA, Chen EK, Xenakis L, Carbone EG, Burgette LF, Chandra A. Building older adults' resilience by bridging public health and aging-in-place efforts. *RAND Corp*; 2018. [online]. Available at: www.rand.org/pubs/tools/TL282.html

Aldrich N, Benson WF. Disaster preparedness and the chronic disease needs of vulnerable older adults. December 15, 2007. www.ncbi.nlm.nih.gov/pmc/articles/PMC2248769.

Al-rousan M, Rubenstein LM, Wallace RB. Preparedness for natural disasters among older US adults: a nationwide survey tala *Am J Public Health.* 2014;104(3):506–511.

American Red Cross. Emergency preparedness for older adult. www.redcross.org/get-help/how-to-prepare-for-emergencies/older-adults.html. Accessed March 2021.

American Red Cross Scientific Advisory Council, in partnership with the American Academy of Nursing. *Closing the gaps: advancing disaster preparedness, response and recovery for older adults, 25 evidence-informed expert recommendations to improve disaster preparedness, response and recovery for older adults across the United States.* January 2020. www.redcross.org/content/dam/redcross/training-services/scientific-advisory-council/253901-03%20BRCR-Older%20Adults%20Whitepaper%20FINAL%20 1.23.2020.pdf

Bell SA, Kullgren JT, Solway E, Malani P. Supporting the health of older adults before, during and after disasters. *Health Affairs Blog*; November 27, 2019. www.healthaffairs.org/do/10.1377/hblog20191126.373930/full.

Brasher K. An Age-Friendly Approach to Disaster Recovery; Central Hume Primary Care Partnership. May 2020. https://extranet.who.int/agefriendlyworld/wp-content/uploads/2020/05/AgeFriendlyApproachToDisasterRecovery.pdf

Capacity-Building Toolkit for including Aging & Disability Networks in Emergency Planning; National Association of County Health Officials, 2019. www.phe.gov/ASPRBlog/Lists/Posts/ViewPost.aspx?ID=342. Accessed March 2021.

Core Emergency Preparedness Rule Elements. www.cms.gov/Medicare/Provider-Enrollment-and-Certification/SurveyCertEmergPrep/Core-EP-Rule-Elements.   Accessed March 2021.

Cornell VJ, Cusack L, Arbon P. Older people and disaster preparedness: a literature review. *Aust J Emerg Manag.* 2012;27(3):49–53.

Disaster preparedness and planning for populations with dementia and other vulnerabilities: an annotated bibliography, Alzheimer's Association, August 2016, www.alz.org/media/Documents/spotlight-disaster-preparedness.pdf. Accessed March 2021.

Gibson MJ, Hayunga M. We can do better: lessons learned for protecting older persons in disasters. AARP Public Policy Institute; 2006. https://assets.aarp.org/rgcenter/il/better.pdf.

Kailes J, Enders A. Moving beyond "special needs": a function-based framework for emergency management and planning. *JDPS*; 2007; 17(4):230–237.

Levac J, Toal-Sullivan D, O`Sullivan TL. Household emergency preparedness: a literature review. *J Comm Health.* 2012;37(3): 725–733. https://doi.org/10.1007/s10900-011-9488-x

Lisa C, Ford E, Okoro C. Natural disasters and older US adults with disabilities: implications for evacuation. March 14, 2007 https://doi.org/10.1111/j.1467-7717.2007.00339

Morrow BH. Identifying and mapping community vulnerability. International hurricane center, Florida International University. *Disasters.* 1999;23(1):1–18.

National Center for Disaster Medicine & Public Health. Caring for older adults in disasters: a curriculum for health professionals. (n.d.). www.usuhs.edu/ncdmph-learn/KnowledgeLearning/2015-OAC.htm. Accessed March 2021.

National Council on Disability. Effective communications for people with disabilities: before, during, and after emergencies. 2014. www.ncd.gov/publications/2014/05272014. Accessed March 2021.

Older Adults and Disasters – American Psychological Association. www.apa.org/pi/aging/resources/older-adults-disasters.pdf. Accessed March 2021.

Older persons in emergencies: an active ageing perspective, World Health Organization, 2008.

Ready.Gov preparedness training for community-based organizations. https://community.
    fema.gov/opentraining. Accessed March 2021.
Regina et al. Improving Disaster Resilience Among Older Adults. *Rand Health Q*uarterly
    2018;8(1):3.
Resources for emergency health professionals. https://emergency.cdc.gov/health-profession
    als.asp. Accessed March 2021.
Shih R, Acosta J, Chen E, Carbone E, Xenakis L, Adamson D, Chandra A. Improving disaster
    resilience among older adults: insights from public health departments and aging-in-
    place efforts. *Rand Health Quarterly* 2018;8(1):AugPMC6075802.
U.S. Administration for Community Living Emergency Preparedness Resources. https://acl.
    gov/emergencypreparedness. Accessed March 2021.

16 Safety Considerations for Older Workers

Sara J. Czaja PhD and Joseph Sharit PhD

CONTENTS

16.1 INTRODUCTION

A critical component of any workplace is the safety, health, and well-being of workers. With the aging of the population and of the workforce, employers and organizations need to ensure that the needs and characteristics of older workers are included in health and safety programs and plans. Workforce aging raises several issues about the age-related impact of work on health. Some of these issues are unique to older adults, and some apply to workers of all ages: for example, the impact of aging on work capacity and susceptibility to occupational diseases and hazards; how best to design jobs and workplaces to maximize the productivity and health and safety of older workers. As noted by a National Academy of Sciences report,[1] there are several important reasons for focusing on the health and safety needs of older workers. An overarching reason is that to maximize employment opportunities, health and safety and productivity, and satisfaction of older workers, it is important to understand what types of tasks older workers are best suited for and how to optimize workplace and job accommodations and support programs to best accommodate employees as they age. This will also help to maximize benefits to organizations and the economy.

The literature on aging and work is evolving as the nature of jobs and organizations are changing. However, an evidence base regarding aging and work is available that can provide guidance regarding strategies to help maximize the health, safety, and productivity of older workers.

What defines an older worker? It is always difficult to define or provide an age cutoff for an "older worker" as there is no consensus about the term – it varies according to context and occupation. Policy makers, business/industry, older adults, and researchers have different perspectives on the meaning of the term "older worker." For example, in the general literature, someone of age 65 and older is generally referred to as an older adult. The U.S. Department of Labor defines an older

DOI: 10.1201/9781003197843-16

worker as someone 55 years or older. The Age Discrimination in Employment Act protects individuals aged 40 and older from discrimination in the workplace based on age. Similarly, a report by the Institute of Medicine[1] focusing on the health and safety of older workers adopted the age of 45 as the cutoff for defining an older worker. This cutoff is likely related to concerns regarding physical workplace demands relative to normative age-related changes in physical abilities. Overall, although chronological age is important with respect to work abilities, the most important factors are the individual's health, their attitudes and preferences about work, and their abilities relative to the job demands.

Until recently, older workers made up a small segment of the U.S. labor force. Now, that picture is changing and the trend towards "early retirement" where workers permanently leave the workforce in their mid-60s is altered, resulting in an "aging" workforce. Currently workers aged 55+ make up about 40% of the workforce[2] and in the next decade, workers aged 65–74 and 75+ are projected to have the fastest rates of labor force growth. Specifically, between 2014 and 2024, the labor force growth rate of people aged 65–74 is expected to be about 55% and for those 75 and older, about 86%; this is in comparison to slower rates of growth for other age groups (Figure 16.1). Employment rates for older workers are increasing for both men and women but tend to be slightly higher for males. Reasons for why many older people are working longer are varied and include concerns about retirement income, healthcare benefits, or a desire to remain socially and productively engaged. Many employers also need to retain older workers. Given current demographic and workforce trends, in many occupations there are not enough younger workers to replace

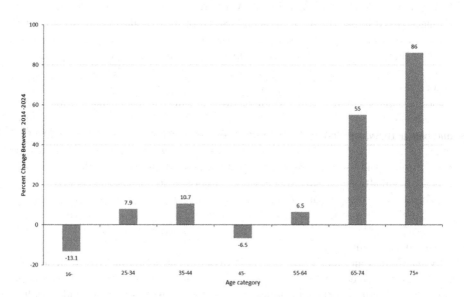

FIGURE 16.1   Projected Changes in the Workforce Between 2014 and 2024 as a Function of Age.

Source: 3

older workers currently in the workforce, which could result in labor shortages. This trend has been particularly compelling during the COVID-19 pandemic.

Older adults are employed in many occupations across a variety of industries but tend to be more concentrated in the service sector, including educational and health services, the management and professional occupations, or sales and retail trade. The rate of part-time work is higher among workers aged 65+.[4] Older workers are also increasingly seeking other employment options such as contract work, opportunities to start second (or third) careers, or to start a new business. It is important to note that because of the continual deployment of technology into the workplace, the nature of jobs and structures of organizations are also changing, which has implications for the employment of older workers. Many workers in fact are in less physically demanding jobs (e.g., more service sector work and less agriculture, manufacturing, and mining work); older adults thus may desire as well as be able to continue to work into their later years.

However, though jobs are becoming less physically demanding, there is evidence that they are becoming more cognitively demanding, which may create challenges for older workers given age-related cognitive changes. Potential problems with skill obsolescence are an additional challenge confronting many older workers. In fact, recent data from a report by the Pew Research Center[5] indicate that about 54% of adults who are currently in the labor force say that it will be essential for them to get training and develop news skills throughout their work life in order to keep up with changes in the workplace, and about 35% of workers perceive that they do not have the education and training they need to advance in their jobs. This chapter advocates that worker training programs need to include health and safety training commensurate with the demands of today's jobs.

Increasing numbers of workers are also working from remote locations such as their home or satellite workplaces. Opportunities to engage in telework may be particularly beneficial for older workers as telework is amenable to part-time work, flexible work schedules, and working within the home, which may help to offset problems that many older adults have with driving, mobility limitations, or caregiving responsibilities. Working in home settings may, however, result in health and safety challenges. For example, poorly designed workplaces in home environments, lack of a suitable chair and desk, or inadequate lighting can result in musculoskeletal disorders due to improper postures or eyestrain. Thus, in today's workplaces health and safety practices and programs such as worker safety training programs may need to extend beyond the traditional workplace into the home and satellite workplaces. Working from home may also limit opportunities for social interaction with friends and colleagues and thus, as underscored in the COVID-19 pandemic, exacerbate problems with social isolation. Social isolation is linked to negative mental, physical, and cognitive health consequences; consequently, it is important to consider the social implications when designing these types of tasks.

In this chapter we will briefly review some findings regarding older workers and outline some recommendations for maximizing the health and safety and productivity of an aging workforce. At the outset it is important to note that older adults, however defined, are not a homogeneous group. Aging is associated with variability and there are vast interindividual and intraindividual differences among older adults. For example, people in their 90s are typically different with respect to functional abilities than those in their 60s. Further, there may be differences among cohorts due to life experiences and historical events. Interindividual differences also reflect differences that occur within an age group: not all 65-year-olds are alike. People vary in terms of language, education, and skill level, which has vast implications for work. Further, the term older worker is likely to be redefined in the future as the perceptions of what it means to be old are changing.

16.2   CHARACTERISTICS OF OLDER WORKERS: MYTHS AND REALITIES

There are many myths about older workers, and unfortunately both managers and some older adults themselves believe those myths. Prominent myths include the notion that older workers are less productive than younger ones, are less interested in work, and less willing to learn new skills, particularly technology skills, and have higher rates of injury and absenteeism. With respect to age differences in productivity the available literature indicates that there is little relationship between age and job productivity, measured through studies of work output and supervisor ratings. Also, the relationship between age and job performance is complex and depends on the nature of the job and the performance metric (e.g., speed vs. accuracy). For example, although some cognitive abilities such as processing speed and working memory decline with age, age-related increases in knowledge, particularly job-related knowledge, more than compensate for these declines, enabling older workers to remain productive. There are also contextual cues in work environments that can aid performance. In addition, older adults generally have lower rates of absenteeism and lower rates of turnover.

Older workers also have lower accident rates than younger workers. Reasons for the lower rate of accidents among older worker may be due to several factors such as experience gathered in the workplace, increased caution and awareness of relative physical limitations, changes in choice of work roles or careers, or changes in job demands such as the increased emphasis on knowledge work. However, older workers have on average more slips, trips, and falls than younger workers, and generally workplace injuries are more costly on average for older than younger workers. Data[6] from a recent review indicate that occupations for which older workers were at the greatest risk for injury included agriculture, forestry, fishing, and hunting; construction; mining, quarrying, and oil and gas extraction; and transportation and warehousing. Injuries that occurred in these occupations were likely due to a mismatch of physical ability in older workers and occupational job demands. Other recent data[7] indicate that employees ages 55–64 are nearly twice as likely as the youngest adult workers to have experienced work-related injuries or illnesses, with back pain being the most prevalent injury or illness. Back pain can result from either activity or inactivity. The use of force, like lifting or moving heavy objects, can cause back pain; an inactive job or a desk job can also contribute to back pain, especially with poor posture or sitting in a chair with inadequate back support.

When accidents involving older workers do occur, older workers often require more time to recover and incidents affecting older workers are more likely to be fatal

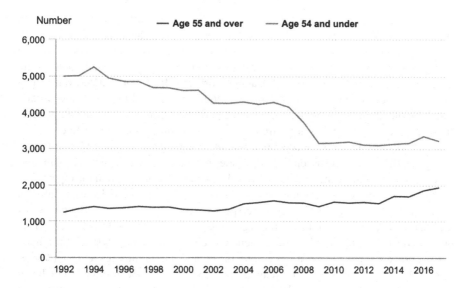

FIGURE 16.2   Fatal Occupational Injuries by Age.

Source: 8

(Figure 16.2). Recent data from the U.S. Bureau of Labor Statistics[8] indicate that while fatal occupational injuries for workers overall declined by 17% from 1992 to 2017, older workers incurred 56% more fatal work injuries in 2017 than in 1992. This trend was especially pronounced for workers in the oldest group, those age 65 and over. This clearly underscores the need for employers to identify strategies to adapt work conditions of work to protect workers of all ages as well as explore opportunities for preventative programs that can maintain or build the health of employees through their working life. As will be noted in the following pages, applications of technology, such as the use of exoskeletons, may also decrease the injury potential for workers of all ages.

There are several ways older workers differ from younger workers, which can have important implications for worker health and safety. Changes in vision that occur with age such as declines in acuity, color vision, and heightened sensitivity to glare have implications for the design of written instructions and manuals and lighting requirements. For example, levels of illumination need to be higher for older adults and potential sources of glare need to be minimized. Many older adults also experience some decline in audition that has relevance to work settings and health and safety. High frequency alerting sounds such as beeps or alarms on equipment may be difficult for older adults to detect. Changes in audition may also make it more difficult for older workers to detect auditory alarm signals or communicate in noisy work environments. Aging is also associated with changes in motor skills, including slower response times, declines in the ability to maintain continuous movements, disruptions in coordination, loss of flexibility, and greater variability in movement.

There are also age-related declines in strength and endurance. These age-related physiological changes have implications for worker health and safety. High-risk jobs for older workers are jobs that present exposure to work risks such as biomechanical risk factors. Jobs that involve lifting or carrying or are highly physically demanding or highly paced pose health risks for aging workers. Many work tasks, even if they do not involve strenuous effort, impose meaningful physical demands on the worker. For example, light assembly tasks may have periodic lifting or carrying components to them. Even if these tasks are infrequent or do not subject the worker to excessive forces, awkward postures, or repetitive motions, they may place the older worker at risk given the increased susceptibility of older workers to such stresses. A main concern with respect to lifting and carrying is the occurrence of back pain or injury.

As previously noted, sedentary work per se can also impose risks, especially on older workers. From a physiological perspective, the advantages of sedentary work are that they take the weight off the legs, stabilize the upper body posture, reduce energy consumption, and place fewer demands on the circulatory system. Prolonged sitting, however, can lead to a slackening of the abdominal muscles and to a curvature of the spine especially if the seating arrangement is poorly designed, which can accelerate the deterioration of the intervertebral discs.[9] Older workers are also more likely to have a chronic condition such as arthritis or hypertension. The prevalence of chronic diseases and comorbidities in older adults has safety implications and impact how and when workers can physically perform their duties. Safety risks for older workers with comorbid conditions are exacerbated for some types of tasks, especially under hazardous work conditions.

Recently, there has been increasing recognition of work-related mental health and psychosocial issues among older workers. Certain workplace situations may have disparate effects on older workers' mental health, such as ageism, increasing physical and cognitive demands, and pressure to retire.[1]

16.3 ENHANCING THE HEALTH AND SAFETY OF OLDER WORKERS

Workplace safety is a broad topic that encompasses many domains, such as protection from chemical and biological agents, noise and lighting issues, vibration, radiation, temperature extremes, unsafe equipment and work areas, poorly designed jobs, and ergonomic hazards (e.g., poorly designed workplaces). Psychosocial factors, such as ageism, also play an important role. There are numerous strategies that can be used to maximize the health and safety of aging workers and make the work environment "age friendly." In this regard it is important to note that the environment in the workplace includes both the physical setting and the structural arrangements of work, such as number of hours worked per week, type of work shift, availability of healthcare and other benefits, training opportunities, and whether employment is full or part-time.

In general, the disciplines of human factors, ergonomics, and occupational safety promote holistic, human-centered approaches that consider physical, cognitive, social, organizational, environmental, and other relevant factors with the goal of making tasks, jobs, products, environments, and systems compatible with the needs, abilities, and limitations of people. From a person-environment fit perspective, if the job has demands that the worker cannot meet, unsafe practices can be incurred, which can result in a higher risk for injury and poor health outcomes. There are numerous guidelines in these literatures that can be used to enable older workers to stay on the job, avoid injury, and facilitate return after illness of injury. Generally, these recommendations take the form of changes in workplace and job design, worker training, and provision of tools. In all cases the emphasis should be on promotion of worker health and safety. Here we summarize some of these recommendations. It should be noted that many of these recommendations benefit workers of all ages.

- Workplace flexibility: To the extent possible, allow workers to have some flexibility in their schedule, work conditions, work organization, work location, and work tasks.
- Match tasks to worker abilities: Avoid jobs that are physically demanding (e.g., heavy lifting, highly paced work, highly repetitive tasks), and allow for self-directed rest breaks.
- Avoid prolonged, sedentary work: For workers who traditionally sit all day, consider sit/stand workstations, and build activity breaks into the job. Invest in ergonomically designed chairs that provide adequate back support.
- Provide work environments that adhere to ergonomic guidelines: Ensure that the workplace is as adjustable as feasible; furthermore, any adjustment mechanisms should be easy to identify and use. If adjusting the workplace is not feasible, consider adjusting the worker's position relative to the workplace. Consider placement of work-related objects and devices to generate appropriate postures and avoid musculoskeletal strain and potential injuries.
- Design of the ambient environment: Provide sufficient levels of illumination, minimize sources of glare, and avoid high brightness ratios between the task area and the surrounding area. Minimize ambient background noise and ensure that warning signals can be detected.
- Manage hazards: Minimize the potential for slip/trip hazards by focusing on the characteristics of walking and standing surfaces and footwear, and physical hazards such as unguarded dangerous equipment or falling objects. Ensure that workers have the proper safety equipment and are trained in their use.
- Provide health promotion and lifestyle interventions: Provide opportunities for physical activity, healthy meal options, risk factor reduction and screenings, coaching, and onsite medical care. Accommodate medical self-care in the workplace and time away for health visits.
- Function allocation: Consider automating tasks that are physically demanding or hazardous, or that require high speeds of routine information processing.
- Provide aiding and support tools: Technological applications such as decision and memory aids can help offset age-related changes in cognition. Other technology aids such as exoskeletons can help reduce the physical demands of jobs.
- Invest in worker training: Help older employees adapt to new technologies and changes in jobs and work environments. Invest in worker safety programs.

- Advocate aging workforce training for managers and supervisors: Raise awareness of the benefits of an aging workforce, characteristics of older workers, and strategies to maximize worker health and safety, productivity, and satisfaction.

REFERENCES

[1]   Institute of Medicine. *Health and safety needs of older workers*. Washington DC: National Academies Press; 2004. https://doi.org/10.17226/10884

[2]   Schramm J, Figueiredo C. *The US essential workforce ages 50 and older*. Washington DC: AARP Public Policy Institute; 2020. Fact Sheet, 655, September. https://doi.org/10.26419/ppi.00111.001

[3]   Toossi M. Labor force projections to 2024: the labor force is growing, but slowly. *Monthly Labor Review*, U.S. Bureau of Labor Statistics, December 2015. https://doi.org/10.21916/mlr.2015.48

[4]   Toosi M, Torpey E. Older workers: labor force trends and career options. U.S. Bureau of Labor Statistics, May, 2017.

[5]   Pew Research Center. Changes in the American workplace. Report, October. Washington DC: Pew Research Institute; 2016. www.pewresearch.org

[6]   Stoesz B, Chimney K, Deng C, Grogan H, Verena Menec V, Piotrowski C, Shooshtari S, Turner N. Incidence, risk factors, and outcomes of non-fatal work-related injuries among older workers: a review of research from 2010 to 2019. *Safety Science*. 2020;126:104668. https://doi.org/10.1016

[7]   Allen K. Workers 55–64 report highest rates of work-related illness or injury. Washington DC: American Association of Retired Persons; 2020. www. aarp.org

[8]   Smith S, Pegula S. Fatal injuries to older workers. *Monthly Labor Review*, U.S. Bureau of Labor Statistics; January 2020. www.bls.gov/opub.../fatal-occupational injuries to older workers

[9]   Sharit J, Czaja SJ. *Job design and re-design for older workers*. In: Hedge JW, Borman WC, eds. *The Oxford handbook of work and aging*. Oxford: Oxford University Press; 2012, pp. 454–482.

[10] Czaja SJ, Boot WR, Charness N, Rogers WA. *Designing for older adults: principles and creative human factors approaches*. 3rd ed. Boca Raton, FL: CRC Press; 2019.

17 Mental Health
Depression, Anxiety, and Social Isolation

Alexandra Woods, PhD and Jo Anne Sirey, PhD

CONTENTS

Mental health difficulties such as depression and anxiety are not a normal part of aging. However, there are many factors that make older adults vulnerable to developing psychiatric conditions. This is an important public health problem as mental health disorders are associated with impairments in physical, cognitive, and social functioning, all of which pose threats to safety in late life. Individual and community-level interventions can reduce the impact of mental health conditions and their related safety risks.

17.1 DEPRESSION

Major depressive disorder is common among older adults, with a prevalence of 1–5% in community samples of adults aged 65 and older (1). While everyone's moods change day to day, low mood may be considered clinically significant depression when it persists or interferes with daily life. It is thought to have multiple causes that involve biological, psychological, and social factors. Biological factors often reflect an imbalance in brain chemistry that can be brought about by genetic predispositions,

DOI: 10.1201/9781003197843-17

age-related cognitive changes, medication side effects, or co-occurring medical illnesses such as cancer, heart disease, diabetes, or Parkinson's disease. Psychological factors that impact depression may include variables such as personality, coping skills, beliefs, and behaviors. Social factors that play a role in late-life depression are commonly loss of loved ones, retirement, lack of social support, declining independence, and financial difficulties.

Approximately half of older adults with depression are experiencing this condition for the first time in late life, while the other half have experienced their first episode of depression at a considerably younger age (2). Those with late-onset depression are typically less likely to have a family history of depression and may be more likely to have vascular risk factors, cognitive deficits, and other structural brain changes that impact mood regulation.

Depression is often underdiagnosed and undertreated, particularly among older adults. Common myths about late-life depression are that the symptoms will go away on their own, one is too old to get help, getting help is shameful or a sign of weakness, or that it is not safe to tell a doctor how one really feels. Sometimes these beliefs reflect generational or cultural norms that are difficult to challenge. However, depressive symptoms are not a normal part of aging. It is not the usual response to loss or the physical ailments of growing older. In fact, contrary to common perception, major depression is less frequent among older adults than among younger adults (1).

17.2   SYMPTOMS OF DEPRESSION

17.2.1   KEY SYMPTOMS

- Depressed or sad mood
- Decreased interest or pleasure in activities

17.2.2   OTHER SYMPTOMS

- Significant changes in appetite or weight
- Sleep disturbances
- Restlessness or sluggishness
- Fatigue or loss of energy
- Lack of concentration or indecision
- Feelings of worthlessness or inappropriate guilt
- Thoughts of death or suicide

Although most individuals experience some form of the above symptoms from time to time, a diagnosis of depression is made if five or more of the nine symptoms are present during the same two-week period and represent a change from previous functioning (3). For a diagnosis of depression, at least one of the symptoms must be either (1) depressed mood or (2) loss of interest or pleasure. It is important to note that this loss of interest or pleasure is not due to inability to engage in activities, but rather a lack of motivation or enjoyment. In addition, the episode is not attributable to the physiological effects of a substance, another medical condition, or a psychotic disorder.

Older adults may be less likely than younger adults to endorse the cognitive and affective symptoms of depression, such as low mood or guilt. They are more likely to report somatic symptoms including sleep disturbances, fatigue, and sluggishness, as well as loss of interest in living and hopelessness about the future (4).

17.3   RISK FACTORS

- Family history of depression
- Medical illness, particularly cardiovascular disease, diabetes, dementia, stroke, and Parkinson's disease
- Disability
- Stressful life events such as bereavement and caregiving
- Poor social support
- Financial strain

17.4   TREATMENTS FOR DEPRESSION

Depression is a treatable condition. Appropriate treatment such as medication, psychotherapy, or a combination of both have been shown to relieve symptoms for many older adults.

Pharmacologic treatments: Depression can be treated by a group of medications called antidepressants. A class of antidepressants called selective serotonin reuptake inhibitors (SSRIs) are advised as the first-line agents for treating depression in older adults (5). Studies show that response rates to antidepressants in depressed older adults are similar to those of younger adults (6). However, the benefits of antidepressants seem to decline with increasing age, which may be a result of coexisting medical conditions, brain changes, or a suboptimal dosage. Pharmacotherapy may be more complicated among older adults due to potential interactions between antidepressants and other medications. Some older adults may need help adhering to their medication regimen because of cognitive status or medical comorbidities. Involvement of family members or other support persons can help improve medication adherence and depression outcomes (7).

Psychotherapy interventions: Psychotherapy has been found to be as effective as antidepressant medication in treating depression among older adults (6). This offers a promising alternative for those with poor tolerance for antidepressants. A number of evidence-based manualized treatments have been adapted for late-life depression.

- *Behavioral activation* seeks to break the relationship between low mood and low activity level by reengaging the individual in pleasurable and meaningful activities.
- *Cognitive therapy* teaches a set of tools to challenge negative beliefs about one's self, others, and the world that have developed as a result of past experiences.

- *Problem-solving therapy* provides training in skills to overcome barriers to effective problem solving.
- *Brief psychodynamic therapy* focuses on increasing awareness of how past experiences impact current functioning and relationships.
- *Life review therapy* was designed specifically for older adults with mild to moderate depression and aims to promote greater acceptance of one's life by recounting the life story and resolving past conflicts.

Electroconvulsive therapy (ECT): ECT is a medical treatment for individuals with severe depression that has not responded to other treatments. It involves a brief electrical stimulation of the brain while the patient is under anesthesia. ECT is used more often in older adults than in any other age group (8). Improvement is seen in more than 80% of patients in most trials, but ECT should be used with caution in older adults due to the risk of adverse effects such as cardiac complications and memory loss.

17.5 ANXIETY

Occasional anxiety is an expected part of life. It is a normal reaction to stress and can even be beneficial in some situations by alerting us to danger, motivating us prepare, and helping us pay attention. Anxiety may arise when facing a stressful problem, navigating a life change, or making an important decision. However, anxiety disorders involve more than temporary worry or fear. For a person with an anxiety disorder, the anxiety is out of proportion to the situation and interferes with normal functioning. Anxiety disorders are common among older adults, with prevalence estimates ranging between 1.2 and 15% in community-dwelling older adults and almost twice as high in clinical settings (9). Similar to major depressive disorders, anxiety disorders are thought to be caused by a combination of biological, psychological, and social factors. The majority of anxiety disorders have been found to develop between childhood and young adulthood, with fewer than 1% developing an anxiety disorder after the age of 65 (10). The detection of anxiety disorders in late life may be complicated by medical comorbidity, cognitive decline, unique changes in life circumstances, and differences in the reporting of symptoms. Anxiety symptoms and disorders have been found to be associated with increased disability, mortality, and use of health services, as well as decreased cognitive functioning (11).

There are several types of anxiety disorders, including generalized anxiety disorder, panic disorder, specific phobias, agoraphobia, social anxiety disorder, and separation anxiety disorder (3). The prevalence rate of generalized anxiety disorder is estimated to be 1.2–7.3% among older adults (12). The other categories of anxiety disorders are found at substantially lower rates in older adults (9). The remainder of this section therefore focuses on generalized anxiety disorder, and the term "anxiety" will refer to generalized anxiety rather than other specific types of anxiety.

17.5.1 Symptoms of Anxiety

- Feeling nervous, anxious, or on edge
- Not being able to stop or control worrying

- Worrying too much about different things
- Trouble relaxing
- Being so restless that it is hard to sit still
- Becoming easily annoyed or irritable
- Being afraid, as if something awful might happen

A diagnosis of generalized anxiety disorder (GAD) is made if the presence of anxiety and worry about a variety of topics occurs more often than not for at least six months and is clearly excessive (3). Excessive worry means worrying even when there is no specific threat present or the worry is disproportionate to the actual risk. The worry is experienced as very challenging to control and is accompanied by at least three of the following symptoms: edginess or restlessness, fatigue, impaired concentration or feeling as though the mind goes blank, irritability, increased muscle aches or soreness, or difficulty sleeping. For a diagnosis of GAD, these symptoms must make it hard to carry out day-to-day activities or cause problems in relationships, at work, or in other important areas of life. The symptoms must also be unrelated to any other medical conditions, not explained by a different mental disorder, and not caused by the use of a substance such as medication, alcohol, or recreational drugs.

Compared to younger individuals, older adults may report more somatic symptoms of anxiety such as restlessness or trouble relaxing, rather than the cognitive symptoms of worry thoughts (9). The content of older adults' worry also tends to focus more on health, disability, and fear of being a burden on others, with fewer concerns about work, finances, and family (12).

17.5.2 RISK FACTORS

- Being female
- Having several chronic medical conditions
- Being single, divorced, or separated (compared to being married)
- Lower education
- Impaired subjective health
- Stressful life events
- Physical limitations in daily activities
- Adverse events in childhood

17.5.3 TREATMENTS FOR ANXIETY

Anxiety among older adults can be treated using medication, psychotherapy, or a combination of both.

Pharmacologic treatments: Antidepressant medications, particularly SSRIs, are widely used to treat anxiety and are considered the preferred agents for use in older adults (13). Antihistamines and beta-blockers, a type of medication used for heart conditions, are sometimes used to control the physical symptoms of anxiety. They are typically taken only as needed or immediately before an anxiety-provoking event. A type of medication called benzodiazepines can provide immediate relief for acute anxiety, but is known to negatively affect cognitive functioning and psychomotor performance. To reduce risk of falls, impairments in driving, and other adverse effects in older adults, benzodiazepines are recommended to be used at the lowest possible dose and only as an adjunct to another primary agent such as an SSRI (13).

Psychotherapy interventions: Psychotherapy can significantly improve anxiety symptoms among older adults and may be a preferable alternative to medication (14). The effects of behavioral interventions on older adults are comparable to those reported in younger patients.

- *Cognitive therapy* identifies and restructures patterns of thinking that fuel anxiety, particularly distorted perceptions of risk or threat.
- *Behavioral relaxation training* involves exercises that counteract the body's physiological stress response.
- *Biofeedback* helps to achieve a more relaxed state at will by watching your own brain-wave patterns on an electroencephalograph and learning how to control the waves.
- *Supportive-expressive therapy* explores conflicts that have developed in early learning experiences and how these are represented in current situations and relationships.

17.6 SOCIAL ISOLATION

Social relationships play an important role in wellness, with older adults more commonly endorsing social engagement rather than physical health when describing successful aging (15). Social isolation may be defined by a lack of belonging, social contacts, quality relationships, fulfilling relationships, and/or engagement. While some older adults maintain strong social connections through neighborly socializing, participation in religious communities, and volunteering (16), many older adults are at risk for reduced social contact due to factors such as loss of loved ones, decreased mobility, health conditions such as hearing impairment, changes in work and family roles, or changes in economic status (17). Other groups who may be at particularly high risk are older adults who live in rural areas and those who are not skilled at using social media. Older adults who live alone may go days without seeing or communicating with another person.

Social isolation has been shown to have severely detrimental effects on both physical and mental health. Isolated older adults are more likely to engage in unhealthy behaviors such as poor diets, tobacco use, and heavy alcohol use (18). Loneliness among older adults is associated with greater functional decline such as difficulty with bathing, eating, and general mobility, and a 45% increased risk of death (19). Social isolation is also strongly associated with depressive symptoms, underlying a significant proportion of depressed mood and accounting for the excess of depression seen in widowhood (20). Furthermore, depressed older adults with greater social support are more likely to benefit from treatment and sustain improvements in mood over time (21).

Given the impact of social isolation on health outcomes, it is recommended that primary care settings, community health clinics, and hospitals incorporate assessment of social isolation into standard care (17). By screening and identifying isolated individuals, at-risk older adults can be referred to community resources to increase social support. Though there is a lack of evidence-based interventions for social isolation, one approach may involve one-on-one befriending protocols in which volunteers make phone calls or in-person visits to socially isolated older adults (22).

17.6.1 MENTAL HEALTH AND INCREASED SAFETY RISKS

Suicide: As noted earlier, one of the nine symptoms of depression is thoughts of death or suicide. Older adults have historically shown the highest suicide rates (23). Older white men have the highest rate of suicides at almost four times the United States' overall age-adjusted rate (24). Risk factors for suicide may include previous suicide attempts, specific behaviors suggestive of suicide planning, lack of compliance with medical or psychiatric treatment, or changes in mental status, behavior, or mood. Protective factors that lower one's risk for suicide include strong coping skills, religious or spiritual beliefs against suicide, responsibility to children or beloved pets, social supports, and a positive therapeutic relationship with a therapist. When assessing suicide risk, it is important to note the difference between passive and active suicidal ideation. Passive suicidal ideation may involve thinking about one's death or that one would be better off dead. This may manifest as a wish to not wake up, to be killed in an accident, or to develop a terminal disease. In contrast, active suicidal thoughts involve an actual plan or intention to end one's life. For those with active suicidal ideation who are at imminent risk of harm, admission to an inpatient hospital is generally indicated. For those with moderate or low risk of suicide, a safety plan such as the sample below is recommended (25).

SAFETY PLAN

Step 1: Warning signs:
1. Suicidal thoughts and feeling worthless and hopeless
2. Urges to drink
3. Intense arguing with girlfriend

Step 2: Internal coping strategies - Things I can do to distract myself without contacting anyone:
1. Play the guitar
2. Watch sports on television
3. Work out

Step 3: Social situations and people that can help to distract me:
1. AA Meeting
2. Joe Smith (cousin)
3. Local Coffee Shop

Step 4: People who I can ask for help:
1. Name Mother   Phone 333-8666
2. Name AA Sponsor (Frank)   Phone 333-7215

Step 5: Professionals or agencies I can contact during a crisis:
1. Clinician Name Dr John Jones   Phone 333-7000
   Clinician Pager or Emergency Contact # 555 822-9999
2. Clinician Name   Phone
   Clinician Pager or Emergency Contact #
3. Local Hospital ED City Hospital Center
   Local Hospital ED Address 222 Main St
   Local Hospital ED Phone 333-9000
4. Suicide Prevention Lifeline Phone: 1-800-273-TALK

Making the environment safe:
1. Keep only a small amount of pills in home
2. Don't keep alcohol in home
3.

The Substance Abuse and Mental Health Services Administration (SAMHSA) developed a resource for conducting a suicide risk assessment called the SAFE-T Pocket Card: Suicide Assessment Five-Step Evaluation and Triage (26), which can also be downloaded via SAMHSA's Suicide Safe mobile application.

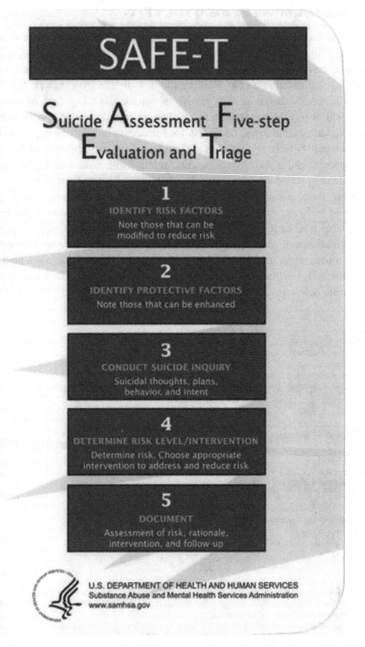

Falls: Mental health conditions have been shown to increase the risk of fall-related injuries for older adults (27). The relationship between depression and falls may occur through several possible mechanisms. First, older adults with depression are more likely to develop a sedentary lifestyle and have decreased levels of physical exercise than older individuals without depression (28). This is partly explained by the association of depression with decreased motivation and energy. The consequences of a lack of physical activity include muscle weakness, problems with balance, and decreased bone strength, which are known risk factors for fall injuries (29). It also appears that there are direct physiologic links between depression and inactivity within the nervous system. Depression may involve psychomotor slowing and cause impairments in attention, concentration, planning, and organization, which in turn impact reaction time and stability (30). Furthermore, the use of antidepressants has been shown to be associated with falling, independent of other factors including depression and dementia (29). Older adults with depression may have greater difficulty adhering to recovery regimens after suffering a fall.

Older adults with anxiety are also significantly more likely to suffer a fall than those who do not report elevated levels of anxiety (31). There are a number of possible ways in which the symptoms of anxiety may increase the risk of falls, such as increased dizziness, gait stiffness caused by increased muscle tension, hypertension, deconditioning due to avoidance or restriction of activity, and hypervigilance leading to distraction (31). One compelling theory suggests that anxiety alters the way an individual perceives risk and produces a bias for threatening stimuli in the environment (32). This leaves less attention for moving in a stable and safe manner, which may result in falls. Anxiety may also increase arousal and lead to stiffened movements that negatively impact balance (33). There is also evidence that medications used to treat anxiety are independent risk factors for falls due to their possible physiological and cognitive side effects (34). The fear of falling is high among older adults with and without a history of falls and may lead to reduced physical and social activity, lower perceived physical health status, and lower quality of life (35).

Given the increased risk of falls among older adults with mental health conditions, fall prevention strategies should be developed specifically for these individuals. An older adult with existing risk factors for falling may consider psychological treatment as an alternative intervention for depression or anxiety when weighing the risks and benefits of medication. Other possible interventions include exercise programs, the use of hip protectors, discussion of fall risk with prescribers of antidepressant medications, and fall prevention training for staff who work in mental health care settings.

Self-Neglect: Self-neglect refers to behavior that threatens one's own health and safety through the inability to perform basic tasks such as maintaining personal hygiene and household tasks, obtaining necessary services, or managing finances (36). This can result in injuries, hazards, and lack of medical care which is a threat to the older adult's safety and independence. Depression has been shown to be associated with self-neglect and may even double the risk of self-neglect in community-dwelling older adults (37). Depression and anxiety are both associated with poorer medical care outcomes, and this may be linked to poor adherence to treatment recommendations (38). Depressive self-neglectors have more untreated medical conditions than self-neglectors without depression, which may be attributed to apathy, placing less value on one's health, viewing services negatively, or poor decision making (39). Given the heightened risk of self-neglect among depressed older adults, mental health clinicians can play an important role in identifying and managing self-neglect alongside family members, home health care workers, and medical providers.

Elder Abuse: Elder abuse is an intentional act or failure to act that causes or creates risk of harm to an older adult. It often occurs at the hands of a caregiver or trusted person. It may involve abuse that is physical, sexual, emotional, financial, or the neglect of an older adult's basic needs. Psychological distress is an obvious consequence of elder abuse, and clinical depression, depression severity, and loneliness also have been identified as risk factors for elder abuse (40). In particular, older adults reporting higher levels of depressed mood, somatic complaints, and interpersonal problems are significantly more likely to have confirmed abuse from a perpetrator (41). Depression may also be associated with several established elder abuse risk factors such as functional disability (42). A depressed elder abuse victim may have difficulty leveraging elder abuse services due to low mood and a heightened sense of hopelessness. The treatment of depression may therefore improve not only one's risk for abuse, but also the ability to leverage elder abuse services after mistreatment has occurred. Screening and behavioral interventions for depression should therefore be delivered in conjunction with abuse resolution services (43).

17.7 CONCLUSION

Mental health problems often cause impairments that impact health and safety among older adults. Disorders such as depression and anxiety are treatable with medication and psychotherapy, and can be mitigated by community-level interventions. Treatment and education among older adults, caregivers, and health care professionals are important for reducing safety risks of older adults with mental health conditions.

REFERENCES

1. Hasin, D. S., Stinson, G. R., & Grant, B. F. (2005). Epidemiology of major depressive disorder: results from the National Epidemiologic Survey on Alcoholism and Related Conditions. *Archives of General Psychiatry, 62*, 1097–1106.
2. Fiske, A., Wetherell, J. L., & Gatz, M. (2009). Depression in older adults. *Annual Review of Clinical Psychology, 5*, 363–389.
3. American Psychiatric Association. (2013). *Diagnostic and statistical manual of mental disorders* (5th ed.). Washington, DC.
4. Christensen, H., Jorm, A. F., Mackinnon, A. J., Korten, A. E., Jacomb, P. A., Henderson, A. S., & Rodgers, B. (1999). Age differences in depression and anxiety symptoms: a structural equation modelling analysis of data from a general population sample. *Psychological Medicine, 29*, 325–339.
5. Alexopoulos, G. S., Katz, I. R., Reynolds 3rd, C. F., Carpenter, D., & Docherty, J. P. (2001). The expert consensus guideline series. Pharmacotherapy of depressive disorders in older patients. *Postgraduate Medicine, 110*, 1–86.
6. Kok, R. M., & Reynolds, C. F. (2017). Management of depression in older adults: a review. *JAMA, 317*, 2114–2122.
7. Unützer, J., & Park, M. (2012). Older adults with severe, treatment-resistant depression. *JAMA, 308*, 909–918.
8. Kelly, K. G., & Zisselman, M. (2000). Update on electroconvulsive therapy (ECT) in older adult. *Journal of the American Geriatrics Society, 48*, 560–566.

9. Bryant, C., Jackson, H., & Ames, D. (2008). The prevalence of anxiety in older adults: methodological issues and a review of the literature. *Journal of Affective Disorders, 109*, 233–250.

10. Kessler, R. C., Berglund, P., Demler, O., Jin, R., Merikangas, K. R., & Walters, E. E. (2005). Lifetime prevalence and age-of-onset distributions of DSM-IV disorders in the National Comorbidity Survey Replication. *Archives of General Psychiatry, 62*, 593–602.

11. Gonçalves, D. C., & Byrne, G. J. (2012). Interventions for generalized anxiety disorder in older adults: systematic review and meta-analysis. *Journal of Anxiety Disorders, 26*, 1–11.

12. Wolitzky-Taylor, K. B., Castriotta, N., Lenze, E. J., Stanley, M. A., & Craske, M. G. (2010). Anxiety disorders in older adults: a comprehensive review. *Depression and Anxiety, 27*, 190–211.

13. Clifford, K. M., Duncan, N. A., Heinrich, K., & Shaw, J. (2015). Update on managing generalized anxiety disorder in older adults. *Journal of Gerontological Nursing, 41*, 10–20.

14. Pinquart, M., & Duberstein, P. R. (2007). Treatment of anxiety disorders in older adults: a meta-analytic comparison of behavioral and pharmacological interventions. *The American Journal of Geriatric Psychiatry, 15*, 639–651.

15. Depp, C. A., & Jeste, D. V. (2006). Definitions and predictors of successful aging: a comprehensive review of larger quantitative studies. *The American Journal of Geriatric Psychiatry, 14*, 6–20.

16. Cornwell, B., Laumann, E. O., & Schumm, L. P. (2008). The social connectedness of older adults: A national profile. *American Sociological Review, 73*, 185–203.

17. Nicholson, N. R. (2012). A review of social isolation: an important but underassessed condition in older adults. *The Journal of Primary Prevention, 33*, 137–152.

18. National Academies of Sciences, E., and Medicine,. (2020). *Social isolation and loneliness in older adults: Opportunities for the health care system*. National Academies Press.

19. Perissinotto, C. M., Cenzer, I. S., & Covinsky, K. E. (2012). Loneliness in older persons: a predictor of functional decline and death. *Archives of Internal Medicine, 172*, 1078–1084.

20. Golden, J., Conroy, R. M., Bruce, I., Denihan, A., Greene, E., Kirby, M., & Lawlor, B. A. (2009). Loneliness, social support networks, mood and wellbeing in community-dwelling elderly. *International Journal of Geriatric Psychiatry, 24*, 694–700.

21. Woods, A., Solomonov, N., Liles, B., Guillod, A., Kales, H. C., & Sirey, J. A. (2021). Perceived Social Support and Interpersonal Functioning as Predictors of Treatment Response Among Depressed Older Adults. *The American Journal of Geriatric Psychiatry, 29*, 843–852.

22. Blazer, D. G. (2020). Social isolation and loneliness in older adults – a mental health/ public health challenge. *JAMA Psychiatry, 77*, 990–991.

23. World Health Organization (WHO). (2014). *Preventing suicide: A global imperative*. Retrieved from Luxembourg.

24. National Institute of Mental Health. (2021). Suicide. Retrieved from www.nimh.nih.gov/health/statistics/suicide

25. Stanley, B., & Brown, G. K. (2012). Safety planning intervention: a brief intervention to mitigate suicide risk. *Cognitive and Behavioral Practice, 19*, 256–264.

26. Jacobs, D. (2009). SAFE-T: Suicide Assessment Five-Step Evaluation And Triage. Retrieved from https://store.samhsa.gov/product/SAFE-T-Pocket-Card-Suicide-Assessment-Five-Step-Evaluation-and-Triage-for-Clinicians/sma09-4432

27. Finkelstein, E., Prabhu, M., & Chen, H. (2007). Increased prevalence of falls among elderly individuals with mental health and substance abuse conditions. *The American Journal of Geriatric Psychiatry, 15*, 611–619.

28. Roshanaei-Moghaddam, B., Katon, W. J., & Russo, J. (2009). The longitudinal effects of depression on physical activity. *General Hospital Psychiatry, 31*, 306–315.

29. Tinetti, M. E., Speechley, M., & Ginter, S. F. (1988). Risk factors for falls among elderly persons living in the community. *New England Journal of Medicine, 319*, 1701–1707.

30. Manning, K. J., Alexopoulos, G. S., McGovern, A. R., Morimoto, S. S., Yuen, G., Kanellopoulos, T., & Gunning, F. M. (2014). Executive functioning in late-life depression. *Psychiatric Annals, 44*, 143–146.

31. Hallford, D. J., Nicholson, G., Sanders, K., & McCabe, M. P. (2017). The association between anxiety and falls: A meta-analysis. *Journals of Gerontology Series B: Psychological Sciences and Social Sciences, 72*, 729–741.

32. Nieuwenhuys, A., & Oudejans, R. R. (2012). Anxiety and perceptual-motor performance: toward an integrated model of concepts, mechanisms, and processes. *Psychological Research, 76*, 747–759.

33. Young, W. R., & Williams, A. M. (2015). How fear of falling can increase fall-risk in older adults: applying psychological theory to practical observations. *Gait & Posture, 41*, 7–12.

34. Huang, A. R., Mallet, L., Rochefort, C. M., Eguale, T., Buckeridge, D. L., & Tamblyn, R. (2012). Medication-related falls in the elderly. *Drugs & Aging, 29*, 359–376.

35. Jung, D. (2008). Fear of falling in older adults: comprehensive review. *Asian Nursing Research, 2*, 214–222.

36. Papaioannou, E. S. C., Räihä, I., & Kivelä, S. L. (2012). Self-neglect of the elderly. An overview. *The European Journal of General Practice, 18*, 187–190.

37. Abrams, R. C., Lachs, M., McAvay, G., Keohane, D. J., & Bruce, M. L. (2002). Predictors of self-neglect in community-dwelling elders. *American Journal of Psychiatry, 159*, 1724–1730.

38. DiMatteo, M. R., Lepper, H. S., & Croghan, T. W. (2000). Depression is a risk factor for noncompliance with medical treatment: meta-analysis of the effects of anxiety and depression on patient adherence. *Archives of Internal Medicine, 160*, 2101–2107.

39. Burnett, J., Coverdale, J. H., Pickens, S., & Dyer, C. B. (2006). What is the association between self-neglect, depressive symptoms and untreated medical conditions? *Journal of Elder Abuse & Neglect, 18*, 25–34.

40. Dong, X., Chen, R., Chang, E. S., & Simon, M. (2013). Elder abuse and psychological well-being: A systematic review and implications for research and policy-A mini review. *Gerontology, 59*, 132–142.

41. Roepke-Buehler, S. K., Simon, M., & Dong, X. (2015). Association between depressive symptoms, multiple dimensions of depression, and elder abuse: A cross-sectional, population-based analysis of older adults in urban Chicago. *Journal of Aging and Health, 27*, 1003–1025.

42. Hays, J. C., Landerman, L. R., George, L. K., Flint, E. P., Koenig, H. G., Land, K. C., & Blazer, D. G. (1998). Social correlates of the dimensions of depression in the elderly. *The Journals of Gerontology Series B: Psychological Sciences and Social Sciences, 53*, P31–P39.

43. Sirey, J. A., Solomonov, N., Guillod, A., Zanotti, P., Lee, J., Soliman, M., & Alexopoulos, G. S. (2021). PROTECT: a novel psychotherapy for late-life depression in elder abuse victims. *International Psychogeriatrics, 33*, 521–525.

18 Suicide Prevention

Kimberly Williams

CONTENTS

Suicide among older adults is a pressing public health problem. Older adults are at higher risk for suicide due to a number of important risk factors. Yet, suicide prevention in older adults is a neglected area.

Providers often lack basic knowledge and training about older adult suicide. As a result, at-risk older adults, who are not likely to self-disclose suicidal ideation, are often not identified. Therefore, implementation of suicide prevention efforts is key. Effective prevention initiatives, which must be prioritized at local, state, and federal levels, deploy a multifaceted approach at individual, group, and population levels. Providers must understand their key role in preventing suicide among older adults and implement practices to reduce risk, identify older adults at risk, and appropriately intervene with next steps to keep older adults at risk safe. In order to further combat suicide among older adults, major public policy changes will be required.

18.1 SUICIDE INCIDENCE AND RISK FACTORS

In many countries, suicide rates among older adults are higher than or as high as young adults. In the United States, older adults die by suicide at a rate nearly 50% greater than younger adults 15–24 years old. The incidence of suicide is particularly high among older, white males, who complete suicide at a rate three times that of the general population. White men over 85 are the population at highest risk; they have a suicide rate four times higher than the nation's overall rate of suicide.

There are several important risk factors for suicide in older adults. These include mental illnesses (particularly depression), chronic physical illnesses, prior suicide attempts, chronic severe pain and declining role function (e.g., loss of independence,

DOI: 10.1201/9781003197843-18

sense of purpose), marked feelings of hopelessness, social isolation, family discord or losses, inflexible personality or marked difficulty adapting to change, access to lethal means, substance use problems, and recent onset of dementia. Older adults are also at risk if they are experiencing a sense of low belonging and perceived burdensomeness. According to Thomas Joiner's Interpersonal-Psychological Theory of Suicide, when someone experiences a lack of social connectedness – specifically, a low sense of belonging – and a sense of burdensomeness paired with acquired capability (a lower level of fear of death and habituated experiences of violence), their risk can become severely elevated. The more risk factors among an older adult, the greater the likelihood of suicide.

The factors that protect older adults from suicide include a sense of meaning and purpose in life, a sense of hope or optimism, social connectedness, skills in coping and adapting to change, and care for mental and physical health problems. Social connectedness, in particular, is a key protective factor against suicide including in terms of the escalation of thoughts of suicide to action. Connectedness also acts as a buffer against hopelessness and psychological pain.

Suicide attempts by older adults are more likely to result in death than attempts by younger adults. Older adults are also less likely to report suicidal ideation compared to younger adults. Older adults who attempt suicide are often frailer, more isolated, more likely to have a plan, and more likely to use more deadly methods. The most common means of suicide among older adults are firearms, poisoning, and suffocation. Older adults are nearly twice as likely to use firearms as a means of suicide than people under age 60.

18.2 PREVENTION STRATEGIES

Prevention of suicide in older adults is a public health imperative and requires many different approaches. Multilayered prevention initiatives that combine universal, selective, and indicated prevention strategies, which are aimed at different stages of suicidality, are the most effective in reducing suicide among older adults.

Universal prevention, which targets the entire population, aims to reduce risk factors for suicide. These strategies include providing information and enhancing skills of older people and their caregivers. Recommendations include implementing depression screenings, limiting access to means of suicide, providing education on suicide risk and protective factors, and developing media guidelines for safe reporting of suicide in older adults.

Selective prevention is aimed at reducing risk in groups with an increased risk for suicide but who may not display signs of suicidal thoughts or behavior such as older adults who experience life transitions or losses that can make them vulnerable to depression and suicide. Selective prevention strategies aim to reduce risk factors for suicide or to improve resilience. Interventions at this level include making systematic screening tools available to staff in medical and nonmedical settings, providing systemic outreach for assessment and support for older adults at risk, increasing provider awareness of the losses that are important to older people, such as retirement, loss of driver's license, and loss of important body functions (e.g., sight, mobility), and improving the treatment of sleep problems, pain, or other physical symptoms that decrease quality of life.

Indicated prevention strategies aim to reduce risk in older adults who have survived a suicide attempt or are at high risk for suicide. Recommended interventions at this level include training professionals to detect, intervene, and manage depression and suicide risk, taking action to ensure the safety and effective treatment of older adults who are at imminent risk for suicide, caring contacts at follow-up such as homes visits, regular phone calls, postcards, or connection to alarm central for immediate help when needed to reassure the older adult that their existence is meaningful, and disseminating practice guidelines for the detection and management of suicide in later life. Because depression is considered to be present in up to 90% of all suicides among older adults, the detection and effective treatment of depression are key to reducing suicides. Routine screening for depression can be done with many instruments, such as the PHQ-9 and the Geriatric Depression Scale (GDS). A variety of psychotherapy interventions (e.g., cognitive behavioral therapy, behavioral therapy, and problem-solving treatment) and antidepressant medications are effective at treating symptoms of depression and can reduce suicidal ideation in some older adults.

18.3   ASSESSING AND ADDRESSING RISK FOR SUICIDE

Providers and even the general public can play an important role in preventing suicide by identifying at-risk older adults and taking appropriate follow-up actions to keep them safe. If providers think an older adult may be suicidal, they should ask the older adult if they are thinking about suicide, as older adults are not likely to report suicidal ideation. Studies show that asking at-risk individuals if they are suicidal does not increase suicides or suicidal thoughts. In fact, studies suggest acknowledging and talking about suicide may reduce rather than increase suicidal ideation.

A common screening tool for depression, the PHQ9, can also identify individuals with thoughts of suicide. The last question of the screening tool asks, "In the last two weeks, have you had any thoughts of hurting or killing yourself?" If an older adult responds positively to this question, there are additional follow-up questions to explore suicidality that can be asked, such as with the P4Screener:

- **Past** suicide attempts: "Have you ever attempted to harm yourself in the past?"
- **Plan**: "Have you had thoughts about how you might actually hurt yourself?"
- **Probability** of completing suicide: "How likely do you think it is that you will act on these thoughts about hurting yourself or ending your life some time over the next month?"
- **Preventive** factors: "Is there anything that would prevent or keep you from harming yourself?"

Positive answers to any of these questions should be discussed with a qualified mental health provider to determine the degree of urgency and steps for further assessment and intervention. Older adults who have suicidal ideation with a plan and intent to act should not be left alone but should be supervised for their safety until emergency services are in place.

Depending on an older adult's level of suicidal ideation or intent, different levels of action can be taken. For an older adult showing low risk (i.e., suicidal thoughts without the endorsement of a prior suicide attempt or suicide plan or intent to self-harm), suggested actions include: showing support, connecting with or referral to the older adult's primary care provider or a mental health provider for an evaluation and/or treatment, urging that the older adult remove means, and identifying possible coping strategies. For an older adult with a moderate risk of suicide (i.e., active suicidal thoughts with or without prior suicide attempt and have either formulated or are contemplating forming a suicide plan) the previously mentioned actions should be taken, consultation/supervision should be sought, as needed, and referral to a mobile crisis team for evaluation should be considered. Mobile crisis teams comprise mental health professionals who can meet with at-risk older adults in their homes or elsewhere in the community. For an older adult at high risk of suicide (i.e., active suicidal thoughts with a prior suicide attempt or suicide plan as well as clear intentions to harm themselves or complete suicide), a supervisor/consultant should be called immediately and emergency medical services (e.g., EMS or 911) should be considered.

18.4   ROLE OF PROVIDERS

Providers working with older adults including aging service, mental health, addiction, and primary care providers can all take action to prevent suicide in older adults. Aging service providers are in a distinct position to identify older adults at risk for depression or suicide because of their delivery of services in the home and in community-based settings. Aging service providers can be trained as community gatekeepers to identify warning signs and refer older adults who are at risk for depression or suicide. Community members who come into contact with older adults such as police, barbers, and educators as well as family members and friends can also be trained as community gatekeepers so they can intervene when an older adult is at risk of suicide. Aging service providers can also introduce depression education and screening into senior centers, senior companion programs, and other aging service programs. They can also provide systematic outreach to assess and support high-risk older adults (e.g., recently widowed, socially isolated older men) in improving life conditions and addressing issues that can reduce stress.

Mental health and addiction service providers can take a variety of actions to reduce the risk of suicide in older adults. They can screen for suicidal ideation among older adults receiving treatment for mental health and substance use issues. Providers can also implement evidence-based practices for depression to reduce the symptoms of depression, improve functioning, improve overall health outcomes, and ensure that older adults receive effective care. Mental health and addiction service providers can also offer assertive help and follow-up after a suicide attempt.

Primary care is regarded as a critical point for preventative intervention. Most older adults who die by suicide have seen their primary care provider within 30 days of taking their own lives but have not been identified as a suicide risk. Primary care providers can implement routine standard screening for depression and suicidal ideation. They can implement collaborative depression care management interventions to optimize the diagnosis and treatment of late-life depression. Evidence-based depression care management models include Improving Mood, Promoting Access to Collaborative Treatment (IMPACT) and Prevention of Suicide in Primary Care Elderly (PROSPECT). These primary care collaborative treatment strategies deliver support from depression care managers (nurses, psychologists, or social workers) who offer education about treatment options, brief psychotherapy, and provide close monitoring of depressive symptoms and medication side effects as well as follow-up of patients. Primary care providers can also enhance the treatment of pain, sleep problems, or other physical symptoms that can decrease an older adult's quality of life and increase suicidal thoughts. They can also develop and use registries to identify and monitor older adults after a suicide attempt.

18.5 PUBLIC POLICY AND SOCIETAL RECOMMENDATIONS

Reduction of suicides among older adults in the United States will take changes beyond greater provider awareness, practice improvements, and public health initiatives. Major changes in public policy are also needed including expansion of, and improved quality of, mental health services for older adults, increased integration of physical and mental health services, increased coordination between physical health, mental health, and aging services, and increased use of technology to improve access. Insurance coverage needs to be modified to include the need for nontraditional services. Greater control of access to lethal means of suicide, especially guns, is also needed.

In addition, the implication of Thomas Joiner's Interpersonal-Psychological Theory of Suicide is that the prevention of suicide cannot be accomplished solely through improved identification of risk, timely intervention, and access to help. Older adults also need a place of belonging and a sense that becoming dependent does not mean becoming an intolerable burden. In our ageist society where many Americans, even older Americans, tend to regard the aging process with disdain and older adults as lacking value, this is a tremendous challenge to overcome. The kind of societal change that is needed around shifting cultural norms to help overcome older adult isolation and feelings of burdensomeness requires a systemic, multipronged undertaking on the part of both the public and private sectors.

BIBLIOGRAPHY

Berman J, Furst, L. *Depressed older adults.* New York: Springer; 2010.

Center for Mental Health Services. *The treatment of depression in older adults evidence-based practices (EBP) KIT.* Rockville, MD: Substance Abuse and Mental Health Services Administration, US Dept of Health and Human Services; 2011.

Conwell Y, Van Orden K, Caine ED. Suicide in older adults. *Psychiatr Clin North Am.* 2011;34(2):451–468.

Dube P, Kurt KB, et. al. The p4 screener: evaluation of a brief measure for assessing potential suicide risk in 2 randomized effectiveness trials of primary care and oncology patients. *Prim Care Companion J Clin Psychiatry.* 2010;12(6):1–15.

Dazzi T, Gribble R, Wessely S, Fear NT. Does asking about suicide and related behaviours induce suicidal ideation? What is the evidence? *Psychol Med.* 2014 Dec;44(16):3361–3363. doi: 10.1017/S0033291714001299. Epub 2014 Jul 7. PMID: 24998511.

Erlangsen A, Nordentoft M, et al. Key considerations for preventing suicide in older adults: consensus opinions of an expert panel. *Crisis.* 2011;32(2):106–109.

Friedman, M. Elder suicide: a public health challenge of the elder boom. *Huffpost Health Living.* 2012.

Holm AL, Salemonsen E, Severinsson E. Suicide prevention strategies for older persons: an integrative review of empirical and theoretical papers. *Nurs Open.* 2021;8(5):2175–2193.

Injury Prevention & Control: Data & Statistics (WISQARSTM). *Fatal injury reports.* www.cdc.gov/injury/wisqars/fatal.html. Accessed September 25, 2021.

Joiner TE. *Why people die by suicide.* Cambridge, MA: Harvard University Press; 2005.

Klonsky DE, May, AM. The three-step theory (3ST): a new theory of suicide rooted in the "Ideation-to-Action" framework. *Int J of Cogn Ther.* 2015;8(2):114–129.

Lapierre S, Erlangsen A, et al. A systematic review of elderly suicide prevention programs. *Crisis.* 2011;32(2):88–98.

Luoma JB, Martin CE, Pearson JL. Contact with mental health and primary care providers before suicide: a review of the evidence. *Am J Psychiatry,* 2002;159(6):909–916.

National Association of State Mental Health Program Directors. Weaving a community safety net to prevent older adult suicide. August 2018. www.nasmhpd.org/sites/default/files/TAC-Paper-10-Prevent-Older-Adult-Suicide-508C.pdf Assessed October 9, 2021.

Substance Abuse and Mental Health Services Administration. Older Americans' behavioral health issue brief 4: preventing suicide in older adults. https://acl.gov/sites/default/files/programs/2016-11/Issue%20Brief%204%20Preventing%20Suicide.pdf Accessed October 8, 2021.

19 Social Isolation and Loneliness

Caitlyn Foy, DOT

CONTENTS

19.1   SOCIAL ISOLATION AND LONELINESS

A 2020 report from the National Academies of Sciences, Engineering, and Medicine (NASEM) found that more than one-third of adults aged 45 and older feel lonely, and nearly one-fourth of adults aged 65 and older are considered to be socially isolated.[1] Older adults are at increased risk for loneliness and social isolation because they are more likely to face factors such as living alone, the loss of family or friends, chronic illness, and hearing loss. Being socially isolated or lonely can affect their health, and these same health conditions can further increase the likelihood of experiencing social isolation or loneliness

Social isolation, loneliness, living alone, and *social withdrawal* are distinct phenomenon.

Social isolation is the objective reduction or lack of social connections, including a lack of contact with friends, family, or others. *Loneliness* is a subjective, emotional feeling which may or may not be accompanied by social isolation.[1,2] In some cases,

DOI: 10.1201/9781003197843-19

social isolation can lead to loneliness, while in others, loneliness can occur without social isolation.

Social isolation has been linked to poor health and higher health care utilization in numerous studies.[1,2] Poor health can be both a contributing factor and a consequence of social isolation. Social isolation is associated with a significant increased risk for premature mortality from all causes, as well as higher rates of depression, increased risk of heart disease and stroke, and cognitive decline.

An individual who feels lonely can feel alone regardless of their level of social contact. Lonely people experience emotional pain. Losing a sense of community and connection can change one's perspective on the world. For example, chronic loneliness can make people feel threatened and mistrustful of others. In the same way that physical pain can activate stress responses in the body, emotional pain can do the same. If this occurs for a long period of time, it can lead to chronic inflammation (overactive or prolonged release of factors that damage tissues) and reduced immunity (ability to fight off disease). Chronic diseases and infectious diseases can be more likely to develop as a result.

Living alone means living in a one-person household. The US Census Bureau defines living alone as living and eating separately from any other persons in the building. The likelihood of living alone increases with age. According to the Institute on Aging (2010), nearly one-third of older adults in America lives alone, outside of a nursing home or hospital.

Social withdrawal is avoiding people and activities one would usually enjoy. In social withdrawal, one excludes themselves from social opportunities. For some people, this can progress to a point of social isolation, where they may even want to avoid contact with family and close friends. Social withdrawal can contribute to anxiety and depression. Different types of social withdrawal have different underlying emotional, motivational, and psychological causes.

Aging itself, independent of other factors, does not cause social isolation or loneliness. However, many of the risk factors that can cause or exacerbate social isolation or loneliness, such as death of loved ones, worsening health and chronic illness, new sensory impairment, retirement, or changes in income, are more likely to affect people over 50.

It is vital to understand how loneliness and isolation impact an aging population so that we can improve the quality of education, social services, and advocacy efforts needed to improve this population's overall quality of life and reduce generational health and wellness disparities.

Older adults face increased risks for loneliness and social isolation as they are more likely to have smaller social networks, live alone, and have chronic illnesses that can limit travel outside the home and socialization. The 2020 report from the National Academies of Sciences, Engineering, and Medicine stated that 24% of community-dwelling Americans (individuals who live independently) aged 65 and older are considered to be socially isolated. The same report found that 35% of adults aged 45 and older and 43% of adults aged 60 and older feel lonely.[1] Further, older adults living in long-term care facilities report significantly higher rates of loneliness than older adults aging in place.[3-5]

Lack of social connectivity is a growing public concern. It is a health risk comparable to other mortality risk factors, including lack of physical activity, substance abuse, obesity, injury, and violence.[8] Feelings of loneliness can lead to increased risk of depression, anxiety, alcoholism, suicidal thoughts, aggressive behaviors, and impulsivity. It is essential for health care providers to educate older adults and their families on signs, symptoms, and risk factors of social isolation and loneliness. Education about and identification of risk factors can better prepare older adults and families to implement strategies in their daily lives and routines to help prevent negative consequences from social isolation.

19.2   RISK FACTORS FOR SOCIAL ISOLATION

While many age groups and generations are at risk for social isolation, which may lead to loneliness, older adults are at a higher risk due to a myriad of individual, community, and societal factors. Individual risk factors include:[1,10]

- Living alone
- Loss of family and friends
- Chronic disease (i.e., coronary artery disease, hypertension, cancer, and cerebral vascular accidents)
- Sensory impairments (i.e., hearing loss, vision changes)
- Decreased independence in activities of daily living and self-care
- Cognitive impairments
- Language barriers
- Psychiatric disorders (i.e., depression, anxiety, substance abuse, and dementia)
- Demographics (i.e., age, race, ethnicity, sexual orientation, and/or gender identity)
- Mobility and other physical impairments
- Obesity
- Sedentary lifestyle
- Caregiving burdens

Community risk factors may pertain to inadequate community mobility, including inaccessible transport and lack of universal design and enforcement of Americans with Disabilities Act standards. They can also relate to one's social support system, including the loss of a partner or friend, retirement, relocating for family or work, and migration of children. Societal risk factors include limited education, ageism, stigma, marginalization, limited socioeconomic resources, and attitudinal barriers.[4,8,11]

Living alone is another substantive risk factor related to social isolation. There is little to no research on the prevalence of older adults living alone. However, a 2016 study researching older adults who have little to no support found that the likelihood of being at high risk for elder orphan status is up to 22.6%.[12] *Elder orphans* (older adults without a spouse or child on whom they can depend) and *Solo agers* (older adults who live alone and do not have children) may experience loneliness as well as complex medical and psychosocial issues that can impact their personal life satis-faction and independence. A growing population of elder orphans and solo agers can impact communities, and the U.S. health care system must create a support system for these individuals. This public health concern poses a financial burden as well as an increased demand for resources.[12] Baby boomers must plan and explore options to secure a support system before issues impacting independence develop.

More research is needed to understand the effects of loneliness and isolation on disempowered populations (those that face discrimination and exclusion due to power disparities and inequities related to economic, political, and sociocultural factors).[13] Current research suggests that lesbian, gay, bisexual, transgender, queer, or questioning (LGBTQIA+) individuals and immigrants are especially affected due to barriers to care, discrimination, and stigma. Older LGBTQIA+ community members have a higher risk of depression, public condemnation, discrimination, poor mental health outcomes, and ageism within LGBTQIA+ communities, which face a higher rate of isolation than the general population, regardless of age.[14,15] Immigrants may also face stressors such as language barriers, loss of family ties, disparities in the community, and poor access to health care.[1,15] In some cultures, such as indi-genous, first nations, or religious groups, removal from tradition, community, or other valued factors (i.e., worship locations, family) may be particularly damaging to mental health and encourage loneliness.[16] In general, it is well established that disempowered populations face greater mental health challenges due to stigma, bias, and restricted access to resources. These factors faced by disempowered populations, particularly older adults in these communities, compound health care disparities and reduce access, services, and public attention.[15]

The COVID-19 pandemic has brought many social issues to the surface including, fear, panic, discrimination, division, isolation, as a result of mourning the loss of over 3 million deaths worldwide.[17] Although not everyone has had the ability or priv-ilege to shelter in place during the pandemic, many older adults have been isolated in their homes or on lockdown at long-term care facilities, unable to see or speak with loved ones. The COVID-19 pandemic has had an outsize impact on the health and well-being of older adults, especially those prone to depression, loneliness, and feeling disconnected from friends and neighbors. It's the responsibility of society to take practical steps to better manage mental health, especially during difficult times, and look for new strategies and methods to reconnect older adults with their communities.

The multifaceted nature and potential for layering these risk factors contribute to global challenges in reducing social isolation for older adults and potentially thereby reducing loneliness.

19.3 HEALTH CONSEQUENCES OF SOCIAL ISOLATION AND LONELINESS

Current research demonstrates the significant impact loneliness and isolation play on mental and physical health, which can reduce lifespan and quality of life. Older adults should be made aware of the negative impacts social isolation and loneliness can have on their health. Loneliness, for example, can increase the risk of chronic illness,[18] physical inactivity,[19,20] and early death.[9] Social isolation and loneliness can negatively impact one's quantity and quality of sleep. Sleep disturbances are associated with increased mortality rates, negative impacts on physical health conditions, and behavioral risk factors, such as lower physical activity, poor medical treatment compliance, increased tobacco and alcohol consumption, and poor diet.[21]

19.4 CHRONIC HEALTH CONDITIONS

- Social isolation can significantly increase a person's risk of cardiovascular disease, stroke, diabetes, cognitive decline, dementia (50% increased risk), depression, anxiety, and suicide.[1]
- Poor social relationships (characterized by social isolation or loneliness) were associated with a 29% increased risk of heart disease and a 32% increased risk of stroke.[1]
- Loneliness among heart failure patients was associated with a nearly fourfold increase in the risk of death, 68% increased risk of hospitalization, and 57% increased risk of emergency department visits.[1]

19.5 ASSESSMENTS

Older adults (and their family, friends, and caregivers), particularly those who are socially isolated, should be aware of the signs and symptoms of loneliness to allow for early intervention, potentially reducing complications and health consequences.[1-6] The following are early signs and symptoms that a person may be at risk for loneliness:

- Lacking pleasure in activities that once brought joy and satisfaction (anhedonia)
- Spending significant amounts of time alone
- Neglecting personal care
- Feeling anxious or distressed when thinking about social interactions
- Feeling self-doubt, hopelessness, and worthlessness when trying to engage socially
- Canceling plans with other people
- Withdrawing from meaningful roles and group activities
- Experiencing social burnout after making failed attempts to engage with others
- Feeling relationships are too superficial, unfulfilling, and lacking depth and intimacy
- Withdrawing and feeling lonely whether or not there are people near
- Struggling to make connections with other people and maintain engaging relationships

Feelings of loneliness from social isolation can develop into more severe and life-threatening concerns, including depression and possible thoughts or intents of suicide. Emergency intervention is needed if a person is at risk of hurting themselves or contemplating suicide. This includes contacting a regional suicide interventionist, local authorities, and an immediate referral to a health care professional or primary care physician.

Nearly all adults aged 50 or older interact with the health care system in some way. Their health care providers have a unique opportunity to identify socially isolated older adults, given that these appointments may be an older adult's only ongoing social engagement. Despite this, only 13% of adults reported that they have been asked about social isolation during primary care examinations (Tung et al., 2021).[22] Health care professionals should encourage and educate older adults to self-identify risk factors that could lead to social isolation, feelings of loneliness, or depression. Sometimes when a person feels depressed or lonely, they may not be self-aware. It is crucial for family members, friends, and caregivers to also understand the warning signs. Health care professionals can assess and monitor older adults' social isolation and depressed feelings through a variety of screenings and assessments, as listed in Table 19.1.

TABLE 19.1

Screens and Assessments to Help Identify Signs of Loneliness and Social Isolation

Loneliness

Title and Citation	Brief Description	Results
Becks Depression Inventory-II (BDI®-II)[23]	• A self-report 21-item measure for adults ranging from 13 to 80 • Measures characteristic attitudes and symptoms of depression	Respondents are categorized into one of four levels of risk of depression: minimal, mild, moderate, and severe
De Jong Gierveld Loneliness Scale[24]	• A self-report 11-item measure • Assesses both emotional and social loneliness	Respondents are categorized into one of four levels: not lonely, moderate loneliness, severe loneliness, and very severe loneliness Results give insight into if the respondent is missing a wider social network or a more intimate relationship
UCLA (University of California, Los Angeles) Loneliness Scale (Version 3)[18]	• A self-report (or interview) 20-item measure for teens through older adults • Assesses how often a person feels disconnected from others	Scores can range from 20 to 80, with higher scores signifying greater loneliness

Social Isolation

Title and Citation	Brief Description	Results
Berkman–Syme Social Network Index[25]	• A self-report questionnaire for adults ranging from 18 to 65 years of age	Respondents are categorized into one of four levels of social connection: socially isolated, moderately isolated, moderately integrated, and socially integrated

TABLE 19.1   (Continued)
Screens and Assessments to Help Identify Signs of Loneliness and Social Isolation

Social Isolation

Title and Citation	Brief Description	Results
	• A composite measure of four types of social connections: marital status, sociability, church group membership, and membership in other community organizations	
Social and Community Opportunities Profile [SCOPE] – short form[26]	• A self-report (or interview) 48-item questionnaire on opportunities for social and community participation • Assesses eight areas; leisure, housing, safety, work, finances, health, education, and relationships	Respondents answers will give insight into perceived opportunities for social and community participation, satisfaction with opportunities, and respondent's subjective well-being
Steptoe Social Isolation Index[9]	• A self-report 5-item measure • Screens for social isolation through queries related to marital/cohabitation status, monthly contact with kids, family, and friends, and group participation	Respondents are categorized as socially isolated or not if they identify with two or more of the five items

19.6   STRATEGIES TO PREVENT SOCIAL ISOLATION AND LONELINESS

Feelings of loneliness in older adults can be combated by creating a plan to improve social participation and increasing resilience factors, such as meaningfulness, grit, self-care, external connection, good health, autonomy, and a positive outlook.[27] General approaches to building and strengthening resilience, and one's quality of life, include:[28]

- Increasing altruistic behaviors and flexibility
- Fostering a good sense of humor
- Increasing opportunities for social participation
- Strengthening self-efficacy, self-esteem, self-determination
- Motivating individuals through interpersonal interactions and relationships
- Making improvements to one's lifestyle (healthy diet, exercise, and sleep routines)
- Engaging in meaningful and fulfilling activities

Identifying and implementing initial changes in an individual's routine may be difficult at first. This section will offer strategies to implement large and small routine changes to improve social participation and resiliency to reduce loneliness.

19.6.1 LIFESTYLE CHANGES

Improving one's lifestyle, including promoting physical activity, a healthy diet, good sleep hygiene, and social participation, can strengthen resilience and increase quality of life.

Daily moderate exercises have many physical and mental health benefits. The Centers for Disease Control (2021)[29] report the benefits of exercise including maintaining independent living, reducing the risk of falls and bone fractures, improving stamina and muscle strength, and helping maintain healthy bones, muscles, and joints while better controlling joint swelling and pain associated with arthritis. Exercising releases endorphins and serotonin, which can improve mood and reduce symptoms of anxiety and depression. Additional health benefits can be gained through greater amounts of physical activity by increasing duration, intensity, or frequency. However, excessive exercise at high levels of activity is not recommended due to the risk of injury. Older adults with a history of cardiovascular disease, osteoporosis, arthritis, hypertension, diabetes, or other chronic or progressive conditions should consult a physician for safe exercise parameters.

Exercise groups can be advantageous for physical activity, cognitive and emotional well-being, and an additional outlet for social participation. The YMCA, for example, offers a variety of programming and wellness activities for older adults, including physical fitness, social activities, and other educational experiences. Silver Sneakers is a health and fitness program designed for adults aged 65 and older that offers virtual and in-person exercise classes for older adults of all fitness levels. Older adults can find additional exercise training and groups through online searches, visiting their public library, community recreation centers, and local gyms. It is advisable to seek out gyms and trainers who have certification in training older adults or who have completed certification from interest groups like the Arthritis Foundation to ensure the highest safety and quality.

Diet can have a significant impact on older adults' health and wellness and their ability to participate in social events outside of the home. Better diet quality has been related to higher overall participation in leisure activities, including social and cognitive activities, in community-dwelling older adults.[30] The National Council on Aging recommends staying hydrated, reading nutrition fact labels, following recommended portion and serving sizes, and prioritizing lean proteins, fresh fruits and vegetables, whole grains, and low-fat dairy in one's diet.[31]

It is a misconception that older adults need less sleep compared to middle adulthood.[32,33] Though older adults may have difficulties falling and staying asleep, they are recommended to have between seven and nine hours of sleep each day. Lack of proper rest and sleep can impact an individual's physical and mental health and their ability to participate in other aspects of their daily lives. Sleep hygiene (sleep routines and habits) can be improved by:

- Following a consistent sleep routine each night
  - Create a more soothing ambiance by listening to music, taking a shower or bath, using calming scents like lavender, dimming the lights, etc.
  - Try relaxation techniques before bed like meditation, yoga, praying, visual imagery, and breathing exercises
- Exercising at the same time each day
  - Avoid vigorous physical activity right before bed
- Keeping the bedroom limited to sleeping and sex
  - Use other rooms for projects, reading, watching TV, using the computer, etc.
- Avoiding the use of electronic devices at least one hour before bed
  - Wearing glasses to block blue light before bed
- Avoiding late afternoon and evening napping, which can disturb one's sleep cycle
- Avoiding large meals, alcohol, caffeine, and other bladder irritants late in the day
- Wearing incontinence underwear or briefs at night if urgency or general incontinence is an issue

Regular exercise, a healthy diet, and good sleep hygiene will positively impact a person's health and wellness. These factors are also associated with better physical and emotional health, which play a significant role in a person's endurance, motivation, and stamina to engage in community and other social events.

19.7   MEANINGFUL AND FULFILLING ROLES AND ACTIVITIES

Full or part-time work, volunteering, and social and leisure activities can be meaningful ways to occupy one's time. Studies link leisure participation with successful aging through improved self-satisfaction and self-efficacy.[34]

19.7.1   WORK AND VOLUNTEERING

According to Pew Research Center (2016), 51% of participants in America stated their jobs are central to their identity, while 47% said their job is just what they do to earn an income.[35] For those who draw satisfaction from work, it may be a priority and goal to maintain employment well into their older adulthood. For those who do not, it may still foster social interaction, create an outlet for self-expression, generate self-respect, improve or help maintain cognitive health and skills, engender pride in one's appearance, and prevent the daily routine from becoming mundane. All of these benefits can help increase resiliency in older adulthood.

Older adults may consider an *encore career* or a second vocation beginning in the latter half of life. This may lead to a greater sense of fulfillment, satisfying a social or public purpose, or offer financial supplementation. For many reasons, the older adult workforce (65 years and older) has been on a steady incline.[36] The U.S. Bureau of Labor Statistics projects the participation rate for workers aged 65–74 to be 30.2% in 2026, compared with 17.5% in 1996, and more than double at 10.8% in 2026 for ages 74 and older, compared to 1996.[37]

While some older adults may be financially prepared to retire, they may not feel ready due to the meaning, purpose, and value their job and career bring to their daily satisfaction and self-identity. Factors believed to help individuals adjust to retirement, which also supports resiliency and socialization, include:[38]

- Having a partner
- Volunteering
- Utilizing community resources (recreation center, public library, etc.)
- Maintaining a professional identity
- Engaging in physical activity
- Exploring or practicing spirituality

Volunteering provides an opportunity for older adults who may not need the income associated with work to reap the benefits from working (as described above) but with fewer stressors. Research has proven the physical and psychological benefits of participation in volunteer activities for elders living in the community.[39] Older adults can seek volunteer opportunities locally through special interest groups like nature centers, animal rescues, museums, art organizations, and religious and faith-based communities. They can volunteer their time as mentors or consultants in areas related to past professional experiences. The American Association of Retired Persons (AARP) and other associations offer resources for those considering retirement to learn about volunteering opportunities, leisure activities, and social events (see Table 19.2).

19.8   ROUTINE PLANNING

Self-satisfaction and purpose can be enhanced by establishing and maintaining routines and roles (aspects of one's identity that relate to one's environment, other individuals, occupations, activities, etc., such as sister, lawyer, gardener, or church member).[40] Structured routines can create a sense of security, decrease stress, improve sleep, and have a positive impact on one's emotional and physical health. Routines and roles are especially valued during times of transition, like coping with loss, retirement, moving, empty-nester syndrome, (grand)child-rearing, or health changes. It is vital to balance daily activities and prioritize those that positively impact wellness.

Older adults desiring social participation may need to be proactive in scheduling social calls and visits. Social participation opportunities have become more accessible than ever now through technology like social media, video chatting, online support groups, and virtual clubs and organizations. Identifying potential support systems and social networks is essential. Families and health care providers can share resources to help older adults increase opportunities for social participation and help them maintain and develop valued roles. Seeking out social opportunities can sometimes make adults feel vulnerable or insecure. It is recommended that individuals set small goals and not get discouraged easily if social situations do not go as planned. Making new friends and acquaintances at any age can be a challenge. Although virtual social participation can be done from anywhere, it does not replace the need for in-person

TABLE 19.2
Resources for Older Adults and Their Families

Services and General Information for Older Adults

National Institute on Aging (NIA)
www.nia.nih.gov/
Health aging information, Alzheimer's updates, and topics of interest for older adults, caregivers, and health care practitioners

National Council on Aging (NCOA)
www.ncoa.org/
Equitable aging supports, advocacy, and health information for older adults, caregivers, and health care practitioners

Administration for Community Living (ACL)
https://acl.gov/
Support, advocacy efforts, and strategies for older adults, people with disabilities, and caregivers living in the community

American Association of Retired Persons (AARP)
www.aarp.org/
Support and Information to empower people over 50 to live how they choose to live

Americorps Seniors
https://americorps.gov/serve/americorps-seniors
Volunteering information to help and serve others

Government of Canada – Programs and Services for Older Adults
www.canada.ca/en/employment-social-development/campaigns/seniors.html

Age UK
www.ageuk.org.uk/
Health and well-being, work, money, and support services information for older adults and their families

Information for Caregivers of Older Adults
Eldercare Locator
https://eldercare.acl.gov/Public/index.aspx
Community supports for older adults and their families

Family Caregiver Alliance
www.caregiver.org/
Services information for caregivers of older adults with physical and cognitive impairments

Alzheimer's Association
www.alz.org/
Resources, help, and support for older adults experiencing Alzheimer's and their families

Social Visiting for Older Adults
DOROT
www.dorotusa.org/
Bringing generations together through social connection

Elder Network
www.elder-network.org/
Services include respite care, home companions, and transportation for older adults

Connect2Affect (AARP Foundation)
https://connect2affect.org/
Help and strategies for older adults or friends and family of older adults experiencing social isolation

contact, human touch, and the need to feel connected through face-to-face, intimate gatherings. This has to be taken into account while planning social participation.

19.8.1 ROLE AND LEISURE EXPLORATION

As people age, their interests and abilities may change. It is important to adapt valued roles and interests to fit the person in their current life stage, which may include finding new roles and leisure activities. Some roles that may be explored are:

- Caregiver
- Mentor
- Social/community group member
- Amateur hobbyist
- Entertainer
- Student and lifelong learner
- Volunteer
- Employee
- Friend
- Family member
- Religious or spiritual group participant
- Home maintainer
- Partner/significant other

Older adults may require support and encouragement to nurture their creative and social sides. When individuals experience feelings of loneliness, dysphoria, or low self-esteem, they may find it challenging to envision taking on new roles, trying new activities and meeting new people. Health care providers or family members may assist the older adult in identifying new leisure activities and may use resources such as Model of Human Occupation's Assessment of Occupational Functioning Modified Interest Checklist.[41] Another resource is AARP's Virtual Community Center, which offers interactive online events and classes designed for learning, self-improvement, and fun for older adults to try from the comfort of their own home.

19.8.2 EXPANDING COMMUNITY AND SOCIAL NETWORKS

In addition to older adults seeking community engagement outside of the home, there are also ways to bring the community to the individual, reducing social isolation.[2] Friendly visiting programs like DOROT, based in New York City, and Elder Network, based in Minnesota, offer programs and services provided by community members of all generations. Many friendly visiting programs offer in-person or remote social interactions, giving respite to support caregivers.

Continuing Care Retirement Communities (CCRC) or LifePlan community models, although more expensive options, offer a continuum of care for residents. If a resident needs additional support or more assistance over time, many CCRC/ LifePlan communities have options for Independent Living, Assisted Living, Memory

Care, or Skilled Nursing. These options can lower the cost of living, facilitate social connections, and provide opportunities for leisure activities.

Baby boomers (people born between 1946 and 1964) are redefining what old age looks like and making strides to reinvent their *second act*. They are developing alternatives to a typical retirement by moving to urban areas, college towns, participating in cohousing and shared housing, successfully changing the way society thinks about retirement and aging. This generation is demonstrating resistance to traditional retirement community models, creating opportunities for innovative approaches.[42]

Intergenerational housing models are a unique alternative for older adults living in communities that promote ties with younger generations. This housing model improves the health and well-being of older adults while providing significant benefits for all residents involved, no matter their age. For example, in some instances, college students live alongside older adults, which reduces housing costs and enhances communication skills and feelings of empathy for both populations. The college students experienced improved academic performance, improved self-worth, increased appreciation for older adults, and reduced anxiety.[43]

Another way to bring the community to the individual is through virtual connections. Those who cannot participate socially in person, such as older adults living alone or with mobility issues, can foster relationships and join communities online. For older populations, social technology – including email, Facebook, online video conferencing, and instant messaging – can lower the levels of loneliness and improve one's sense of connectedness.[44] It is important to consider that some groups may not have adequate access to broadband, which could limit the availability of social media tools and applications. Refer to Chapter 12 for more information on the benefits of social media and application recommendations to encourage social connectivity online.

19.9 FOLLOWING UP WITH A PHYSICIAN

Mental health stigma continues to be a barrier for older adults in need of and seeking mental health support and treatment. Older adults with depression do not often seek out mental health treatment. Further, African American older adults are more likely to internalize stigma and even less likely to seek treatment than their white counterparts.[45] As older adults may be resistant to express mental health concerns, they can benefit from support from close friends, family members, and health care providers. If an older adult or a family member suspects symptoms of depression like dysphoria, insomnia or hypersomnia, significant weight changes, fatigue, feelings of worthlessness or excessive guilt, indecisiveness, inability to concentrate, and/or recurrent thoughts of death or suicide,[46] they should immediately reach out to a physician for further guidance, assessment, and treatment. Older adults should be encouraged to speak honestly, create open lines of communication, bring written questions to the physician, feel comfortable taking notes, and ask follow-up questions to better understand the physician's recommendations and their options for treatment.

19.10 CONCLUSION

Older adults have a higher risk of social isolation, which can lead to feelings of loneliness. Over time, these feelings can result in health consequences, negatively affecting physical, socio-emotional, and mental well-being, and diminishing overall quality of life. Warning signs of social isolation and a state of loneliness should be understood by older adults, caregivers, and health professionals to screen for and monitor older adults' well-being. Older adults feeling socially isolated and lonely can benefit from increasing contact with social supports, creating more engaging interactions, exploring leisure and special interest groups and activities, and connecting with others (in-person and virtually).

REFERENCES

[1] National Academies of Sciences, Engineering, and Medicine. *Social isolation and loneliness in older adults: opportunities for the health care system.* Washington, DC: National Academies Press; 2020. doi.org/10.17226/25663

[2] Marczak J, Wittenberg R, Doetter LF, et al. Preventing social isolation and loneliness among older people. *Eurohealth (Lond).* 2019;25(4):3–5. https://apps.who.int/iris/handle/10665/332493

[3] Gardiner C, Laud P, Heaton T, Gott M. What is the prevalence of loneliness amongst older people living in residential and nursing care homes? A systematic review and meta-analysis. *Age and Ageing.* 2020;49(5):748–757. doi:10.1093/ageing/afaa049

[4] Jansson AH, Muurinen S, Savikko N, et al. Loneliness in nursing homes and assisted living facilities: prevalence, associated factors and prognosis. *J Nurs Home Res Sci.* 2017;3:43–49. doi.org/10.14283/jnhrs.2017.7

[5] Simard J, Volicer L. Loneliness and isolation in long-term care and the COVID-19 pandemic. *J Am Med Dir Assoc.* 2020;21(7): 966–967. doi:10.1016/j.jamda.2020.05.006

[6] Cornwell EY, Waite LJ. Social disconnectedness, perceived isolation, and health among older adults. *J Health Soc Behav.* 2009;50(1):31–48. doi:10.1177/002214650905000103

[7] Coyle CE, Dugan E. Social isolation, loneliness and health among older adults. *J Aging Health.* 2012;24(8):1346–1363. doi:10.1177/0898264312460275

[8] Holt-Lunstad J, Smith TB, Baker M, Harris T, Stephenson D. Loneliness and social isolation as risk factors for mortality: a meta-analytic review. *Perspect Psychol Sci.* 2015;10(2):227–237. doi:10.1177/1745691614568352

[9] Steptoe A, Shankar A, Demakakos P, Wardle J. Social isolation, loneliness, and all-cause mortality in older men and women. *Proc Natl Acad Sci USA.* 2013;110(15):5797–5801. doi:10.1073/pnas.1219686110

[10] Taylor HO, Taylor RJ, Nguyen AW, Chatters L. Social isolation, depression, and psychological distress among older adults. *J Aging Health.* 2018;30(2):229–246. doi:10.1177/0898264316673511

[11] Milton C. Social isolation and loneliness among older people: advocacy brief. Geneva, Switzerland: World Health Organization; 2021. www.who.int/publications/i/item/9789240030749

[12] Carney MT, Fujiwara J, Emmert BE, Liberman TA, Paris B. Elder orphans hiding in plain sight: a growing vulnerable population. *Curr Gerontol Geriatr Res.* 2016. doi:10.1155/2016/4723250

[13]   Glossary. National Collaborating Centre for Determinants of Health. https://nccdh. ca/glossary/entry/marginalized-populations. 2022. Accessed February 8, 2022.

[14]   Hoy-Ellis CP, Ator M, Kerr C, Milford J. Innovative approaches address aging and mental health needs in LGBTQ communities. *Generations.* 2016;40(2):56–62. https://pubmed.ncbi.nlm.nih.gov/28366982/

[15]   Kern S, Reitz M, Seruya F, et al. Marginalized populations. In: Fratantoro CA, ed. *Occupational therapy in community and population health practice.* 3rd edition. Philadelphia, PA; FA Davis; 2020:457–487.

[16]   Viscogliosi C, Asselin H, Basile S, et al. Importance of Indigenous elders' contributions to individual and community wellness: results from a scoping review on social participation and intergenerational solidarity. *Can J Public Health.* 2020;111(5):667–681. doi:10.17269/s41997-019-00292-3

[17]   The true death toll of COVID-19: estimating global excess mortality. World Health Organization. www.who.int/data/stories/the-true-death-toll-of-covid-19-estimating-global-excess-mortality. 2022. Accessed February 2, 2022.

[18]   Russell D. UCLA loneliness scale (version 3): reliability, validity, and factor structure. *J Pers Assess.* 1996;66(1): 20–40. doi:10.1207/s15327752jpa6601_2

[19]   Hawkley LC, Thisted RA, Cacioppo JT. Loneliness predicts reduced physical activity: cross-sectional & longitudinal analyses. *Health Psychol.* 2009;28(3): 354–363. doi:10.1037/a0014400

[20]   Biordi DL, Nicholson NR. Social isolation. In: *Chronic illness: impact and intervention.* 8th ed. Jones & Bartlett Learning; 2013:85–115.

[21]   Das A. Loneliness does (not) have cardiometabolic effects: a longitudinal study of older adults in two countries. *Soc Sci Med.* 2019;223:104–112. doi.org/10.1016/j.socscimed.2018.10.021

[22]   Tung EL, De Marchis EH, Gottlieb LM , Lindau ST , Pantell MS. Patient experiences with screening and assistance for social isolation in primary care settings. *J Gen Intern Med.* 2021;36(7):1951–7. doi:10.1007/s11606-020-06484-9

[23]   Beck AT, Steer RA, Ball R, Ranieri W. Comparison of Beck depression inventories – IA and II in psychiatric outpatients. *J Pers Assess.* 1996;67(3): 588–597. doi:10.1207/s15327752jpa6703_13

[24]   de Jong-Gierveld J, Kamphuis F. The development of a Rasch-type loneliness scale. *Appl Psychol Meas.* 1985;9(3):289–299. doi:10.1177/01466216850090030

[25]   Berkman LF, Syme SL. Social networks, host resistance, and mortality: a nine-year follow-up study of Alameda County residents. *Am J Epidemiol.* 1979;109(2):186–204. doi:10.1093/oxfordjournals.aje.a112674

[26]   Huxley P, Evans S, Madge S, et al. Development of a social inclusion index to capture subjective and objective life domains (Phase II): psychometric development study. *Health Technol Assess.* 2012;16(1):iii–241. doi: 10.3310/hta16010

[27]   Kamalpour M, Watson J, Buys L. How can online communities support resilience factors among older adults. *Int J Hum Comput Interact.* 2020;36(14):1342–1353. doi:10.1080/10447318.2020.1749817

[28]   Resnick B. Resilience in older adults. *Top Geriatr Rehabil.* 2014;30(3):155–163. doi:10.1097/TGR.0000000000000024

[29]   Physical activity. Centers for Disease Control and Prevention. www.cdc.gov/physicalactivity/basics/pa-health/index.htm. 2021. Accessed June 24, 2022.

[30]   Bloom I, Edwards M, Jameson KA, et al. Influences on diet quality in older age: the importance of social factors. *Age and Ageing.* 2017;46(2):277–283. doi:10.1093/ageing/afw180

[31] Healthy eating tips for seniors. National Council on Aging. www.ncoa.org/article/
      healthy-eating-tips-for-seniors. February 23, 2021. Accessed October 26, 2021

[32] 10 myths about aging. National Institute on Aging. www.nia.nih.gov/health/10-
      myths-about-aging. 2021. Accessed October 26, 2021.

[33] Newsom R, DeBanto J. Aging and sleep. Sleep Foundation. www.sleepfoundation.
      org/aging-and-sleep. October 23, 2020. Accessed October 26, 2021.

[34] Adams KB, Leibbrandt S, Moon H. A critical review of the literature on social
      and leisure activity and wellbeing in later life. *Ageing Soc.* 2010;31(4):683–712.
      doi:10.1017/S0144686X10001091

[35] How Americans view their jobs. Pew Research Center. www.pewresearch.org/soc
      ial-trends/2016/10/06/3-how-americans-view-their-jobs./2016. Accessed October
      26, 2021.

[36] Sewdas R, de Wind A, van der Zwaan L, et al. Why older workers work beyond the
      retirement age: a qualitative study. *BMC Public Health.* 2017;17(1):672. doi:10.1186/
      s12889-017-4675-z

[37] Labor force participation rate for workers age 75 and older projected to be over
      10 percent by 2026. U.S. Bureau of Labor Statistics. www.bls.gov/opub/ted/2019/
      labor-force-participation-rate-for-workers-age-75-and-older-projected-to-be-over-
      10-percent-by-2026.htm. May 29, 2019. Accessed October 29, 2021.

[38] Barbosa LM, Monteiro B, Murta SG. Retirement adjustment predictors – a system-
      atic review. *Work Aging Retire.* 2016;2(22):262–280. https://doi.org/10.1093/wor
      kar/waw008

[39] Lum TY, Lightfoot E. The effects of volunteering on the physical and mental health
      of older people. *Res Aging.* 2005;27(1):31–55. https://doi.org/10.1177/016402750
      4271349

[40] American Occupational Therapy Association. Occupational therapy practice frame-
      work: domain and process. 4th edition. *Am J Occup Ther.* 2020;74(2). https://doi.
      org/10.5014/ajot.2020.74S2001

[41] Modified interest checklist. Model of Human Occupational Therapy: University of
      Illinois at Chicago. www.moho.uic.edu/productDetails.aspx?aid=38. 2022. Accessed
      February 10, 2022.

[42] Greenfield EA, Scharlach A, Lehning AJ, Davitt JK. A conceptual framework for
      examining the promise of the NORC program and village models to promote aging
      in place. *J Aging Stud.* 2012;26(3):273–284. doi.org/10.1016/j.jaging.2012.01.003

[43] Suleman R. Bhatia, F. Intergenerational housing as a model for improving older-
      adult health. *BCMJ.* https://bcmj.org/articles/intergenerational-housing-model-
      improving-older-adult-health. 2021;63(4):171–173.

[44] Chopik WJ. The benefits of social technology use among older adults are mediated
      by reduced loneliness. *Cyberpsychol Behav Soc Netw.* 2016;19(9):551 –556.
      doi:10.1089/cyber.2016.0151

[45] Conner KO, Copeland VC, Grote NK, et al. Mental health treatment seeking among
      older adults with depression: the impact of stigma and race. *Am J Geriatr Psychiatry.*
      2010;18(6):531–543. doi:10.1097/JGP.0b013e3181cc0366

[46] Depression. U.S. Department of Health & Health Services. www.mentalhealth.
      gov/what-to-look-for/mood-disorders/depression. March 1, 2021. Accessed March
      10, 2022.

20 Caregiver Safety and Self-Care

Lauren Sponseller

CONTENTS

20.1 CAREGIVING ROLES

20.1.1 RESPONSIBILITIES

Individuals who enter the role of caregiver are often unpaid family members, friends, or neighbors who provide assistance and care to an individual with an acute or chronic condition.[1] There are various responsibilities that caregivers are involved in, including assisting with activities of daily living (ADLs), decision making, and managing medical care devices.[1-2] Caregivers may provide assistance in daily tasks such as bathing, dressing, and feeding, and these tasks may become more complex and involved if the care recipient resists help or has multiple medications with varying schedules that must be closely managed daily.[1] Caregivers may need to make decisions on behalf of the individual regarding financial, medical, and legal matters if they are not able to do so due to increasing cognitive decline.[3] When managing an individual's medications, caregivers are responsible for noting any potential side effects of medications as well as early signs and symptoms of conditions such as dementia.[1] In addition, caregivers become routinely involved in the care of devices and medical procedures, for example, providing skin care around a central line or gastrostomy tube feedings.[1-2]

DOI: 10.1201/9781003197843-20

20.1.2 CHALLENGES

Caregivers may face a wide range of challenges throughout the time they are providing care. They frequently report feeling unprepared and not having sufficient knowledge to provide the proper and necessary care, as they often are unsure of the type and amount of care they need to provide and worry about making the right decisions for care recipients.[1] Further, caregivers may be unaware of available resources in their communities and how to access them.[1] They also experience difficulties when interacting with healthcare professionals, as many caregivers receive little guidance from healthcare staff due to disagreements surrounding the needs of the patient upon discharge.[1] Discharge has been described by family caregivers as an abrupt and upsetting event, where they are left feeling abandoned because healthcare providers did not support or prepare them for the individual's transition home.[1] There is also the potential for decline in caregiver health in those who perceive themselves as burdened, as research has shown higher risks for fatigue, sleep disturbances, lower immune functioning, increased blood pressure, and higher risks for cardiovascular disease.[1] Social relationships may become impacted as well, and this can occur both between the caregiver and care recipient as well as for the caregiver and their relationships with family members and friends. The relationship between caregiver and care recipient may change and the caregiver may become more focused on specific caregiving tasks and experience a deep sense of loss as the individual's cognitive condition declines.[4] Relationships with family members and friends may become neglected or more challenging to maintain as the caregiver has decreasing resources of energy and time.[4]

20.2 SAFETY WITH CAREGIVING

20.2.1 MENTAL

Family caregivers can experience significant emotional symptoms and lower quality of life as the patient's illness progresses, and they must continuously change and adapt to the demands of new patient needs.[2] Caregiver burden appears to be a precursor to depressive symptoms, and caregiving activities that require high levels of time and energy often result in deprivation of basic necessities such as sleep and exercise.[1] These responsibilities also lead to less time to recover from illness, which may compromise the ability of caregivers to continue to provide adequate services. If caregivers are to continue in providing appropriate care, relief from the distress and demands of maintaining the required care must be prioritized.[1] When considering how to remedy this, effective and adaptive coping strategies may play a protective role in reducing the caregiver's distress.[5] Problem-focused strategies, which are based on one's ability to manage the environmental event, can help to reduce caregiver burden. Some examples of this include using strategies focusing on communication skills, anger management, and deep breathing to promote positive responses to negative situations.[6] Problem-focused coping strategies are beneficial in helping caregivers gain more control over stressful situations, their own behaviors, and their time management, allowing them to feel less burdened in all aspects of their lives.[6]

20.2.2 Biomechanical

Caregiving requires a large responsibility, and few caregivers have received formal training in safely executing activities involving greater physical exertion. For example, assisting with transfers or helping care recipients who may be at risk for falling and could injure themselves during these tasks.[4] Without formal training, this task could be performed in an unsafe manner and put both the caregiver and their care recipient at risk of harm. Also, physical stress and strain may occur when assisting with walking, lifting heavy wheelchairs into cars for appointments, helping care recipients into and out of bed, or lifting care recipients if they fall.[2] Family caregiver education and training are critical to improve caregivers' skill and confidence in their abilities to assist care recipients with transfers and mobility to reduce the chance of injury during caregiving activities.[7–8] Physical and occupational therapists support this role by providing caregivers with safe patient handling education and making appropriate recommendations for equipment based on the caregiver's ability and care recipient's level of function and deficits. They also educate caregivers on proper body mechanics, cueing techniques, and appropriate mobility aids that can support appropriate caregiving roles.[7]

20.3 SELF-CARE

20.3.1 Self-Awareness/Support

Caregiving is an essential role with many demands. However, many internal facets impact our ability to be successful caregivers while maintaining occupational balance and well-being. Recent research has explored the impact that self-awareness and support have on caregiving ability and individuals' overall well-being.[9–11] Various tactics have been found to improve self-awareness including reflection, identification of triggers, and emotional regulation. All of these tactics allow caregivers to "act rather than react."[10] Reflection or journaling was found to remind caregivers of the good that comes from their role and details about their own well-being as well. "Writing, listing accomplishments, and reflecting on specific topics had positively affected participants' awareness of positive aspects in life" and reminding them of the good that can come out of caregiving.[11] Journaling also "showed promise in giving caregivers greater awareness of their own well-being and encouraged social sharing and could become part of a solution" (ibid.).

Self-awareness as a caregiver is a great trait. However, external support can make all the difference. Support for caregivers can come in many forms. This can include emotional support, trained/professional support, and community support. Encouraging individuals to utilize these supports "empowered caregivers to share their identities and experiences as a caregiver within their social networks and provided caregivers reassurance with regard to their own mental health."[11] Although emotional support is critical for the well-being of the caregiver, physical support from other parties can be the most beneficial thing. There is a need for support in the "recruitment of competent help ... the verification of qualifications and references of people hired to help, and support in finding professional experts available for the needs of elderly and informal caregivers."[9] Nonprofessional support is also helpful. Small acts like helping with the individuals' meal or safety was shown to convey respect and support for the caregivers (ibid.).

It is important to find ways in which caregiver stress can be minimized, as higher levels of stress is associated with lower life satisfaction in caregivers.[12] Research supports that participating in in-person support groups moderated by professionals has many positive effects for caregivers of individuals with dementia. This includes decreasing depression and stress; it also promotes feelings of preparedness in being able to manage memory loss. Cognitive-behavioral interventions that teach caregivers on how to reframe their thoughts has shown to decrease experienced anxiety, stress, and depression. Overall, occupational therapists can promote caregiver well-being through providing education, mindfulness, and stress management techniques and teaching coping skills. It is imperative that therapists allow caregivers to ask questions and communicate their needs and show that they are listening and that they are heard.[13]

20.4 CAREGIVER BURNOUT

Oftentimes, the health concerns and emotional management of individuals who care for their loved ones with chronic neurological conditions aren't given attention. Although caregivers may feel grateful to be able to care for their loved ones and improve their personal relationship with them, there are many physical and emotional challenges that they may face.

Hale and Marshall[14] noted that caregivers of individuals who have more complex needs are at a greater risk of experiencing caregiver burnout and individual health challenges, and about 20% of caregivers are financially strained. Therefore, it is important for health professionals to take a multidisciplinary approach and educate caregivers on how they can take care of themselves so they can improve the quality of life of themselves and their loved ones.

Sullivan and Miller[4] discussed specific ways in which caregivers can avoid burnout. They suggest that actively asking doctors questions about their loved one's conditions may increase caregivers' understanding of the new role they may be taking, as well as make them feel more confident. It is also important for caregivers to take care of themselves and attend to their emotional, spiritual, and physical health, and practice living a healthy lifestyle (gaining adequate sleep, exercise, a balanced diet) to be able to provide care for their loved one. This includes acknowledging one's emotions (even if they are unwanted) and noting that it is common to feel sad, hopeless, anxious, and fearful at times. It's important that caregivers continue to engage in activities and interests that make them happy, as well as maintain connections with family and friends. Further, connecting with individuals who are going through a similar experience may provide a sense of comfort, community, and additional support. It is imperative that caregivers accept help from others at times and allow themselves to have "holidays" to avoid burnout and refuel one's tank. Utilizing other professional agencies such as meal/medicine delivery services, transportation services, and adult day care centers can help decrease caregiver burden. Further, each state has its own organizations for specific diseases which can seek for assistance. Caregivers from the United States search for "The National (specific disease) Association of America" online to find these.

Other ways for caregivers to prevent burnout include maintaining a relationship with their loved one where they are able to express their emotions in a healthy manner. This can be supported through utilizing extremal resources, such as counseling or family therapy if complications come up. Caregivers can also motivate their loved one to remain as independent as possible for as long as they can prevent the risk of caregiver burnout. This can be done by ensuring that the individual has other professionals (such as a home health aide) on their team and implements environmental adaptations, routines, and assistive technology that can promote safe independence (i.e., grab bars, nonslip rugs).[14,4]

REFERENCES

[1]   Reinhard SC, Given B, Petlick NH, Bemis A. Supporting family caregivers in providing care. In: Hughes RG, ed. *Patient Safety and Quality: An Evidence-Based Handbook for Nurses*. Rockville, MD: Agency for Healthcare Research and Quality; 2008: 341–404. www.ncbi.nlm.nih.gov/books/NBK2665/

[2]   Sherman DW. A review of the complex role of family caregivers as health team members and second-order patients. *Healthcare (Basel)*. 2019;7(2). doi: 10.3390/healthcare7020063

[3]   Lee JJY, Barlas J, Thompson CL, Dong YH. Caregivers' experience of decision-making regarding diagnostic assessment following cognitive screening of older adults. *J Aging Res*. 2018. doi: 10.1155/2018/8352816

[4]   Sullivan AB, Miller D. Who is taking care of the caregiver? *J Patient Exp*. 2015;2(1):7–12. doi: 10.1177/237437431500200103

[5]   Lavarone A, Ziello AR, Pastore F, Fasanaro AM, Poderico C. Caregiver burden and coping strategies in caregivers of patients with Alzheimer's disease. *Neuropsychiatic Dis Treat*. 2014;10:1407–1413. doi: 10.2147/NDT.S58063

[6]   Ghane G, Farahani MA, Seyedfatemi N, Haghani H. Effectiveness of problem-focused coping strategies on the burden on caregivers of hemodialysis patients. *Nurs Midwifery Stud*. 2016;5(2). doi: 10.17795/nmsjournal35594

[7]   Powell-Cope G, Pippins KM, Young HM. Teaching family caregivers to assist safely with mobility. *Am J Nurs*. 2017;117(12):49–53. doi: 10.1097/01.NAJ.0000527485.94115.7e

[8]   Agency for Integrated Care. Body mechanics: positioning, moving, and transfers. 2020. Accessed March 01, 2022. www.aic.sg/resources/Documents/Brochures/Caregiving%20Support/Body%20Mechanics%20EN.pdf

[9]   Freitas S, Silva AC, Teixeira HJ, Teixeria JC. The needs of informal caregivers of elderly people living at home: an integrative review. *Scand J Caring Sci*. 2012;27(4):792–803. doi: 10.1111/scs.12019

[10] Stjernsward S, Hansson L. A qualitative study of caregivers' experiences, motivation, and challenges using a web-based mindfulness intervention. *Community Ment Health J*. 2020;56(3):416–425. doi: 10.1007/s10597-019-00477-7

[11] Bosch L, Kanis M, Dunn J, Stewart KA, Kröse B. How is the caregiver doing? Capturing caregivers' experiences with a reflective toolkit. *JMIR Ment Health*. 2019;6(5). doi: 10.2196/13688

[12] García-Castro FJ, Hernández A, Blanca M. Life satisfaction and the mediating role of character strengths and gains in informal caregivers. *J Psychiatr Ment Health Nurs.* 2021. doi: 10.1111/jpm.12764

[13] Piersol CV, Canton K, Conner SE, Giller I, Lipman S, Sager S. Effectiveness of interventions for caregivers of people with Alzheimer's disease and related major neurocognitive disorders: a systematic review. *Am J Occup Ther.* 2017;71(5). doi: 10.5014/ajot.2017.027581

[14] Hale D, Marshall K. Managing caregiver stress. *Home Healthcare Now.* 2022;40(1). doi: 10.1097/NHH.0000000000001037

21 Opioid Use and Abuse in Older Adults

Tracy Offerdahl-McGowan

CONTENTS

21.1 SOME BASICS ABOUT PAIN

Acute pain serves an important purpose as it is a necessary and protective function for survival. It allows us to live in a world full of physical pitfalls and dangers. It immediately alerts a person that something is physiologically or anatomically wrong. This type of pain generally responds well to treatment with opioids.

Chronic pain is more difficult to define, as it may serve no purpose similar to acute pain, and it may produce no useful symptoms, and little to no autonomic nervous system signs (i.e., little/no "fight or flight" response). It may be defined as pain that persists ≥ 3 months beyond the "usual" timeframe of acute disease or the healing of an injury. In addition, it may not be easily treated with routine or typical pain management methods.

The treatment of acute and chronic pain is a complicated endeavor and it is a very subjective problem to evaluate. When opioids became popular as the drugs of choice for many acute and chronic pain scenarios, the current "Baby Boomer" population was right in the middle of it, and now that population of patients is our current group of Medicare enrollees. No matter the age group, the appropriate treatment of pain and the potential for opioid use disorder (OUD) continues to be a timely topic when discussing our older adult population of patients. While opioids have been the mainstay for the treatment of acute pain, palliative care, and end-of-life drug regimens, they have also become a controversial part of the treatment of certain chronic pain syndromes.

The term "opiate" derives its name from chemical structures that are related to compounds derived from the opium poppy, which is still grown legally and illegally all over the world. From a "purist" pharmacology perspective, the natural compounds (morphine and codeine) are "opiates" whereas the semisynthetic and synthetic

TABLE 21.1
Popular Types of Opioids That Are Available in the United States

Natural Opioids	Semi-Synthetic Opioids	Synthetic Opioids
Morphine	Hydromorphone	Fentanyl
Codeine	Hydrocodone	Methadone
	Oxycodone	Tramadol
	Heroin	

compounds are called "opioids." Colloquially, most medical professionals refer to all of the drugs in this class as "opioids," and we will use that terminology here. In addition, the term "narcotic" generally refers to any drug that causes narcosis (sleep) but it is typically a term referring to opiates/opioids and it tends to refer to substances with abuse or addiction potential (Table 21.1).

All opioids work to control pain because they decrease a person's perception of pain. They also produce a euphoric effect (sense of well-being) in most people, which may be part of the "lure" to misuse or abuse an opioid. Opioids work because they "plug" into the opioid receptors mu, kappa, and delta that are located in the brain, the spinal cord, and in some tissues. Opioids cause adverse drug effects for the same reason, by binding to these opioid receptors.

Common adverse drug effects associated with opioids are many, and these can be more severe in older adults. These include:

• Sedation
• Change in mental status or full delirium
• Falls resulting in fractures or other injuries
• Miosis (constriction of the pupil)
• Constipation
• Nausea and vomiting
• Respiratory depression leading to overdose and death

In general, the most common adverse drug events seen in older adults are central nervous system (CNS) impairment and falls/fractures. When we add common comorbidities or multimorbidities seen in adult patients such as visual impairment, hearing impairment, cardiovascular disease, musculoskeletal conditions, dementia, and stroke, the addition of opioid adverse drug effects becomes increasingly problematic and dangerous.

21.2 APPROPRIATE TREATMENT OF PAIN

The American Geriatric Society (AGS) supports the use of opioids in acute pain management; however, the evidence for using opioids to treat chronic pain is lacking in all populations, including older adults. In addition, most of the studies evaluating opioid use in older adult populations lacks those who are "frail" and those with multimorbidities. The AGS has put together a set of questions in order to determine if opioids should be used in older adult patients with moderate to severe pain.

- How is this type of pain typically treated?
- Are there any alternatives to treat this type of pain that are equivalent or better for pain control, side effects, and quality of life?
- Does the patient have comorbidities or multimorbidities that can increase the risk of opioid-related adverse drug effects?
- Is the patient or their caregiver likely to manage the opioid treatment plan appropriately and responsibly?
- Does the practitioner/prescriber feel capable treating this patient and their type of pain?
- Does the patient need a specialized pain management physician?
- Would the patient benefit from multidisciplinary or co-management of their pain?
  - Considerations should involve evaluation of medical, behavioral, or social circumstances.

In general, the overall of risk of OUD in older adult patients with no current or past substance abuse is low; however, it is not possible to predict those that will go on to misuse opioids, and OUD is a disease that affects people in every socioeconomic class and every age group.

From 1999 to 2020, there were more than 932,000 deaths due to a drug overdose in the United States. In 2020, there were > 90,000 deaths due to drug overdose, representing a 31% increase in deaths compared to the year before and approximately 75% of those deaths involved an opioid. In The number of opioid deaths continues to increase yearly. We are losing the battle in this opioid crisis.

The older adult population is no exception, and the misuse and abuse of opioids has steadily increased in this patient population in the last 25 years (Chang and Compton 2016).In general, older adult patients have a decrease in both the breakdown (metabolism) and renal elimination of opioids. The result of this is a potentially dangerous increase in blood levels of the opioid(s) which may result in exaggerated adverse drug effects. Opioid misuse or abuse poses other risks in older adults including cardiac problems, motor vehicle accidents, poor surgical outcomes, falls, fractures, and death.

With the approval of drugs like long-acting oxycodone (OxyContin™), oxycodone + acetaminophen (Percocet™, Endocet™), and hydrocodone + acetaminophen (Vicodin™, Lortab™), the availability of potent oral opioid analgesics became available to inpatient and outpatient treatment plans for pain. Long-acting oxycodone (OxyContin™) is a particularly controversial drug. It was approved in 1995 and was marketed as a product that would have a lower potential for abuse. Unfortunately, it was quickly determined that opioid abusers could quickly achieve a remarkable "high" by crushing the OxyContin™ and swallowing it or snorting it. This led to a very high level of abuse that has not ceased. There are some that describe this period of the mid-1990s as a "huge manufacturing and marketing success" as well as a "public tragedy" when specifically referring to the approval of OxyContin™. Of note, >75% of heroin users state that the first opioid they used was a prescription opioid.

The Centers for Disease Control and Prevention (CDC) reference three distinct "waves" that led to our current opioid crisis and opioid overdose deaths in America.

- 1990s: There was an increase in deaths due to prescription opioid overdoses, as many more opioid choices were available for doctors to prescribe and for patients to take. The prescription opioids that were the predominant culprits included morphine, codeine, oxycodone, hydrocodone, and methadone. Shockingly, the number of people who admitted to abusing OxyContin™ was approximately 400,000 in 1999 and 1.9 million just four years later in 2003.
- 2010: There was a marked rise in deaths due to heroin overdoses. When patients can no longer get prescribers to write for prescription opioids, or when they no longer get a good enough "high" with prescription opioids, they many times move to the illicit opioid, heroin.
- 2013: There was a marked rise in deaths due to synthetic opioid use. This surge was due to the synthetic opioid, fentanyl. The drug is used intravenously in inpatient settings, as a patch under the brand name Duragesic™, and it can also be illicitly manufactured. Most overdose deaths were due to the illicit fentanyl.

Clinical Pearl: It is important to note that no older adult patient should be denied appropriate treatment of pain (with opioids) if they have moderate to severe pain with or without functional impairment and/or a diminished quality of life, as no patient should needlessly suffer.

There are many reasons why the misuse and abuse of opioids is problematic in older adults. Research has indicated that some people are predisposed or more vulnerable to substance abuse. In addition, the use and abuse of "psychoactive" drugs like opioids happens for many reasons. People use them to find relief from anxiety, stress, boredom, or pain. People use opioids because they find pleasurable changes in their moods like relaxation or euphoria. They also may use opioids to fit into social groups. Older adults may misuse opioids for any or all of these reasons. When there is a blurred line between taking opioids to relieve pain versus taking opioids to improve their mood, a dangerous habit has begun. Although this is typically unintentional, it may start the patient on a dangerous road towards addiction and OUD. It all ends the same, however, with negative physical, psychological, and social consequences.

Clinical Pearl: It is estimated that approximately 30% of opioid overdose deaths involve concomitant use of a benzodiazepine (Liu et al. 2011). The combination of a benzodiazepine and an opioid is particularly dangerous because they both have a longer duration of activity in older adult patients and they are both significant central nervous system depressants. When used together, the risk of respiratory depression and death is increased.

Benzodiazepines are typically used for anxiety, panic disorders, alcohol dependence, seizure disorders, and as muscle relaxers.

Common benzodiazepines include alprazolam (Xanax™), lorazepam (Ativan™), diazepam (Valium™), and clonazepam (Klonopin™).

"Addiction" may be defined as "compulsive use despite harm," as a person's life is not improved by use of the drug. While people with OUD initially misuse the opioid because it causes pleasant feelings or sensations, later on they take the opioid to prevent the unpleasant feelings or sensations of NOT taking it. There begins the genesis of "addiction" and OUD.

Tools have been developed to help identify current or future drug-behavior problems. They are:

- **Opioid Risk Tool (ORT)**: The ORT has a very high sensitivity and specificity to identify people who are likely to develop OUD. It quantifies the risk of developing OUD and includes information about family history of substance abuse, personal history of substance abuse (including alcohol and nicotine in some cases), history of preadolescent sexual abuse, and certain psychological diagnoses (attention-deficit hyperactivity disorder (ADHD), obsessive compulsive disorder (OCD), bipolar disorder, depression, and anxiety). While some practitioners find great value in using this tool, some find that it tends to stigmatize patients and it is of little value.
- **Urine Drug Screening:** This is a simple tool that is used regularly to check for the presence of other contraindicated or illegal substances. It is also a way to ensure that the patient is actually taking the prescribed opioid which helps to decrease drug diversion.
- **Prescription Drug Monitoring Program (PDMP):** An electronic database used by most states that tracks controlled substance **prescriptions** (opioids, benzodiazepines, sleep agents, etc.). It tracks which prescribers are writing the prescriptions, which pharmacies are filling the prescriptions, and how the patient has paid for the prescription. This is a valuable tool that should be utilized every time a prescriber is considering an opioid (or other controlled substance) for a patient and every time a pharmacist is filling a prescription for a controlled substance for a patient. PDMPs can help identify patients who may be misusing prescription opioids or other prescription drugs and who may be at risk for overdose. Some experts believe that tools such as the PDMP may actually force people with OUD to move to illegal drug sources like heroin or "homemade" fentanyl, as these are not tracked by the PDMP.
- **Morphine Milligram Equivalents (MME):** Any dose of an opioid increases the risk of falls, fractures, and death. However, higher doses of an opioid carry an even bigger risk, including death due to overdose. Calculating a patient's total daily dose of an opioid helps practitioners to determine which patients are at highest risk of overdose and therefore will benefit from dosage reduction and closer monitoring. Patients who take ≥ 50 MME/day are two times more likely to overdose than those patients taking lower doses, and patients taking ≥ 90 MME/day are at an even higher risk. This led to the CDC to develop the "CDC Guidelines for Prescribing Opioids for Chronic Pain" and to further recommend that most patients should not take ≥ 90 MME/day of any opioid, except under special circumstances. Morphine is the "prototype" for the opioid group of drugs, which means that all of the other opioids are compared to morphine. This includes potencies and doses.

TABLE 21.2
Opioid Conversions Compared to Morphine

Opioid Doses in mg/day	Conversion Factor*
Morphine	1
Codeine	0.15
Hydrocodone	1
Oxycodone	1.5
Hydromorphone	4
Fentanyl patch (mcg/hour)	2.4

* Any number > 1 indicates that drug is MORE potent than
  morphine, whereas any number < 1 indicates that drug is
  less potent than morphine.

Here is an example of how "MMEs" are calculated.

1. Determine the total daily amount of each opioid the patient is taking.
2. Convert each of the opioids to MMEs and multiply the dose of each opioid by the conversion factor (see Table 21.2)
3. Add them together (if taking more than one opioid).

Example: Mrs. Smith is a 74-year-old patient who has been taking 1–2 tablets of oxycodone 5mg/325mg acetaminophen (Percocet™) every 6 hours as needed to treat chronic pain associated with severe spinal stenosis.

1. According to Mrs. Smith she takes 2 tablets every 6 hours almost every day. 5mg oxycodone/tablet × 2 tablets = 10mg every 6 hours = 40mg of oxycodone TOTAL per day.
2. Table 21.2 shows that oxycodone has a conversion factor of 1.5. 40mg × 1.5 = 60mg
3. Mrs. Smith is taking 60MME per day and is definitely at risk of accidental overdose. Furthermore, her daughter states that she frequently runs out of medication before she should (i.e., the prescriber writes for a 30-day supply and she is normally out of tablets by day 22 or 23). This indicates that the patient is misusing the medication by taking more than prescribed which means that her MME is higher than 60 on many days, putting her at even higher risk of accidental overdose.

Mrs. Smith, along with any patient taking an opioid (as well as their caregivers, family, and friends) would benefit from having naloxone (Narcan™) on hand in case of accidental overdose. This lifesaving medication is now approved as a nasal spray that can temporarily stop or even reverse the effects of an opioid overdose, including heroin. Some states have implemented naloxone access laws that provide civil or

TABLE 21.3
Drugs Approved for Medication-Assisted Treatment of OUD

Buprenorphine	Naltrexone	Methadone

Subutex™ – sublingual tablet
Sublocade™ – subcutaneous
    injection
	Vivitrol™ – intramuscular
	    injection
		Dolophin™ – oral tablet
		Methadose™ – oral
		    concentrate

Combination Products
Buprenorphine + naloxone
Suboxone™ – sublingual tablet or sublingual/buccal film
Bunavail™ – buccal film
Cassipa™ – sublingual film
Zubsolv™ – sublingual tablet

criminal immunity to licensed healthcare clinicians and lay responders who administer naloxone to reduce overdose.

21.3 TREATMENT OF OPIOID USE DISORDER (OUD)

Despite alarming statistics, the treatment of OUD is not easily accessible to all patients, as it is a highly regulated (federal and state levels) and restrictive endeavor. In the U.S. there are three drugs approved for use in medication-assisted treatment programs (Table 21.3).

For the greatest benefit, medication-assisted treatment should be used along with behavioral therapies. Even after successful treatment of OUD and abstaining from opioids for months or years, users are vulnerable to relapse. This happens despite feeling very motivated to remain drug-free. While relapse is likely a multimodal problem, one of the explanations pertains to likely long-term changes in the brain resulting in a "reward" system and a "wanting" of the opioid.

21.4 CONCLUSION

The subject of pain management with opioids in older adult patients is a complicated one. Older adults typically have medications and multimorbidities that contribute to severe adverse drug effects from opioids, including CNS depression and falls and fractures. Healthcare and medical professionals, along with the families and caregivers of older adult patients on opioids, need to be aware of the risks of opioid use and the potential for opioid use disorder. Early intervention will help decrease the morbidities and mortality associated with the use of these agents in our older adult patients.

BIBLIOGRAPHY

Baird CA. Substance-related and addictive disorders. In: Jones JS, Fitzpatrick JJ, Rogers VL. Psychiatric-mental health nursing. 2nd ed. New York: Springer LLC;2017:291–325.

Bechara A, Berridge KC, Bickel W, et al. A neurobehavioral approach to addiction: Implications for the opioid epidemic and the psychology of addiction. *Psychol Sci Publ Int.* 2019;20(2):96–127. doi: 10.1177/1529100619860513. Accessed October 22, 2021.

Brott NR, Peterson E, Cascella M. Opioid, risk tool. Updated May 19, 2021. In: StatPearls [Internet]. Treasure Island (FL): StatPearls; 2021 Jan–Available from: www.ncbi.nlm. nih.gov/books/NBK553147/. Accessed October 13, 2021.

Centers for Disease Control and Prevention. 2022. Data resources: Analyzing prescription data and morphine milligram equivalents (MME). Available from: www.cdc.gov/opioids/ data-resources/. Accessed October 29, 2021.

Centers for Disease Control and Prevention: Drug Overdose Deaths. Available from: www.cdc. gov/drugoverdose/deaths/index.html. Accessed December 30, 2022.

Centers for Disease Control and Prevention: opioid data analysis and resources. Available from: www.cdc.gov/opioids/data/analysis-resources.html. Accessed October 2, 2021.

Chang YP, Compton P. Opioid misuse/abuse and quality persistent pain management in older adults. *J Gerontol Nurs.* 2016;42(12):21–30.

Dowell D, Haegerich TM, Chou R. CDC Guideline for prescribing opioids for chronic pain – United States, 2016. *JAMA.* 2016;315(15):1624.

Gladden M, O'Donnell J, Mattson C, et al. Changes in opioid-involved overdose deaths by opioid type and presence of benzodiazepines, cocaine, and methamphetamine – 25 states, July–December 2017 to January–June 2018. *Morb Mortal Wkly Rep.* 2019;68(34):737–744. Accessed October 29, 2021.

Gloth FM. Pharmacological management of persistent pain in older persons: Focus on opioids and non-opioids. *J Pain.* 2011;12(3), Suppl 1:S14–S20. doi: 10,1016/ J.Jpain.2010.11.006. Accessed September 1, 2021.

Gressler LE, Martin BC, Hudson TJ, et al. Relationship between concomitant benzodiazepine-opioid use and adverse outcomes among US veterans. *Pain.* 2018;159(3):451–459.

Heins SE, Castillo RC. Changes in opioid prescribing following the implementation of state policies limiting morphine equivalent daily dose in a commercially insured population. *Med Care.* 2021;59(9):801–807. doi: https://doi.org/10.1097/MLR.0000000000001 587. Accessed October 2, 2021.

Heins SE, Frey KP, Alexander GC, et al. Reducing high-dose opioid prescribing: state-level morphine equivalent daily dose policies, 2007–2017. *Pain Med.* 2020;21(2):211–215.

Jassal M, Egan G, Dahri K. Opioid prescribing in the elderly: a systematic review. *J Pharm Tech.* 2020;36(1):28040. doi: 10.1177/8755122519867975. Accessed October 18, 2021.

Karaca-Mandic P, Meara E, Morden NE. The growing problem of cotreatment with opioids and benzodiazepines. *BMJ* 2017;356:j1224.

Khan M, Mahmood Laila, Fisher D, et al. Assessment of pain: Tools, challenges, and special populations. In: Moore RJ. *Handbook of pain and palliative care.* 2nd ed. Cham, Switzerland: Springer Nature AG; 2018:175–199. https://doi.org/10.1007/ 978-3-319-95369-4.

Larsson C, Hansson EE, Sundquist K. Chronic pain in older adults: prevalence, incidence, and risk factors. *Scand J Rheumatol.* 2017;46(4):317–325. doi: 10.1080/ 03009742.2016.1218543. Accessed October 29, 2021.

Lee BL, Zhao W, Yang KC, et al. Systematic evaluation of state policy interventions targeting the US opioid epidemic, 2007–2018. *JAMA Netw Open.* 2021;4(2):e2036687. doi: 10.1001/jamanetworkopen.2020.36687. Accessed October 29, 2021.

Liu EY, Tamblyn R, Filion KB, et al. Concurrent prescriptions for opioids and benzodiazepines and risk of opioid overdose: Protocol for a retrospective cohort study using linked administrative data. *BMJ Open.* 2021;11:e042299. doi: 10.1136/bmjopen-2020-042299. Accessed October 11, 2021.

Mattson CL, Tanz LJ, Quinn K, et al. Trends and geographic patterns in drug and synthetic opioid overdose deaths – United States, 2013–2019. *Morb Mortal Wkly Rep.* 2021;70(6):202–207. doi: http://dx.doi.org/10.15585/mmwr.mm7006a4external iconexternal icon. Accessed October 29, 2021.

Perry BL, Odabas M, Yang KC, et al. New means, new measures: Assessing prescription drug-seeking indicators over 10 years of the opioid epidemic. Society for the Study of Addiction. 2021. https://doi.org/10.1111/add.15635. Accessed September 28, 2021.

Prescription Drug Monitoring Program. Updated February 12, 2021. Centers for Disease Control and Prevention, National Center for Injury Protections and Control. www.cdc.gov/opioids/providers/pdmps.html. Accessed August 23, 2021.

Rowland FN, Fallon B. Chronic pain in the elderly: The use of opioids and selected adjuvants. *Top Geriatr Rehabil.* 1993;8(4):27–37.

Suryadevara U, Holvert R, Averbuch R. Opioid use in the elderly. *Psychiatr Times.* 2018;35(1).

Trescot AM, Helm S, Benyamin R, et al. Opioids in the management of chronic non-cancer pain: An update of American Society of the Interventional Pain Physicians' (ASIPP) Guidelines. *Pain Phys.* 2008;Opioids Special Issue:11:S5–S62.

U.S. Department of Health and Humans Services, Center for Disease Control and Prevention. Calculating total daily dose of opioids for safer dosage. www.cdc.gov/drugoverdose/pdf/calculating_total_daily_dose-a.pdf. Accessed September 1, 2021.

United States Food and Drug Administration: Information about medication-assisted therapy. www.fda.gov/drugs/information-drug-class/information-about-medication-assisted-treatment-mat. Accessed January 15, 2022.

Webster LR, Webster RM. Predicting aberrant behaviors in opioid-treated patients: Preliminary validation for the opioid risk tool. *Pain Med.* 2005;6(6):432–442.

West NA, Severtson SG, Green JL, et al. Trends in abuse and misuse of prescription opioids among older adults. *Drug Alcohol Depend.* 2015;1:117–121.

Wilson N, Kariisa M, Seth P, et al. Drug and opioid-involved overdose deaths – United States, 2017–2018. *MMWR Morb Mortal Wkly Rep.* 2020;69(11):290–297.

Weimer MB, Wakeman SE, Saitz R. Removing one barrier to opioid use disorder treatment: Is it enough? *JAMA.* 2021;325(12):1147–1148. doi: 10.1001/jama.2021.0958. Accessed October 18, 2021.

Yaksh T, Wallace M. Opioids, analgesia, and pain management. In: Brunton LL, Hilal-Dandan R, Knollmann BC. *Goodman & Gilman's: The pharmacological basis of therapeutics.* 13the ed. New York: McGraw Hill Education;2018: https://accessmedicine-mhmedical-com.libsalus.idm.oclc.org/content.aspx?bookid=2189&sectionid=170269577. Accessed October 22, 2021.

22 Reducing Harm Related to Substance Use by Older Adults

Michelle Knapp, DNP, NP-P, PMHNP-BC, FIAAN, Donna E. McCabe, DNP, GNP-BC, PMHNP-BC and Madeline A. Naegle, Ph.D, CNS-PMH, BC, FAAN

CONTENTS

22.1 OVERVIEW

Substance use is a long-standing human behavior. People in the 21st century, however, can benefit from a greater body of research, information, and reflection on popular culture about substances used and their health implications. The largest population of users of all substance are between 26 and 55; as people age, the general trend is to use fewer nonprescription substances and in lesser amounts. Patterns of use in America are tracked through Substance Abuse Mental Health Services (SAMHSA) and global trends are tracked through the World Health Organization (WHO). Available data apply most accurately to early and middle age adults; trends of use among adults over 55 are less commonly tracked. The 21st century is unique in that people everywhere are living longer, and longevity often means continuing patterns and lifestyles established early in life. Adults 65 and older are now reporting longer lifetime substance use in greater amounts and frequency than in prior generations.[1] Cultural trends and traditions vary greatly, but certain patterns persist: men are more likely to use all substances, and alcohol is the substance most widely used around the globe. Its harmful use accounts for 5.3% of all deaths annually and is a causal factor in more than 200 disease and injury conditions.[2] With sedentary lifestyle, tobacco use, and poor diet, harmful alcohol use increases risks for noncommunicable (chronic) disease.[3] The percentage of older adult alcohol users is lowest in Portugal and highest in Norway, and its use is currently increasing among women and extending well into old age. In the United States, more than one in ten adults over the age of 65 report "binge drinking," that is, consuming five or more (men) and four or more (women) drinks at a given time.[4]

The primary substances used by older American adults are alcohol (43%), cannabis (4.1%), stimulants, opioids (1.3%), and tobacco products (14%). Nearly nine in ten (89%) of adults 65 and older report currently taking prescription medicine as compared to 75% of persons 50–64; use of other drugs and alcohol continues strong in both cohorts.[5] Older adults are also more likely than their younger counterparts to be taking multiple prescription medications. More than half of adults 65 and older (54%) report taking four or more prescription drugs compared to 32% of adults 50–64 years old and about one in ten adults 30–49 (13%) and 18–29 (7%).[6] "Polypharmacy," concomitants drugs taken for multiple medical conditions, increases the risk for adverse drug reactions; when multiple psychoactive drugs are taken for pleasure (polysubstance abuse), risks rise.[7]

22.2 AGING INCREASES VULNERABILITY

The potential harm associated with medications, drugs, and other substances derives from the nature of the drug itself, the mode of use, and amounts of use. For older adults, the normal functional declines of aging, as well as additional physical and psychological symptoms of common health conditions, increase the risks associated with substance use. The most common system changes influence the capacity to combat toxins and metabolize medications and other drugs. Excretory capacity changes and modifies management of body fluids, affecting metabolism. Reduced hepatic reserve also raises vulnerability to the effects of substances. The loss of muscle mass and strength (sarcopenia) plus decreased reflexes and overall decrease in neurons and neurotransmitters can result in less precise balance and increase the risk for falls. Overall nutritional status may decline and weight loss, lower levels of body fluids, and undernutrition mean that smaller amounts of a drug can have greater impact on brain and body functions. Physical and mobility limitations mean that balance and exercise capacities may be impaired. Comorbid medical conditions can also increase the degree and severity of the normal changes of aging.[7]

22.3   LINKS TO CHRONIC CONDITIONS

Over 80% of individuals 65 years and over live with at least one chronic health condition such as cardiovascular disease, diabetes mellitus, or arthritis; 77% have two or more.[8] Co-occurring chronic health conditions increase cognitive, psychological, and physiologic vulnerability to all medications and substances. Regular use of substances can hasten development of such conditions, impair one's ability to participate in self-care, and worsen overall health. Persons with diagnosed SUDs have a higher prevalence of the top 19 major health problems and a higher 10-year mortality rate.[9] Older adults with chronic conditions use many prescribed and over-the-counter medications which can increase the risk of drug interactions. Psychoactive medications can potentiate the effects of other substances leading to greater risk of harm. Common medications that interact with alcohol include those for medical conditions such as metformin for diabetes, statins for cholesterol disorders, and over-the-counter pain medications including ibuprofen and acetaminophen.

Chronic pain, often a primary driver of the opioid use and abuse, afflicted approximately 13.5 million adults aged 65 and older in 2018.[10] Prescriptions for opioid analgesics for chronic pain were written for 26.8% of adults 65+ in 2018, and although there has been a modest decline in opioid use, the management of pain in noncommunicable (chronic) disease poses major care challenges.[11] Older adults may self-medicate with alcohol or misuse combinations of prescribed and illicit drugs to achieve psychoactive effects. Safety can be maximized by addressing psychological and emotional risks which older adults face using substances. These include the financial and personal losses associated with transitions in living or care situations, loss of loved ones, forced retirement, changes to income, poor health status, and/or chronic illness. As reenforced by the SARS 2, COVID-19 epidemic, older people are especially vulnerable to the detrimental effects of social isolation which have been shown to trigger or worsen substance use disorders and other mental health problems.[12] A history of substance use disorders or another current mental illness further increases vulnerability.

Key issues likely to arise for older adults experiencing problematic substance use include health and social inequities, stigma, and discrimination, all of which deter help seeking and contribute to negative health outcomes. Older adults may be less likely to discuss issues of concern with providers or have difficulty finding providers specializing in care of older adults due to poor accessibility and workforce deficits. Health professionals trained in care of older adults and mental health, SUDs, are in short supply around the world.

22.4   HEALTH IMPLICATIONS FOR OLDER ADULTS OF COMMONLY USED DRUGS

Substance use disorders are among the most common psychiatric disorders in older adults. The most frequently used substances include alcohol, tobacco, marijuana, sedative hypnotics, opioid analgesics, and stimulants.[13] Cannabis use has increased and is greatest among those undergoing treatment for mental health issues and/or those who use alcohol.[14] Among older adults, admissions to substance use treatment facilities show an increase of stimulant, marijuana, and opiates use.[1] During the COVID-19 pandemic, opioid overdoses surged, largely due to misuse and diversion of synthetic opioids. As overdoses increased, the disparities in overdose deaths and treatment gaps for at-risk populations, such as Black and Hispanic American men, also increased. A recent study of national prescription drug claims from 2011 to 2016 showed that one in ten adults are at risk of opioid overdose, with older adults relying on high doses of opioids more than younger adults.[15]

22.5 HEALTH AND SAFETY RISKS BY DRUG CLASS

The major substance categories commonly used by older adults are generally classified by intoxication effects: depressants, sedative/hypnotics, opioids, stimulants, and tobacco. Health risks vary depending on the class (see Table 22.1). Physical and mental health conditions may be precipitated or exacerbated depending on substance-specific pathophysiological effects. The risk of death increases with the number of an individual's medications, and polypharmacy in adults 65 and older is a global concern.[16] Drug misuse, taking a medication in ways other than prescribed, can negatively impact the older adult's ability to perform activities of daily living. Misuse

TABLE 22.1
Classification of Substances with Consequences of Toxic Exposure

Substance	Category	Consequences of Use
Alcohol	Depressant	Depression, anxiety, cardiac symptoms, increased sensitivity to pain, weakened immune function, sleep problems, hepatic and renal diseases
Opioids	Depressant	Opioid-induced hyperalgesia, low testosterone, low estrogen, GI problems, skin conditions, secondary infections
Cocaine, amphetamines	Stimulant	Dysphoria, nervousness, insomnia, stroke, GI problems, irregular heartbeat, elevated blood pressure, acute psychosis, chronic psychotic disorders, skin conditions, secondary infections, tooth and gum decay
Caffeine	Stimulant	Dysphoria, nervousness, insomnia, GI problems, irregular heartbeat, elevated blood pressure
Tobacco	Stimulant	Headaches, attentional problems, GI problems, cardiac illness, respiratory illness, tooth staining and decay
Cannabis, synthetic cannabis	Hallucinogen	Nervousness, paranoia, psychosis, adult-onset psychotic disorders, tooth staining and decay
Anabolic-Androgenic Steroids	Steroids	Depression, mood swings, frank paranoia, secondary infection, gynecomastia, enlarged prostate, elevated blood pressure, elevated glucose, decreased high-density lipoprotein

is specifically linked to increased risk for overdose related to benzodiazepines and opioids and is believed to occur in as many as 8.7% of users.[17]

22.5.1 ALCOHOL

While substance use has historically been viewed as a societal burden, positive social sanctions for use have increased over the years. In older adults, use patterns continue from early years into retirement. Bareham and colleagues performed a systematic review of perceptions of alcohol use among older adults and found that most report maintaining a routine, ritualistic, controlled consumption pattern a socially responsible part of life.[18] Alcohol is a depressant and negatively affects several systems. While the metabolism of alcohol cannot be accelerated, it can be slowed secondary to medical conditions and natural aging declines that impair hepatic and kidney metabolism. Acute effects appear at lower doses in older adults and include impaired judgment, coordination problems, and sedation. The National Institute of Alcohol Abuse and Alcoholism[4] recommendations for adults over 65 proscribe that those who do not take medication and are in good health should limit their total alcohol consumption to no more than seven drinks per week and should consume no more than three drinks on any given day. Persons with medical conditions such as Major Depression, or those taking certain medications (e.g., pain killers) should consume less alcohol or abstain completely.

Chronic use in excess, that is, more than one standard drink, defined as 1½ oz. spirits, 4-5 oz. wine, 11oz. beer per day, in persons over 65, can precipitate and/or exacerbate hepatic and renal dysfunction, cardiac and pulmonary conditions, depression, anxiety, and neurological and cognitive disorders. Alcohol elevates suicide risk in older adults with both depression and alcohol use disorders.[19] Chronic alcohol consumption also increases risk of seizures and Wernicke's Korsakoff's syndrome, leading to irreversible cognitive damage. Alcohol and tobacco dependence frequently occur together and alcohol may contribute to the reinforcing nature of tobacco use.[20]

TABLE 22.2
CDC Alcohol Consumption Definitions

Type of Alcohol Consumption	Definition
Heavy Alcohol Drinking	The quantity of alcohol consumed in one week by men or women: • Men –15 or more drinks • Women – 8 or more drinks
Binge Alcohol Drinking	Pattern of drinking alcohol that corresponds to the number of alcoholic drinks consumed usually within a two-hour period (on a single occasion) by men and women Men – 5 or more drinks Women – 4 or more drinks

Centers for Disease Control and Prevention. (May 11, 2021). *What is excessive drinking?* Alcohol and Public Health: Frequently Asked Questions. www.cdc.gov/alcohol/fact-sheets/alcohol-use.htm

---

Heavy drinkers have been shown to smoke cigarettes faster following low- and high-dose alcohol, in comparison to placebo beverages (see Table 22.2).[21]

Older adults are less likely than younger persons to exceed recommended drink limits, but they experience higher rates of alcohol-related harm on lower doses. Alcohol use by older adults is associated with more frequent hospitalizations, increased adverse medication-alcohol interactions, and depressive symptoms.[22] Alcohol use and the development of disease is highest for cancer, supporting recommendations against even moderate use of alcohol. Individuals who consume moderately remain at risk for binge-drinking and subsequent alcohol use disorders. Holahan and colleagues studied the effect of binge drinking on mortality in 446 moderate alcohol drinkers with a mean age of 62.[23] While 372 consumers spread out their drinking evenly, 74 had episodic heavy drinking. Compared with non-binge drinkers, moderate alcohol consumers who binged doubled their chances of dying within the next 20 years.

22.5.2 CANNABIS

A growing number of older adults report cannabis use. Cannabis contains the psychoactive cannabinoid (THC), the effects of which include euphoria, anxiety, impaired judgment, paranoia, and elevated heart rate. While there is increasing interest in the potential for mental health applications, scientific findings are mixed. Some recent evidence ascribes anxiolytic, neuroprotective, antioxidant, anti-inflammatory, anti-depressant, antipsychotic, and hypnotic pharmacological actions due to several phytochemicals commonly found in the marijuana plant.[24] It is unclear whether these effects are consistent in individuals who have SUDs. A recent review found that 9.0% of past-year marijuana users reported problems due to marijuana use, including emotional or physical problems, low activity, engagement in dangerous activities, problems at home or work, or conflict with family or friends.[25] In addition, aging adults who smoke marijuana are found to have higher rates of separation, divorce, and recent major depressive episodes, and there are other social disparities concerns. Dinitto and Choi reviewed data from the National Survey on Drug Use and Health conducted by SAMHSA[26] Findings of the sample of 5,325 adults aged 50 years and older indicated that past-year marijuana users were more likely than non-users to be between the ages of 50 and 64 years, Black or multiracial, unmarried, report better health but have more psychological distress, and use other substances. Marijuana use is now legal in medical conditions and recreational use in more than 40 states. A recent review (2000–2017) found that the greatest increase in marijuana use was among individuals aged 50 and over, with the most marked increase in those 65 or older.[27] Growing tolerance in attitudes toward marijuana use suggest the continuing need for research on this substance and its effects on personal, social, and health outcomes. Common correlates of use in older adults include being male, being unmarried, having multiple chronic diseases, having psychological stress, and using other substances such as alcohol, tobacco, other illicit drugs, and prescription drugs. Patients commonly request medical marijuana prescriptions for anxiety, depression, and pain. As with other substances of abuse, use can exacerbate psychiatric conditions. Evidence remained mixed on the benefits of marijuana use; excess use is not uncommon.[27]

22.5.3 OPIOIDS

Older adults have chronic pain in rates as high as 45–85% and are likely to be prescribed analgesic opioids. In 2018, about 4% of adults aged 65 and over were prescribed opioids.[10] The Mu receptor is activated by opioid analgesics and is responsible for their reinforcing action which can lead to opioid use disorder. The side effects of opioids include sedation and the shutting down of the respiratory center as in overdose. Tolerance to the drug's effects grows and greater doses are needed to achieve the same relief, translating to significant potential for psychologic and physiologic dependence over time. Chronic use can lead to changes in mood; when in withdrawal states, individuals become highly anxious and depressed. Older people who misuse opioids are likely to have comorbid psychiatric and other substance use disorders. Opioids also slow gastrointestinal processes and can cause chronic constipation, precipitating or exacerbating medical problems. Chronic use leads to opioid-induced hyperalgesia, which changes a person's perception of pain. This state of nociceptive sensitization causes a paradoxical response; the individual can become more sensitive to certain painful stimuli. The CDC recommends non-pharmacological and non-opioid-based medications as the first-line treatment for chronic, non-cancer pain.[28]

22.5.4 SEDATIVE/HYPNOTICS

This class of drug works on the Gaba receptor and includes the "Z-drugs" (zolpidem, zaleplon, etc.) and benzodiazepines (alprazolam, lorazepam, clonazepam, diazepam, etc.) Anxiety and sleep disorders are the most common reasons for prescription of sedative/hypnotics to older adults, despite the high risk of older adults experiencing negative side effects.[29] Most Food and Drug Administration–approval studies were conducted in populations which did not include older adults, and long-term follow-up evaluating efficacy was also lacking. Benzodiazepines cause short-term deficits in memory, learning, attention, and visuospatial ability, and are associated with long-term cognitive problems and dementia,[30] falls resulting in fractures, and motor vehicle accidents.[31] Benzodiazepine use is associated with a considerable increase in all-cause mortality, with exposed patients dying at a 1.2–3.7 times higher rate per year compared with unexposed individuals.[32] It is unclear whether the correlation is causal or these drugs are being prescribed more frequently to patients at higher risk of dying. Nonbenzodiazepine sedatives, such as zolpidem and zaleplon, are associated with risks similar to those of benzodiazepines: with notably high risks for falls and accidents, they should not be used in elderly patients as alternative sleep agents.

In 2015, the American Geriatric Society (AGS) published the fourth update of the so-called Beers Criteria.[33] These criteria serve as evidence-based recommendations by the AGS to guide decision making for prescribing contrast to these recommendations, more than 10% of American women and 6% of American men 65–80 years old filled at least one prescription for benzodiazepines in a one-year period, approximately one-third of them receiving benzodiazepines for longer than 120 days in a year.[29]

22.5.5 STIMULANTS

Amphetamines and cocaine are stimulants that act on a variety of neurotransmitters, mainly dopamine and norepinephrine. In clinical scenarios, stimulants are used in older adults to treat narcolepsy and Parkinson's disease, as fall prevention, to augment late-life depression treatment and treat apathy and catatonia.[34] Nonetheless, misuse of amphetamines and cocaine effects emotional regulation, contributes to risky decision making, and enhances reward/reinforcement processes. Stimulants have an "activating" effect on the central nervous system producing elevated heart rate and blood pressure, and chronic use can lead to stroke or myocardial infarction. Older adults who use stimulants are at great risk for dental decay, weight loss, weakness, and subsequent falls. Elevated dopamine release causes acute effects such as paranoia, psychosis, and agitation. Chronic use of stimulants depletes dopamine, causing depression and anxiety, and may be linked to lifelong psychotic disorders.[35] Stimulant use interrupts healthy sleep patterns, and severe sleep deprivation can lead to psychosis.

22.5.6 TOBACCO

Smokers in the United States have roughly three times the mortality rates of similar people who never smoked, and life expectancy for smokers is about ten years less than nonsmokers.[36] The major causes of excess mortality among smokers are smoking-related diseases like cancer and respiratory and vascular diseases.[37] Rates of smoking among individuals with any mental health disorder is high and even higher in those with severe mental illness.[38] While smokers often state that smoking helps to decrease their anxiety, literature suggests that nicotine can also precipitate and exacerbate depression and anxiety.[38] One study suggested that smokers were 1.3 times more likely to have depression and 1.5 times more likely to have anxiety than lifetime nonsmokers. In 2018, about 8% of adults aged 65 and over smoked cigarettes.[36] Although tobacco use by Americans has decreased over time, the prevalence of smoking > 24 cigarettes daily has remained steady among older adults.[39] Findings of a recent study sample of 160,113 participants 70 and older in the NIH-AARP Diet and Health Study found mortality rates directly associated with the number of cigarettes smoked per day.[40] Cigarette use results in more than 480,000 deaths per year in the United States, with high rates for older adults who initiated smoking at a young age.[37]

Use of e-cigarette and smokeless tobacco is increasing and also carries risk. Smokeless tobacco is a known cause of cancer and ventricular arrhythmias and may increase risk of sudden death.[37] Risk of tobacco and electronic nicotine products is also misunderstood. Public health researchers quantify tobacco products on "the continuum of risk," with conventional cigarettes having the highest risk, an intermediate risk for smokeless tobacco products and e-cigarettes, and tobacco cessation having the lowest risk.[41] A meta-analysis of two longitudinal U.S. mortality studies, 1986–2009, found a 12-fold risk increase in mortality related to lung cancer in exclusive cigarette smokers when compared with those who never smoked.[42]

Severe SUDs are chronic illnesses. The etiology and progression of substance use disorders can be understood through a life course lens. Patterns and consequences of substance use change throughout adulthood and into older adulthood. Substance use

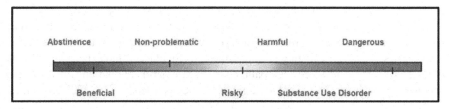

FIGURE 22.1   The Spectrum of Substance Use.

TABLE 22.3
Criteria for Substance Use Disorders

Physiological Changes
• Tolerance to substance
• Withdrawal from substance
Note: if the substance is taken as prescribed, these criteria are not valid.

Continued Use Despite Negative Consequences
• Role failure or reduction of activities at work, home, school, other settings
• Social and relationship problems
• Reducing recreational activities
• Physical hazards
• Physiological harm

Loss of Control Over Use
• Larger amounts and/or longer periods of substance use
• Feeling incapable to cutting back or controlling use of substance
• Increased time spent obtaining, using, or recovering the substance
• Craving the substance
• Compulsion to use the substance

Severity: Established based on the number DSM-V criteria describing use
Mild: 1–3 criteria
Moderate: 4–5 criteria
Severe: 6 or more criteria

Source: APA, 2013.

occurs on a spectrum from abstinence to substance use disorder (see Figure 22.1). *Risky substance use* is any pattern of use (quantity, frequency, duration) that increases an individual's risk of causing harm to self or others. Substance use disorders are found on the far right of the spectrum, caused by repeated problematic exposure. Core features of a substance use disorder include impaired control, social impairment, risky use, and craving (see Table 22.3).[43] The *Diagnostic and Statistical Manual of Mental Disorders* (5th ed.) outlines criteria for all substance use disorders.[44] Most categories of substances include disorders of intoxication, withdrawal, and substance use disorder (mild, moderate, or severe). Importantly, many individuals with severe substance use problems diagnostically qualify for more than one chronic substance use syndrome at any given time (e.g., alcohol-induced intoxication while having an alcohol use disorder [AUD]).

Older adults carry the burden of lifelong exposure to substances. Use of illicit drugs among adults aged 50–59 has more than doubled since the year 2000, and most individuals who use illicit substances late in life started before age 30.[14] Alcohol and tobacco use beginning in adolescence and young adulthood are reported to be a risk factor for substance use disorders in later years.[14] Literatures shows that the presence of an early substance use disorder doubles the odds for chronic and expensive medical illnesses such as arthritis, chronic pain, heart disease, stroke, hypertension, diabetes, and asthma and exacerbates underlying psychiatric disorders.[45]

22.6 INTERVENTIONS AND BEST PRACTICES TO REDUCE HARM

22.6.1 APPROACHES TO CARE

Partnering with older adults around health promotion, medical interventions, and system utilization is essential to achieving health care goals. Care delivery should be **patient-centered**, that is, care planning and delivery are shaped by patient preferences and patient specific needs and outcomes that maximize health. Optimally, providers collaborate with patients and their families in decision making and comprehensive care planning that considers people's emotional, mental, social, spiritual, financial needs. Participation, partnering, collaboration, and information sharing are key operating terms. In assessing for health and safety issues related to substance use, discussions should be focused on maximizing health and autonomous function and away from terms that connote judgement about patterns of substances use. Older adults in particular are deterred from sharing information about use by stigmatizing terms and classification of substance use disorders as "problematic, bad behavior or moral failing." The fact that clinicians avoid obtaining a comprehensive substance use history connotes an unwillingness to explore risky or harmful use of substances despite potential associated health and safety issues. It also suggests an ignorance of the projected rise in persons over 50 with substance use disorders to 3.1 million in early 2021.[11] *Comprehensive assessment*, a component of primary care, should produce a list of all medications, over-the-counter drugs, food supplements, "recreational" drugs, alcohol, tobacco/tobacco products used by the individual.

Advocacy for **Integrated Care models** promises to increase prevention and health promotion, versus illness care, for older adults. The **WHO ICOPE model** is a proactive, macro approach connecting agencies, services, and health providers in the provision of evidence-based interventions to prevent, slow, or reverse declines in the physical and mental capacities of older adults. The ICOPE guidelines apply to primary care/general care settings and require countries to make the needs and preferences of older adults to be central to the design of services and coordination of service delivery. Maximizing health and function by intervening early and providing assistance with hearing and vision losses, cognitive and affective changes, nutrition and mobility is "best practice." A focus on early detection opens the door to health teaching and screening for harmful use of alcohol and misuse of medications and other drugs.[46]

Screening for harmful use of alcohol, tobacco, and other substances parallels techniques widely used to identify potential health disorders. The positive results of a screening test may indicate a problem but are not conclusive and a full assessment is necessary. The United States Preventive Service Task Force (USPTF) (2020) recommends screening for unhealthy drug/substance use in all adults over the age of 18. Numerous **brief screening** instruments can detect **unhealthy alcohol use** with acceptable sensitivity and specificity in primary care settings. One–three-item screening instruments have the best accuracy for assessing unhealthy alcohol use in adults. The NIDA Quick Screen alcohol question[47] is a single alcohol screener which was effective 81.8% of the time. The NIDA screening tools are designed for use with adults 18 years and over. While appropriate for use with older adults, while they are, sensitive and 79.3% specific for identifying unhealthy alcohol use, they are under utilized in primary and geriatric care settings. The NIDA Quick Screen follows the NIDA-recommended consumption limits. The NIDA Quick Screen for Other Substances was adapted from a single question screener.[48] Testing found it was 100% sensitive and 73.5% specific for detection of a drug use disorder. A positive screen for illicit substance use, tobacco, or misuse of prescription drugs identifies risky use and necessitates exploration and referral for further intervention. The Alcohol, Smoking, and Substance Involvement Screening Test (ASSIST) e-questionnaire, also developed by WHO, has been adapted to target psychoactive prescription medications in addition to tobacco, alcohol, and other drugs. It is an eight-item questionnaire which can be administered by a health worker/health professional using paper and pencil or by computer. It takes **about five–ten minutes** to administer.[49] These instruments include the Alcohol Use Disorders Identification Test-Concise (AUDIT-C). AUDIT-C is a brief alcohol screening instrument that reliably identifies persons who are hazardous drinkers or have active alcohol use disorders (mild, moderate, or severe). The AUDIT-C is a modified version of the ten-question AUDIT screener developed by WHO (1998).[50] The AUDIT-C three questions are scored on a scale of 0–12. Each AUDIT-C question has five answer choices (0–4 points). For men, a score of 4 or more is considered positive, optimal for identifying hazardous drinking, or active alcohol use disorders. In women, a score of 3 or more is considered positive. The higher the score, the greater the likelihood that alcohol consumption is affecting the individual's safety.[50]

Smoking places older adults at high risk for chronic lung disease and cancer. Because the habit is long-standing, they may have tried to quit many times without long-term success. Public health and community focused messages can have the effect of inhibiting disclosure due to guilt and fear. **Screening for smoking** in all health settings is recommended and routine public health messages like the 5As: Ask, Advise, Assess, Assist, Arrange[51.] are linked to online resources, over-the-counter nicotine replacement regimens, and hot lines in most states. The USPTF Force (2020) recommends screening persons with smoking histories of 30 years, those who are current smokers, or those who have quit in the last 15 years. The technique used is low-dose computed tomography (LDCT) and it is increasingly available.

22.6.2 SCREENING, BRIEF INTERVENTION, REFERRAL TO TREATMENT (SBIRT)

SBIRT is a universal screening and prevention approach which, when used in medical settings, has demonstrated some success in assisting people to change behavior around drug and alcohol use.[52] It includes (S) screening, brief intervention (BI) including education, and referral to treatment (RT). Treatment approaches range from counseling, intensive outpatient therapy, acute hospitalization, and/or 28-day rehabilitation and potentially, long-term care. Pharmacotherapy is an appropriate modality in conjunction with any of these options. SBIRT can be used in any setting to assess likelihood of problematic substance use (e.g., emergency room, primary care office, hospital clinic, homecare visit). Discussion of abstinence and harm reduction techniques can be integrated into screening and BI, even with a negative screening outcome. Motivational interviewing principles underlie the brief intervention. When a screening outcome is positive, the provider should explore the patient's substance use, consider where the patient's use pattern fits on the continuum, identify the health implications of patterns of use, and establish the individual's interest in discussing the potential for change. Results improve when the patient assents before education is provided, and referral to treatment as indicated are discussed. The empathic discussion should include nonjudgmental, open-ended questions and affirmation of the individual's capacity for behavior change. A brief intervention lasts from 5 to 30 minutes, helps the clinician gauge readiness for change, and be framed in a way that keeps the conversation open for further intervention. Elements of the brief intervention include: raising the subject; providing feedback; enhancing motivation; and providing advice for improved health. The evidence for efficacy of using SBIRT is uneven. Findings suggest, however, that older adults respond positively and make some initial changes. Reenforcement of the message appears to result in better outcomes.[53]

Motivational Interviewing, based on principles of behavior change, is now widely used in health promotion and situations where the need for changing unhealthy habits is evident. Developed in 2008 to assist people with problem drinking, it has proven effective for specific behaviors beyond drinking, like weight loss and exercise. The "change talk" builds on empathic communication through open-ended questions about health and an identified behavior. Optimally, information available to the health care provider like laboratory tests, screening results, or clinical observations are the basis of discussing the need for change. In problem drinking, for example, behavior change must acknowledge the individual's ambivalence about change, evaluate pros and cons, affirm freedom of choice and capacity for change, and reflect back to key points made by provider and client. No advice or "prescription" for action is given without the patient's consent and expressed willingness to consider change. Motivational interviewing has been used with some success to help individuals reduce opioid use, increase self-esteem, decrease depression, and decrease risks of substance use.

22.6.3 HARM REDUCTION APPROACHES TO IMPROVE SAFETY

Harm reduction techniques are evidence-based policies and programs designed to reduce the adverse health, social, and economic consequences of substances for persons still engaging in use.[18] A harm reduction approach can be helpful because it can **encourage, rather than dissuade, individuals to seek adequate care and facilitate access to medical and social services the need for which may be linked to use.** Individuals have the right to choose their own pathways to recovery which do not always require abstinence but still deserve education and steps to increase accessibility which will allow them to consider options about drug use. Understanding the context of use and its associated risks is key to successful integration of harm reduction. Health professionals and others working with older adults should have realistic expectations and work from person-centered perspectives. **This requires that health professionals put health and safety first, recognize the limitations of a medical model, and move away from personal bias about use.** For example, recommendations for ways to use more safely have saved the lives of injection drug users when providers share information on syringe exchange and access to Naloxone. Further, although cannabis use has consequences, there is evidence for its benefit in pain treatment, and for it being a safer choice than chronic use of opioids which carry a greater risk of overdose. Smoking, vaping, or ingesting cannabinoids may be less risky than benzodiazepine use for an older person with insomnia and anxiety.[54] Many older adults correlate alcohol's role to greater well-being and ease of social engagement and do not recognize the risks associated with alcohol use.[45] They have been found to be more motivated to make changes to use if they believe that limiting consumption will enable them to maintain their social "quality of life."[18] Discussing ways to modify drinking patterns can help older adults see the benefits of lighter consumption when providers give examples related to health, activity, and relationships. These can motivate older adults to make healthier decisions. Highlighting risks of drinking for individuals such as interrupted sleep, low mood, and the emergence of cardiac symptoms can link to potential gains of limiting intake. Harm reduction practices also include use of needle exchanges, safe consumption/injection sites, and decreased overall use of any substance. Another approach to harm reduction is explaining safety in the context of timing use, for example, not using marijuana, alcohol, or benzodiazepines before driving, limiting drinking to the company of others and omitting analgesics or other medications which interact with alcohol if drinking.

22.7 FINANCIAL COSTS TO SOCIETY

The financial costs of substance use problems are burdensome to individuals, families and the society. A recent study evaluated data from more than 158 million hospital emergency room and other hospital encounters from the 2017 Healthcare Cost and Utilization Project Nationwide Emergency Department Sample and National Inpatient Sample.[55] The total annual estimated cost attributable to substance use disorders was $13.2 billion; cost ranged from $4 million for inhalant-related disorders to $7.6 billion for alcohol-related disorders.[55] The WHO identifies addressing alcohol use as a keystone to the sustainability[2] of global development goals as treatment to address the cyclical damage created by polypharmacy and substance use lead to adverse outcomes that further increase the financial burden on health care systems.[56] For older adults benzodiazepines and opioids are major contributors to hospitalization and service utilization as their side effects result in increased morbidity and hospitalization.[30,31] While ambulatory care settings are ideal and less expensive places to treat substance use problems, detection of substance use disorders and rates of "best practice" medication-assisted therapies for SUDs remain low.[1] Deficits associated with misdiagnoses and high use of hospital and emergency services reflect limited access to consultation with care providers knowledgeable about substance use disorders.[45] Rates of hospital admissions for those with prior histories of substance use treatment have increased over the past few decades and are magnified by polysubstance use; the costs are rising.[1]

22.8   POLICY IMPLICATIONS

Safety risks for older adults with mental health/SUDs are significant. As the global aging population grows and continues alcohol/drug use patterns from earlier life, risks to health and safety will also increase. The results of the SARS 2, COVID-19 pandemic, the ease of viral transmission, required isolation, severe consequences, and bereavement have highlighted global mental health challenges across the life span. Children, adolescents, and older adults, however, have been impacted most heavily. Rates of self-reported major depression have risen by 27.6%, rates of anxiety disorders by 25.6%;[57] and rates of alcohol use rose in most Western countries. The prevalence of mental health/SUDs in older adults is not new, but in many medical, community, and long-term care settings screening and evaluation are insufficient to clearly identify the extent of these problems, compromising the physical and psychological safety of older adults. Mental health screening for older adults is frequently limited to screening for conditions common in this age group like cognitive decline and depression. Pre-pandemic rates of depression and anxiety were 10% and among persons over 60, persons diagnosed with alcohol use disorders were 5.1% (M), 2.4% (F), and binge drinking rates were 21.5% (M) and 9.1% (F). Rates for anxiety and depression in 2020 rose to 24% among adults over 65.[58] These outcomes call for action to reduce harm and increase the safety of substance use by older adults by **confronting stigma, acting to constrain prescription drug abuse, applying public health strategies to screening, and using innovative technical approaches with history taking and medical records.**

Policy updates by the White House Office of Drug Control Policy 2017[59] launched an initiative to **destigmatize language** in government agencies and documents which would align policies with the 2013 edition of the *Diagnostic and Statistical Manual of Mental Disorders of the American Psychiatric Association*. Labeling and stigmatizing language have been shown to perpetuate negative attitudes among providers and patients and act to inhibit help seeking by older adults. This policy requires the substitution of "severe substance use disorder" for addiction, and "substance user" for addict, among several examples. Providers and others are encouraged to drop the terms "dirty" versus "clean" when supplying data on body fluid analysis and to embrace evidence-based medication-assisted treatment as opposed to viewing abstinence as the only avenue for recovery. Accepting severe SUDs as chronic illnesses may also shift detrimental attitudes.

The opioid pandemic continues to have adverse outcomes for older adults. Between 2010 and 2015, opioid-related hospitalizations and emergency room visits among older adults rose precipitously by 34% and 74% respectively.[60] The prevalence of chronic pain, high numbers of prescriptions for synthetic opioid analgesics, and associated overdose deaths prompted action by the CDC in the form of guidelines for prescribing opioid analgesics. The *Quality Improvement and Care Coordination: Implementing the CDC Guideline for Prescribing Opioids for Chronic Pain,* 2018, addresses several issues specific to managing chronic pain, noting considerations for chronic pain in older adults and other special populations. The evidence-based guidelines recommend that a full interdisciplinary team of specialists be available for consultation and regular review of the patients' registration in a prescription drug monitoring program (PDMP). Dosage recommendations include assessment and use of the lowest effective dose possible for a prescribed opioid analgesic. Low-risk dosing guidelines are aimed at pain management and enhanced function. Evaluation of patient records and evaluation of status every three months, as well as checking patient accessibility to a provider who can provide medication-assisted treatment (MAT) is indicated. Appropriate use of the electronic health record (EHR) is essential to quality assurance.[28]

Policies that reference and emphasize the importance of screening for alcohol and all types of drug use in primary care and geriatric settings would assist in improving safety. Assuring the inclusion of education on screening for all health professionals and community workers remains a goal. The USPTF recommendation for screening for alcohol use has been adopted in some settings and the SBIRT and NIAAA screening tools are used. Uptake, however, has been slow and unhealthy alcohol use and cannabis use are on the rise. While noting some positive outcomes of screening older adults using SBIRT, Thomas Babor recommends the use of SBIRT-Plus as well as the adoption of a public health approach to messaging.[53] An SBIRT-plus approach could tailor the intervention to population needs, increase the use of motivational interviewing in the encounter, and reenforce the intervention on recurrent health/medical visits. He also proposes the use of social marketing of SBIRT, noting that the older population and their care givers often have little access to information about safe levels of alcohol use outside of medical settings. This might include posting flyers in senior centers and assisted living facilities and nursing home to raise awareness of the importance of screening. Social media may also be a way of advertising the health advantages of screening and providing short tools for self-assessment.

EHR has been embraced in many health systems, and advocates for more widespread screening and the use of evidence-based screening tools with older adults suggest EHRs as a potential avenue in service of alcohol, tobacco, and other drug use screening.[53,61] These could be upgraded to include screening information on the type, quantity, and frequency of all drugs used and screening/intervention outcomes. Many older adults can respond to digitized screening tools, with paper and pencil screenings in common spoken languages available on request.[62] Policy derived from best practice guidelines and linked to hospital accreditation are two ways this could be enacted.

The need to address safety related to substance use by older adults is a real and present one which will increase as the number of adults living longer grows. While the number of older adults diagnosed with an SUD remains low in relation to other age groups, a large number of older adults are consuming alcohol in amounts which detract from their health, and a high number are binge drinking. The use of prescription and over-the-counter medications is highest for those over sixty-five and drugs used in combination pose the great risk of mortality and hospitalization.

RESOURCES

Ashley Kirzinger Follow @AshleyKirzinger on Twitter, Tricia Neuman Follow @tricia_ neuman on Twitter, Juliette Cubanski Follow @jcubanski on Twitter, and Mollyann Brodie Follow @Mollybrodie on Twitter Published: 2019 Aug 09.

Bakken MS, Engeland A, Engesæter LB, Ranhoff AH, Hunskaar S, Ruths S. Risk of hip fracture among older people using anxiolytic and hypnotic drugs: a nationwide prospective cohort study. *Eur J Clin Pharmacol.* 2014;70(7):873–880. doi:10.1007/s00228-014-1684-z

Barker MJ, Greenwood KM, Jackson M, Crowe SF. Persistence of cognitive effects after withdrawal from long-term benzodiazepine use: a meta-analysis. *Arch Clin Neuropsychol.* 2004;19(3):437–454.doi:10.1016/S0887-6177(03)00096-9

Blow FC, Brockmann LM, Barry KL. Role of alcohol in late-life suicide. *Alcohol Clin Exp Res.* 2004;28(Suppl 5):48S–56S. doi:10.1097/01.alc.0000127414.15000

CDC Guideline for Prescribing Opioids for Chronic Pain. Clinical Tools. www.cdc.gov/opio ids/providers/prescribing/clinical-tools.html

Center of Excellence for Integrated Health Solutions. www.thenationalcouncil.org/integrated-health-coe/resources/

Chhatre S, et al. Trends in substance use admissions among older adults. *BMC Heal Serv Res.* 2017;17(1). doi:10.1186/s12913-017-2538-z

Han BH, Sherman S, Mauro PM, Martins SS, Rotenberg J, Palamar JJ. Demographic trends among older cannabis users in the United States, 2006–2013. *Addict.* 2017; 112(3):516–525. doi:10.1111/add.13670

Han B, Moore A, Ferris R, Palamar J. Binge drinking among older adults in the United States, 2015–2017. *J Am Geriat Soc.* 2019 July 31;67(10). https://doi.org/10.1111/jgs.16071

HIGN Alcohol Use Screening and Assessment for Older Adults. https://hign.org/consultgeri/try-this-series/alcohol-use-screening-and-assessment-older-adults

Huhn AS, Strain EC, Tompkins DA, Dunn KE. A hidden aspect of the U.S. opioid crisis: rise in first-time treatment admissions for older adults with opioid use disorder. *Drug Alcohol Depend.* 2018 Dec 1;193:142–147. doi: 10.1016/j.drugalcdep.2018

KFF Data Note: Prescription Drugs and Older Adults.

Leelakanok N, Holcombe AL, Lund BC, Gu X, Schweizer ML. Association between polypharmacy and death: a systematic review and meta-analysis. *J Am Pharm Assoc (2003).* 2017;57(6):729–738.e10. doi:10.1016/j.japh.2017.06.002

Lehmann S, Fingerhood M. Substance-use disorders in later life. *N Engl J Med.* 2018 Dec 13;379(24): 2351–2360. doi: 10.1056/NEJMra1805981

Lloyd SL, Striley CW. Marijuana use among adults 50 years or older in the 21st century. *Gerontol Geriatr Med.* 2018;4:2333721418781668. doi:10.1177/2333721418781668

Moriarty F, Hardy C, Bennett K, Smith SM, Fahey T. Trends and interaction of polypharmacy and potentially inappropriate prescribing in primary care over 15 years in Ireland: a repeated cross-sectional study. *BMJ Open.* 2015;5(9):e008656. Published 2015 Sep 18. doi:10.1136/bmjopen-2015-008656

NCOA Alcohol & Substance Abuse for Older Adults. www.ncoa.org/older-adults/health/beh avioral-health/alcohol-substance-abuse

*NIA Consumer Health Information: Alcohol Use. www.nia.nih.gov/health/topics/alco hol-use-or-abuse

*NIAAA Rethinking Drinking. Alcohol & Your Health. www.rethinkingdrinking.niaaa. nih.gov/

*NIDA Substance Use in Older Adults Drug Fact. www.drugabuse.gov/publications/substa nce-use-in-older-adults-drugfacts

NIDA Drug Use and Its Consequences Increase Among Middle-Aged and Older Adults. www. drugabuse.gov/news-events/nida-notes/2019/07/drug-use-its-consequences-increase-among-middle-aged-older-adults

NIDA. (2017, Oct 24). Substance Use Disorders Are Associated with Major Medical Illnesses and Mortality Risk in a Large Integrated Health Care System. https://archives.drugab use.gov/news-events/nida-notes/2017/10/substance-use-disorders-are-associated-major-medical-illnesses-mortality-risk-in-large-integrated. Accessed 2021, Nov 1.

SAMHSA Screening, Brief Intervention, and Referral to Treatment (SBIRT). samhsa.gov/sbirt

SAMHSA Treatment Improvement Protocol (TIP) 26: Treating Substance Use Disorder in Older Adults. https://store.samhsa.gov/product/treatment-improvement-protocol-tip-26-treating-substance-use-disorder-in-older-adults/PEP20-02-01-011

SAMHSA Linking Older Adults with Resources on Medication, Alcohol, and Mental Health, 2019 Edition. https://store.samhsa.gov/product/Get-Connected-Linking-Older-Adults-with-Resources-on-Medication-Alcohol-and-Mental-Health-2019-Edition/SMA03-3824

Sarris J, Sinclair J, Karamacoska D, Davidson M, Firth J. Medicinal cannabis for psychiatric disorders: a clinically-focused systematic review. *BMC Psychiatry*. 2020;20(1):24. doi:10.1186/s12888-019-2409-8

Wu LT, Blazer DG. Substance use disorders and psychiatric comorbidity in mid and later life: a review. *Int J Epidemiol*. 2014;43(2):304–317. doi:10.1093/ije/dyt173

Zhong G, Wang Y, Zhang Y, Zhao Y. Association between benzodiazepine use and dementia: a meta-analysis. *PLoS One*. 2015;10(5):e0127836. doi:10.1371/journal.pone.0127836
*Consumer-based information

REFERENCES

1. Chhatre S, Cook R, Mallik E, Jayadevappa R. Trends in substance use admissions among older adults. *BMC Health Serv Res*. 2017;17(1):584. Doi:10.1186/s12913-017-2538-z

2. World Health Organization. *Global Status Report on Alcohol and Health 2018*. Geneva, Switzerland: WHO Press; 2018, p. vii. Accessed 2020 Dec 8.

3. GBD 2015 Risk Factors Collaborators. Global, regional, and national comparative risk assessment 79 behavioural, environmental and occupational, and metabolic risks or clusters of risks, 1990–2015: a systematic analysis for the Global Burden of Disease Study 2015. *Lancet*. 2016;388(10053):1659–1724.

4. National Institute of Alcohol Abuse and Alcoholism. 2020. nih.niaaa.gov. Accessed 2021 Nov 4.

5. Blanco C, Lennon I. Substance use disorders in older adults: overview and future directions. *2021, Gen Jrnl*. 2020–21:44(Winter 4).

6. Kirzinger A, Neuman T, Cubanski J, Brodie M. Data Note: Prescription Drugs and Older Adults, Kaiser Family Fund (KFF). 2019 Aug 9. Accessed 2021 Oct 24.

7. Zwicker D, Fulmer T. Reducing adverse drug effects. In: Boltz M, Capezuti E, Fulmer T, Zwicker D. *Evidence-based geriatric nursing protocols*. 4th ed. New York: Springer; 2012, pp. 324–362.

8. National Council On Aging (NCOA) – Chronic Disease Self-Management Facts – National Council.www.ncoa.org › article › get-the-facts-on-chroni.

9. Bahorik AL, Satre DD, Kline-Simon AH, et al. Alcohol, cannabis, and opioid use disorders, and disease burden in an integrated health care system. *J Addict Med.* 2017;11(1):3–9.

10. Schieber LZ, Guy GP Jr, Seth P, Losby JL. Variation in adult outpatient opioid prescription dispensing by age and sex – United States, 2008–2018. *MMWR Morb Mortal Wkly Rep.* 2020;69:298–302. doi: http://dx.doi.org/10.15585/mmwr.mm6911a5exter nal icon.

11. Dahlhamer J, Lucas J, Zelaya C, et al. Prevalence of chronic pain and high-impact chronic pain among adults – United States, 2016. *MMWR Morb Mortal Wkly Rep.* 2018;67:1001–1006. doi: http://dx.doi.org/10.15585/mmwr.mm6736a2external icon1.

12. Satre DD, Hirschtritt ME, Silverberg MJ, Sterling SA. Addressing problems with alcohol and other substances among older adults during the COVID-19 pandemic. *Am J Geriatr Psychiatry.* 2020; 28(7):780–783. doi:10.1016/j.jagp.2020.04.01

13. Han BH, Palamar JJ. Trends in cannabis use among older adults in the United States, 2015–2018. *JAMA Intern Med.* 2020;180(4):609–611. doi:10.1001/jamainternmed.2019.7517

14. McCormick CD, Dadiomov D, Trotzky-Sirr R, Qato DM. Prevalence and distribution of high-risk prescription opioid use in the United States, 2011–2016. *Pharmacoepidemiol Drug Saf.* 2021;30(11):1532–1540. doi:10.1002/pds.5349

15. Bhagavathula AS, Vidyasagar K, Chhabra M, et al. Prevalence of polypharmacy, hyperpolypharmacy and potentially inappropriate medication use in older adults in India: a systematic review and meta-analysis. *Front Pharmacol.* 2021;12:685518. Published 2021 May 19. doi:10.3389/fphar.2021.685518

16. Maree RD, Marcum ZA, Saghafi E, Weiner DK, Karp JF. A systematic review of opioid and benzodiazepine misuse in older adults. *Am J Geriatr Psychiatry.* 2016;24(11):949–963. doi:10.1016/j.jagp.2016.06.003

17. Bareham BK, Kaner E, Spencer LP, Hanratty B. Drinking in later life: a systematic review and thematic synthesis of qualitative studies exploring older people's perceptions and experiences. *Age Ageing.* 2019;48(1):134–146. doi:10.1093/ageing/afy069

18. Hedegaard H, Bastan BA, Trinidad JP, Spencer M, Warner M. Drugs most frequently involved in drug overdose deaths: United States, 2011–2016. *Natl. Vital Stat. Rep.* 2018;67: 1–14.

19. Green R, Bujarski S, Lim AC, Venegas A, Ray LA. Naltrexone and alcohol effects on craving for cigarettes in heavy drinking smokers. *Exp Clin Psychopharmacol.* 2019;27(3):257–264. doi:10.1037/pha0000252

20. Day AM, Kahler CW, Spillane NS, Metrik J, Rohsenow DJ. Length of smoking deprivation moderates the effects of alcohol administration on urge to smoke. *Addict Behav.* 2014;39(5):976–979. doi:10.1016/j.addbeh.2014.01.023

21. Piasecki TM, Jahng S, Wood PK, Robertson BM, Epler AJ, Cronk NJ, … Sher KJ. The subjective effects of alcohol-tobacco co-use: an ecological momentary assessment investigation. *J Abnorm Psychol.* 2011;120(3):557–571. doi:10.1037/a0023033

22. Holahan CJ, Brennan PL, Schutte KK, Holahan CK, Hixon JG, Moos RH. Late-life drinking problems: the predictive roles of drinking level vs. drinking pattern. *J Stud Alcohol Drugs.* 2017;78(3):435–441. doi:10.15288/jsad.2017.78.435

23. Drysdale AJ, Platt B. Cannabinoids: mechanisms and therapeutic applications in the CNS. (0929–8673 (Print)). 12. The National Institutes of Health. Marijuana and cannabinoids: a neuroscience research summit. Bethesda; 2016.

24. Lloyd SL, Striley CW. Marijuana use among adults 50 years or older in the 21st century. *Gerontol Geriatr Med.* 2018;4:2333721418781668. doi:10.1177/2333721418781668

25. Dinitto DM, Choi NG. Marijuana use among older adults in the U.S.A.: user characteristics, patterns of use, and implications for intervention. *Int Psychogeriatr.* 2011;23(5):732–741. doi:10.1017/S1041610210002176

26. Walsh Z, Gonzalez R, Crosby K, S Thiessen M, Carroll C, Bonn-Miller MO. Medical cannabis and mental health: a guided systematic review. *Clin Psychol Rev.* 2017;51:15–29. doi:10.1016/j.cpr.2016.10.002

27. Centers for Disease Control and Prevention. *Quality improvement and care coordination: implementing the CDC guideline for prescribing opioids for chronic pain.* National Center for Injury Prevention and Control, Division of Unintentional Injury Prevention, Atlanta, GA; 2018.

28. Olfson M, King M, Schoenbaum, M. Benzodiazepine use in the United States. *JAMA Psychiatry.* 2015;72(2):136–142. doi:10.1001/jamapsychiatry.2014.1763

29. Rummans TA, Davis LJ Jr, Morse RM, Ivnik RJ. Learning and memory impairment in older, detoxified, benzodiazepine-dependent patients. *Mayo Clin Proc.* 1993;68(8):731–737. doi:10.1016/s0025-6196(12)60628-4

30. Cumming RG, Le Couteur DG. Benzodiazepines and risk of hip fractures in older people: a review of the evidence. *CNS Drugs.* 2003;17(11):825–837.doi.10.2165/00023210-00004

31. Palmaro A, Dupouy J, Lapeyre-Mestre M. Benzodiazepines and risk of death: results from two large cohort studies in France and UK. *Eur Neuropsychopharmacol.* 2015;25(10):1566–1577.

32. American Geriatrics Society 2015 Beers Criteria Update Expert Panel. American Geriatrics Society 2015 updated Beers Criteria for potentially inappropriate medication use in older adults. *J Am Geriatr Soc.* 2015;63(11):2227–2246. doi:10.1111/jgs.13702

33. Sassi KLM, Rocha NP, Colpo GD, John V, Teixeira AL. Amphetamine use in the elderly: a systematic review of the literature. *Curr Neuropharmacol.* 2020;18(2):126–135. doi:10.2174/1570159X17666191010093021

34. Roncero C, Daigre C, Grau-López L, et al. An international perspective and review of cocaine-induced psychosis: a call to action. *Subst Abus.* 2014;35(3):321–327. doi:10.1080/08897077.2014.933726

35. U.S. Department of Health and Human Services. Smoking cessation: a report of the Surgeon General. Atlanta: U.S. Department of Health and Human Services, Centers for Disease Control and Prevention, National Center for Chronic Disease Prevention and Health Promotion, Office on Smoking and Health; 2020. Published 2021.www.hhs.gov/surgeongeneral/reports-and-publications/tobacco/2020-cessation-sgr-factsheet-key-findings/index.html. Accessed October 17, 2021. 2003;17(11):825–837. doi:10.2165/00023210-200317110-00004

36. Smoking and Tobacco Use. Centers for Disease Control and Prevention. Centers for Disease Control and Prevention. Updated April 28, 2020. www.cdc.gov/tobacco/data_statistics/fact_sheets/health_effects/tobacco_related_mortality/index.htm. Accessed October 17, 2021.

37. Fluharty M, Taylor AS, Grabski M, Munafo MR. The association of cigarette smoking with depression and anxiety: a systematic review. *Nicotine Tob Res.* 2017;19(1):3–13. doi.10.1092/ntr/ntw140

38. Overall Tobacco Trends. American Lung Association. www.lung.org/research/tre nds-in- lung-disease/tobacco-trends-brief/overall-tobacco-trends. Published 2021. Accessed October 18, 2021.

39. Nash SH, Liao LM, Harris TB, Freedman ND. Cigarette smoking and mortality in adults aged 70 years and older: results from the NIH-AARP cohort. Am J Prev Med. 2017;52(3):276–283. doi:10.1016/j.amepre.2016.09.036

40. Gottlieb S, Zeller M. A nicotine-focused framework for public health. N Engl J Med. 2017;377(12):1111–1114. doi:10.1056/NEJMp1707409

41. Fisher MT, Tan-Torres SM, Gaworski CL, Black RA, Sarkar MA. Smokeless tobacco mortality risks: an analysis of two contemporary nationally representative longitudinal mortality studies. Harm Reduct J. 2019;16(1):27. doi:10.1186/s12954-019-0294-6

42. National Institute on Drug Abuse. The science of drug use and addiction: the basics. www.drugabuse.gov/publications/media-guide/science-drug-use-addiction-basics. July, 2018. Accessed October 15, 2021.

43. Diagnostic and statistical manual of mental disorders: DSM-5. 5th ed. American Psychiatric Association; 2013.

44. McLellan AT. Substance misuse and substance use disorders: why do they matter in healthcare? Trans Am Clin Climatol Assoc. 2017;128:112–130.

45. World Health Organization. Framework on integrated, people-centred health services. Report by the Secretariat for the sixty-ninth world health assembly – Provisional agenda item 16.1. 2016. Report No.: A69/39.

46. Smith PC, Schmidt SM, Allensworth-Davies D, Saitz R. Primary care validation of a single-question alcohol screening test. J Gen Inter Med. 2009;24(7):783–788.

47. Smith PC, Schmidt SM, Allensworth-Davies D, Saitz R. A single-question screening test for drug use in primary care. Arch Intern Med. 2010;170(13):1155–1160. doi:10.1001/archinternmed.2010.140

48. World Health Organization. The Alcohol, Smoking and Substance Involvement Screening Test (ASSIST): Manual for Use in Primary care. 2010.73pp.WHO. www. drugabuse.gov/sites/default/files/ pdf/nmassist.pdf).

49. NIDA CTN Common Data Elements. Instrument: AUD-C. Questionnaire.2021. https://cde.drugabuse.gov>instruments

50. Agency for Healthcare Research and Quality. Five Major Steps to Intervention ("The 5A's")2012. Agency for Healthcare Quality and Research. Rockville, Maryland.

51. Knapp M, McCabe DE. Screening and interventions for substance use in primary care. Nurs Pract. 2019;44(8):48–55.

52. Babor T. SBIRT-Plus. Adding populations health innovations to enhance alcohol screening a brief intervention effectiveness. Gen J. 2020–21:44(Winter 4)

53. Yang KH, Kaufmann CN, Nafsu R, et al. Cannabis: an emerging treatment for common symptoms in older adults. J Am Geriatr Soc. 2021;69(1):91–97. doi:10.1111/jgs.16833

54. Peterson C, Li M, Xu L, Mikosz CA, Luo F. Assessment of annual cost of substance use disorder in US hospitals. JAMA Netw Open. 2021;4(3):e210242. Published 2021 Mar 1.

55. Lehmann S, Fingerhood, M. Substance-use disorders in later life. N Engl J Med. 2018 Dec 13;379(24):2351–2360. doi: 10.1056/NEJMra1805981

56. The Lancet: COVID 19 Pandemic Led to Stark Increases in Depressive and Anxiety Disorders Globally in 2020, with Women and Younger People Affected Most. News Release Oct 8, 2021. www.thelancet.com/journals/lancet/article/PIIS0140-6736(21)02143-7/fulltext

57. Koma W, True S, Biniek JF, Orgera K, Garfield R. One in four adults report anxiety or depression amid the COVID-19 pandemic. Oct 9, 2020. Kaiser Family Foundation. Newsletter: Medicare.

58. White House office of National Drug Control Policy. Jan 9, 2017. Changing the Language of Addiction. https://obamawhitehouse.archives.gov/blog/2017/01/13/changing-language-addiction

59. Weiss AJ, Heslin KC, Barrett ML, Izar R, et al. Opioid-related inpatient stays and emergency department visits among patients aged 65 years and older, 2010 and 2015: statistical brief #244. *Healthcare Cost and Utilization Project (HCUP) Statistical Briefs.* Rockville (MD) 2018.

60. Han BH, Masukawa K, Rosenbloom D, Kuerbis A, Helmuth E, Liao DH, Moore AA. Use of web-based screening and brief intervention for unhealthy alcohol use by older adults. *J Subst Abuse Treat.* 2018 Mar;86:70–77. doi: 10.1016/j.jsat.2018.01.002. Epub 2018 Jan 3. PMID: 29415854; PMCID: PMC5808575.

23 Hearing and Aging

Radhika Aravamudhan, Ph.D., CCC-A, FAAA

CONTENTS

23.1 INTRODUCTION

Aging-related changes can impact the outer ear, middle ear, inner ear, and/or the central auditory pathways including the auditory cortex. The entire auditory system undergoes change with aging as do other organs in the body (Weinstein, 2013).

Outer Ear: Aging brings about shrinking of skin, along with changes to pigmentation around the pinna (external ear) and the ear canal. These changes which may not directly affect hearing must be examined by the audiologist to notify the individual of any noticeable changes from previous visits. Audiologist also work with cerumen removal from the ear canal. Cerumen, or ear wax, is created by accumulation of secretions from the sebaceous glands, dust particles, and dislodged hair follicles in the ear canal. Small amounts of cerumen (ear wax) are protective in nature by serving as a water repellent, coating the ear canal, and helping in trapping unwanted material and objects entering the ear. However, when this accumulates and hardens, it can block sound signals reaching the middle ear, leading to hearing loss. There is an increase in sebaceous gland activity as an individual ages and leads to increased wax accumulation.

Clinical Pearl: Audiologists when examining the ear usually evaluate the impact of the wax accumulation on hearing loss especially in aging individuals.

Some of the other common conditions seen in the outer ear are squamous cell carcinoma in the pinna, various forms of dermatitis on the pinna, and external auditory canal and basal cell carcinoma. If the audiologist notices any abnormal growth during ear examination, they will report to the patient and refer for further evaluation both to a dermatologist and the individual's primary care. These forms of changes to the pinna and external auditory canal, while not specific to aging, have an increased prevalence among older adults. Hence, attention to these changes is necessary while evaluating and treating the aging adult.

Clinical Pearl: These conditions in the pinna and external ear canal can affect cerumen management, examination of the ear as well as making ear mold impression for hearing aids. Any observation of a lesion, growth, color change in skin, unusual pigmentation must be reported and referred to dermatologist and primary care physician.

23.2   CERUMEN IMPACTION

Managing cerumen is also another important element in hearing care for older adults. Cerumen impaction, a condition where a significant amount of ear hardens in the ear canal blocking the acoustic signal from reaching the middle ear, can have both medical and audiological significance. Common symptoms of cerumen impaction include pain, ringing sound in the ear (tinnitus), fullness of ear, blocking sensation, and sometimes vertigo. Cerumen impaction also leads most commonly to hearing loss, typically high frequency and conductive hearing loss.

Cerumen causes a constriction in the wall of the ear canal blocking the sound travelling to the middle ear past the constriction. Older adults, especially those with comorbid medical conditions, may be prone to complications from cerumen removal. In addition, older adults may injure or traumatize their ear in attempting to remove wax. Foreign objects including q-tips, hair pins, folded paper and others should never be used to attempt to remove wax. Cerumen should be removed under the care of a health professional.

Clinical Pearl: Cerumen impaction may not always be recognized by the older adult, especially those with underlying cognitive impairment, and should be managed under the supervision of a health care provider.

23.3   MIDDLE EAR

Anatomical and physiological changes to the middle ear also play significant role in the impact on hearing in aging individuals. Some of the common diseases of the middle ear include fluid accumulation, infection, trauma to the middle ear, tumors of the middle ear, other congenital conditions like abnormalities of the small bones of the middle ear, and other vascular abnormalities. Although all these affect hearing, none of these is specifically due to aging. Of these middle ear disorders, aging individuals are more prone to infections of the middle ear. One can expect both chronic and acute ear infections in the older adult. Older adults are more prone to upper respiratory infections, which leads to increased prevalence of middle ear infections.

Clinical Pearl: Evaluation and monitoring of any middle ear infection and co-management of the condition is extremely important in aging individuals to prevent recurrent middle ear infections.

23.4   INNER EAR AND NEURAL PATHWAYS

The process of aging affects the inner ear. Aging often results in changes in speech perception and is responsible for high frequency hearing loss noted in aging adults. This type and pattern of hearing loss noticed in aging individuals is also referred to as *Presbycusis*.

23.4.1   Age-Related Changes Beyond the Cochlea

While hearing is mostly associated with the ear, actual hearing happens due to the ability of the brain to interpret the signals received from the cochlea. Hearing cannot take place without the appropriate neural activity in the brain, in turn the auditory cortex. The signals from the cochlea reach the auditory nerve, then pass through the central auditory pathway through the brain stem, finally reaching the auditory cortex.

Clinical Pearl: While these age-related changes in the central auditory pathway are universal, the extent to which they impact hearing, and the extent of age-related changes are highly variable amongst individuals. All assessments and interventions should be tailored to everyone depending on their extent of behavioral and physiological changes.

23.5   OTHER COMORBID CONDITIONS RELATED TO AGING

While changes related to aging underlies most of the hearing impairment in aging individuals, there are also other medical conditions that can cause hearing loss in older adults. These are:

- Medication side effects
- Noise exposure
- Sudden hearing loss
- Auto immune disorders
- Metabolic conditions
- Meniere's disease, and other infections and temporal bone injuries

These can be mostly discovered in the case history and audiometric patterns and other diagnostic tests.

Clinical Pearl: Audiologists take a detailed case history that includes questions on all the underlying conditions and comorbid conditions like diabetes, heart conditions, vascular diseases, and other causes that impact the general population.

In summary, it is important for hearing health also to be documented in the health records along with other conditions in aging adults. Older adults should have hearing screening performed annually and should be referred for a full diagnostic evaluation by an audiologist if concerns are raised by the patient, caregiver, or medical provider.

23.6   CONSEQUENCES OF AGE-RELATED HEARING LOSS

All the previously mentioned physiological and anatomical changes lead to some form of mostly irreversible hearing loss that impacts quality of life for the aging adult. While there are several types of hearing loss, most of the age-related hearing loss is sensorineural in nature and is mostly seen in the higher frequencies. The fact is some people may not want to admit they have trouble hearing. Having trouble hearing not only impacts day to day activities for the individual, but also makes it difficult to receive medical care, understand and follow medical advice. Another fact to note is that the hearing loss occurs gradually that the aging individual may not be aware of it. This is mostly noticed by friends and relatives before the aging adult realizes it. The hearing loss impacts speech perception both in quiet and noise. The difficulty increases in noisy situations. There are several audiometric patterns seen in age-related hearing loss, but the most common pattern seen is high frequency hearing loss in both ears. The prevalence and severity vary by age and is seen highest in nursing home residents compared to the community-based population. This can be attributed to the increased prevalence of other comorbid conditions in residents of nursing homes than community-based population. Overall hearing thresholds decline progressively with age and in high frequencies. Other risk factors contribute to individual variability in hearing loss and its impact on quality of life.

23.7   IMPACT OF HEARING LOSS ON THE AGING

- Hearing and balance problems
- Depending on the degree of hearing loss, loss of ability to communicate
  - Family members should watch for
    - signs for asking to repeat
    - loud television settings
    - asking several "huh"
    - isolating themselves from group conversations
    - lack of interest in outdoor activities
    - change in their normal behavior in daily activities
- Decreased interest in social life because of difficulty in communicating effectively
- Without assistance from others, it becomes difficult for the individuals with untreated hearing loss to discuss their diagnosis and treatment plans with their care providers

- Cognitive impairment: while hearing loss does not appear to cause cognitive problems, it does amplify the impact of cognitive impairments within the individuals
- Depression: while hearing loss does not cause depression, there is a high correlation between hearing loss and depression in aging individuals

More than reporting hearing loss, aging individuals mostly report difficulty understanding speech both in quiet and noisy situations. They also report increased difficulty understanding female voices and children. Most evidence suggest that when testing in quiet ambience and when the signal is loud enough, most people report ability to hear and understand. However, there is a significant drop in both audibility and speech understanding when there is background noise. It is predicted that changes in auditory thresholds are the primary contributor to individual differences in speech recognition ability. While speech understanding is a big issue, audibility of speech becomes a primary contributor to speech understanding

23.8   COGNITIVE FACTORS

Aging of the brain also has impact on the cognitive aspects of the individual. This age-related decline in cognitive abilities has been associated with understanding speech and has been documented by several studies. However, when a cognitive decline is noted in aging individuals, this does not affect hearing and speech understanding in isolation. Other abilities such as generalized decline in motor and perceptual sensibilities are also noted. Even though there is a correlation between cognitive abilities and speech understanding, the extent of impact continues to be explored in various research studies (Ronnberg J.L. et al., 2016).

In summary, evaluation of the aging adult should have a thorough case history that includes questions on various comorbid conditions; a test battery approach, where tests that can assess the entire auditory system; and speech testing. Thorough drug history also becomes very relevant in the aging adult. Further electrophysiological tests can become a part of test battery if needed to evaluate the auditory system beyond the cochlea and can also include central auditory tests as needed.

23.9   POTENTIAL CONSEQUENCES OF HEARING LOSS EXPERIENCES BY OLDER ADULTS

Hearing problems can impact physical, emotional, and cognitive health. Even though age-related hearing loss does not cause a profound hearing loss in most individuals, even some amount of hearing loss can have an impact on the individual's personal safety and their social and emotional life. Hearing loss leads to:

- Communication gap: they will not be able to communicate effectively with others
- Isolation from large family gatherings or social life
- Low self-esteem

- Strain in family relations due to lack of communication
- Difficulty in living independently
- Interference with their safety
- Failure to participate in telephone conversations
- Failure to participate effectively in their medical care

Other than this, there are also other concerns because of their inability to hear:

- Fire/smoke alarms
- Door knocks
- People calling them

These problems grow over time and can lead to decline in well-being and quality of life for the individual with hearing loss. It is very important to address hearing loss as it impacts all facets of life.

Advice for health care professionals: While communicating with elderly patients, it is important for health care providers and other professionals involved in the care of the individual to follow the following:

- Find a quiet place to talk to help reduce background noise
- Stand in good lighting and use facial expressions or gestures to give clues
- Face the person and speak clearly; maintain eye contact
- Speak slowly and clearly but naturally
- Do not try to speak too loudly, because this reduces the clarity of speech
- Do not hide your mouth, eat, or chew gum while speaking
- Repeat yourself if necessary, using different words
- Only one person should talk at a time
- Ask the patient to repeat what they heard and ensure they understood all the instructions
- Lower the pitch of their voice when you are speaking with the patient

23.10 REFERRAL TO HEARING EVALUATION

As you work with elderly patients, remember to look for sign of hearing loss. They may ask you to repeat what you have said, or look lost, or just nod. A discussion around hearing evaluation on an annual basis can help identify any hearing loss early before it can get in the way of communication. Tinnitus can also be disruptive and cause sleep deprivation and difficulty paying attention to conversation. Dizziness, tinnitus, and hearing loss can be indicators of Meniere's disease and needs follow-up with a full evaluation by an audiologist. Health care providers need to refer these individuals to audiologist for a full evaluation.

23.10.1 REHABILITATION GOAL AND OPTIONS

What happens after the hearing loss is detected? An evaluation starts with a detailed assessment of the individual's day-to-day activities, their need, their environmental factors, and social life. Psychological and social variables are closely linked to the rehabilitation choices that individuals make. Since age-related hearing loss is not a profound loss and the individual still has viable hearing, most often hearing aids are a viable option to make the sounds audible (Weinstein, 2013). The goal of rehabilitation or treatment options should be improving the quality of life for the individual.

- Effective communication being a key to better quality of life, better understanding of the individual's communication needs becomes cornerstone for the treatment to work well.
- Evaluating the impact of hearing loss on an individual becomes essential in effective rehabilitation process.
- Hearing loss is progressive in nature. Any type of rehabilitation plan must include regular follow-up visits to monitor the changes and the benefits of the rehabilitation plan. Both psychological and social variables need to be considered closely to design an effective rehabilitation plan, with the primary goal of improving the quality of life of the individual. (Abrahamson, 1997)

Clinical Pearl: Presbycusis is progressive in nature and follow-ups and regular monitoring of both the hearing loss and the effectiveness of the rehabilitation plan is essential.

Different components of the audiologic rehabilitation plan include:

- Hearing aid selection
- Fitting
- Hearing aid orientation
- Assessment of psychological and social needs
- Counseling on the need of the hearing aid
- Using auditory rehabilitation approaches such as auditory and visual integration training
- Use of assistive listening devices
- Training on identification of difficult hearing situations and changing the environmental variables, and if needed, follow up auditory rehabilitation
- Individual or group sessions to ensure appropriate use of hearing aids. (One of the important aspects of auditory rehabilitation plan in older adults is the training of the person who will be taking of the individual or the significant other.)

23.11   HEARING AID FITTING

Currently there are several options for hearing aids and, with a growing number of over-the-counter (OTC) devices, patients may be confused about their options. Hearing aids are devices that can amplify sounds without distorting the quality of sound, make it easier for the individual to hear and understand the conversation.

- The audiologist can recommend a range of assistive devices, many of which can be purchased over the counter, as well as hearing aids that require prescriptions and need to be programed based on the patient's needs and preferences.

- Although decrease in size of hearing aids is important for aesthetics, the dexterity issues with aging individual make it difficult to use small devices.

While hearings aids are not covered by traditional Medicare, some Medicare Advantage programs include a limited hearing benefit, as do some federal and private health insurance plans. Veterans should check with the VA concerning eligibility for hearing benefits. The Veterans Affairs programs do provide a significant support towards hearing aids with select devices.

23.12   OTC DEVICES

Although OTC devices are available, they will usually work well only for milder hearing losses. It is important to counsel patients on the need to follow up with an audiologist when these OTC devices are not enough. OTC hearing aids are self-fitting and designed for adults with mild to moderate hearing loss.[1]

Sound amplifiers: Personal sound amplification products (PSAPs) are designed for recreational use for individuals with no hearing loss and simply make all environmental sounds louder. These are considered hearing "amplifiers," not true hearing aids. Though they offer limited benefit, they can provide some help with communication needs for mild losses.

Prescription hearing aids: Prescription hearing aids are customized, and an audiologist will test and identify the amount of hearing loss and customize the hearing aids to optimize the patient's hearing. A broad range of features are offered on various hearing aids. As new products become available, individuals with hearing loss should check the features for the products. For instance,

- Rechargeable or replaceable batteries: Rechargeable batteries may be more expensive in the short term, but easier to use and less expensive over time.
- Bluetooth audio streaming: Streaming is helpful when talking on the phone, watching TV, etc.
- Behind the ear or in the ear hearing aid: Behind the ear may be more comfortable, in the ear may be more aesthetically appealing and easier to use.
- Telecoil: A telecoil allows you to connect to your telephone or a loop system that provides direct audio input to sound in large-area listening situations.

The selection of hearing aid or any device must include the factors of dexterity of the individual to ensure appropriate use. Once the selection and fitting are completed, the audiologist should orient the individual to the appropriate use of the device. It is very helpful to have the caregiver, family member, or significant other during these sessions. Because aging also affects the tactile sense as well as visual abilities, it is important that the individual adjusts the settings of dexterity and visual ability in the hearing aid. Use of a magnifier to see the small controls is also an important aspect of the session. It is essential to follow up with a written set of instructions regarding care of the hearing aid and its maintenance. While the patient does not need to know all the choices, the audiologist as they asses the patient's needs, will recommend the features that benefit the patient.

23.12.1 Adaptive or Assistive Listening Devices

When hearing aids are not enough, other assistive listening devices can be combined with the hearing aids to provide specific additional benefits. One such device is a remote microphone that is paired with the hearing aid and can be placed on a table that picks up signals from all the speakers at the table to transmit the signal back to the hearing aid. This helps patients to listen to a group conversation without having to move closer to the speaker. These are also nonbulky and easy to move. Other facilities such as FM radio stations and infrared transmitters too are available, though these do not add up to the cost.

Other than the regular hearing aids, the assistive devices that help with specific situations are as follows.

- Safety devices that have strobes can be placed under pillows and these vibrate.
- Carbon monoxide detectors alert if carbon monoxide exceeds safe levels. Detectors may be hardwired, plug-in, or battery powered.
- Doorbell signalers provide alerts when someone is at the door. These work with or without an existing doorbell system.
- Smoke alarm signalers alert if the smoke alarm is activated.
- Telephone signalers alert when the phone rings. Some signalers plug into the telephone line and electrical outlet. Other signalers attach to the side of the telephone.
- Wake-up alarm signalers alert when the alarm clock goes off. These products range from portable alarm clocks with built-in strobe lights to alarm clocks with a built-in outlet where a lamp or vibrating alert can be plugged in.
- Weather alerts that report upcoming storms or dangerous weather conditions with extra loud sirens; texts explaining the alert; or lights indicating the severity of the alert.
- Text telephone devices (TTY) phones: In the past, wireline telephone and analog cellular networks generally were compatible with TTYs, but digital wireless networks were not. However, now these are available with all phones. It is important to counsel the patients about this service. These help convert the voice into text for the individuals to follow the conversation over the phone.

Although there are new software and technologically advanced devices, each device needs to be selected for the individual patient based on their communication needs and situations when they would be using the device.

Counseling: The important aspect of counseling is that it should emphasize the fact that hearing through a hearing aid takes time to adjust. Patience is an important aspect for the success of using the hearing aids. Wearing the hearing aid longer each day helps in adjusting to using hearing aids. Counseling should also include basic hearing aid operations like how to turn on and off, correct way of wearing the aid, adjusting the volume settings if applicable, charging the aids, identifying the right and left aid, earmold care if applicable, and keeping the aid clean. Along with the instructions, going over what not to do is essential. Also, it is important to ensure that the patient knows what to do and whom to call if there is a problem with the aids.

If the individual does not benefit from hearing aids, there are other options including cochlear implants. Weighing the risk and benefits of the surgical procedure becomes essential when considering these options. Along with this, there are also other assistive listening devices that may be of benefit to many individuals. Each of these options can be evaluated for benefit, risk, and cost of the option. Care should also be taken to not overwhelm the patient with too many options. Giving viable options that increases the quality of life becomes a key component of rehabilitation success.

Interprofessional aspects of aging individual care: As mentioned in the earlier chapters, aging causes changes in all aspects of the individual. If a patient needs multiple appointments with various health care professionals, they may not be able to get to their appointments without assistance from others. Treatment for one of the senses, such as hearing, may impact or be impacted by the ability of other senses such as vision. Thus, an interprofessional approach, with communication between the various providers, becomes important. For example, poor vision and limited dexterity brings added challenges in operating simple devices like hearing aids. Team-based practices with an interprofessional approach also helps with fewer appointments for the individual and improved patient outcomes.

23.13 COMMUNICATION CONSIDERATIONS

Caregivers may hear comments from patients such as

- "I don't like these hearing aids. Everything is too loud but I don't understand anything" – A.K. 85-year-old hearing aid user
- "Hearing aids have helped me to hear my grand kids" – 65-year-old hearing aid user

These are not uncommon comments from hearing aid users. One of the biggest myths with hearing aids is that it eliminates all communication barriers for people with hearing loss. When we think hearing aids would help our patients hear better, it is important to understand that it is not like wearing glasses, where after a few days our vision adjusts to the glasses. Aural rehabilitation and adjustment to listening with hearing aids is an important aspect of fitting patients with hearing aids. Communication is an essential aspect of our daily lives, which is one of the primary reasons for people with hearing loss to seek audiologist services. The ability to maintain meaningful social life with meaningful interactions with their communities, promote independent life and functional communication needs, limit the communication gaps are some of the main goals of people seeing audiologists help (Sweetow and Sabes, 2010). With aural rehabilitation of older listeners, communication and understanding spoken language is the goal.

Pandemic and communication for people with hearing loss: The COVID-19 pandemic isolated people in many ways; it had particular impact on people with hearing loss because of the masks. People with hearing loss rely heavily on visual cues and contextual listening. When visuals cues are eliminated by masks, the individuals with hearing loss become more reliant upon their hearing aids. Even masks with a clear covering have issues like fogging, reducing the clarity of speech, etc. Older people in nursing homes, elderly care facility are further isolated due to limits on visitors. Since individuals with hearing loss rely heavily on face-to-face communication, phone calls become more difficult to continue communication. Video calls help when the technology is available but simple two-way communication is more complicated during a pandemic.

Older adults in a hospital: When older patients are hospitalized without any visitors or care takers to assist them, having a hearing loss makes it extremely difficult for the physicians and other health care providers to communicate the diagnosis and treatment options to the patients. Prior to COVID-19, communication with the patient frequently included family members in discussions with caregivers. As a result of the COVID-19 pandemic, these discussions often have occurred directly with a patient who may have hearing loss. Most of the older patients are not asked if they are able to hear the provider. Most often focus is on the condition for which the patient was admitted to the hospital. Since, audiologists are not a part of emergency room team, most older adults do not get evaluated for hearing abilities when in the hospital. Unless they already own a pair of hearing aids, there is no easy way for the providers to know if the patient heard them or not. These aspects of communication need more attention, especially when the patients may have hearing loss. Holistic management of hearing-impaired patients must incorporate an understanding of their communication as well as psychological and social needs of the individual.

NOTE

1   It is widely believed that the 8/17/222 final ruling by the U.S. Food and Drug Administration establishing OTC hearing aids will greatly increase the nonprescription options available to the public.

REFERENCES

Abrahamson, J. (1997). Patient education and peer interaction facilitate hearing aid adjustment. Supplement to *Hearing Review.* 1:19–23.

Gates, G.A., Mills, J.H. (2005). Presbycusis. *Lancet.* Sep 24–30;366(9491):1111–1120. doi: 10.1016/S0140-6736(05)67423-5. PMID: 16182900.

Peterson, M. (1994). Physical aspects of aging: Is there such a thing as "normal"? Geriatrics. 49;2:45–50

Rönnberg J, Lunner T, Ng EH, Lidestam B, Zekveld AA, Sörqvist P, Lyxell B, Träff U, Yumba W, Classon E, Hällgren M, Larsby B, Signoret C, Pichora-Fuller MK, Rudner M, Danielsson H, Stenfelt S. Hearing impairment, cognition and speech understanding: exploratory factor analyses of a comprehensive test battery for a group of hearing aid users, the n200 study. Int J Audiol. 2016 Nov;55(11):623–42. doi: 10.1080/14992027.2016.1219775. Epub 2016 Sep 2. PMID: 27589015; PMCID: PMC5044772.

Sweetow, R. W., & Sabes, J. H. (2010). Auditory training and challenges associated with participation and compliance. *Journal of the American Academy of Audiology, 21*(9), 586–593.https://doi.org/10.3766/jaaa.21.9.4

Weinstein, B.E. (2013). *Geriatric audiology.* Thieme.

Willott, J. (1996). Anatomic and physiologic aging: A behavioral neuroscience perspective. *Journal of American Academy of Audiology.* 7:141–151.

24 Common Eye Diseases, Their Visual Outcomes, and Strategies for Enhancing Use of Remaining Vision

Erin Kenny, OD, FAAO, Jean Marie Pagani, OD, FAAO, Sarah D. Appel, OD, FAAO, AAO, Marcy Graboyes, ACSW, LSW, MSW and Elise B. Ciner, OD, FAAO

CONTENTS

24.1 INTRODUCTION

Eye diseases and disorders as well as their associated visual impairments are major global health care concerns. By the year 2032, the baby boomer population will have aged into the Medicare ranks and the rapid growth of the population from ages 65 to the mid-80s in the United States (US) will cause dramatic increases in the prevalence and cost of vision impairment. By 2050, the impaired and blind populations are projected to reach 7.3 million and 3.1 million, respectively.[1] These factors highlight the need for early diagnosis and prompt referral of older adults with eye disease to maximize treatment benefits and to minimize vision loss.[2] Visual impairment may be associated with social and functional decline.[3] Adults with vision impairment often have lower rates of workforce participation and productivity and higher rates of depression and anxiety. In the case of older adults, vision impairment can contribute to social isolation, difficulty walking, a higher risk of falls and fractures, and a greater likelihood of early entry into nursing or care homes.

There are numerous potential causes of vision impairment. However, we will highlight the areas of greatest relevance and those encountered most frequently in the aging population, namely, cataracts, diabetic retinopathy, stroke, open angle glaucoma and age-related macular degeneration (AMD). This chapter describes the characteristics of these eye disorders and discusses available treatments. Additionally, a brief overview of the visual changes, strategies, and resources for each disorder are provided with a discussion that emphasizes rehabilitative strategies that enhance the use of remaining vision for individuals who have experienced irreversible vision loss. It should be noted that this chapter is not fully comprehensive of all eye diseases or their visual sequelae and management.

24.2 CATARACTS

Cataract is by far the most common eye disease in the US. A cataract is a cloudiness or opacity in the normally transparent crystalline lens of the eye (Figure 24.1). The clouding associated with cataracts results in reduced visual acuity and contrast sensitivity. While surgery often restores clarity with the removal of the cataract and the insertion of an intraocular lens, some older adults may not be candidates for cataract surgery.

The onset of cataracts in older adults is usually symptomatic and surgery can normally be performed at any stage. Late-stage cataracts may lead to slightly increased risk of complications. Age is not a barrier to consideration of cataract surgery. Surgical intervention should be considered for any patient who has a problem with visual function or performing activities of daily living, and wishes to improve vision. Medical comorbidities are less critical than for other outpatient operations. A study

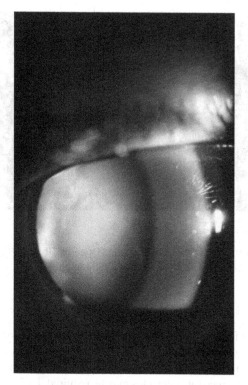

FIGURE 24.1 Cataract.

Pagani, JM. Digital Image of Cataract. 10, May 2022. Author's personal collection taken at The Eye Institute, Philadelphia, PA

assessing the impact of cataract surgery on health-related quality of life in nursing home residents found that, as well as improvement in visual acuity, patients showed significant improvement in psychological distress and social interaction after undergoing cataract surgery.[4] The majority of older adults who are registered as visually impaired due to cataracts had other contributing causes for their visual loss and/or were unwilling or were medically unfit to undergo surgery.[5]

24.2.1 TREATMENT OPTIONS

Standard cataract extraction with intraocular lens implantation, which is performed under local anesthesia, results in improved vision for the majority of patients. Cataract extraction is significantly associated with a lower risk of dementia development[5]. A common complication after cataract surgery is a posterior capsule opacification (PCO) (Figure 24.2). PCO causes a gradual deterioration of visual acuity, which would otherwise improve after a successful cataract extraction. The negative impact of PCO on vision can often be reduced or eliminated with laser treatment.

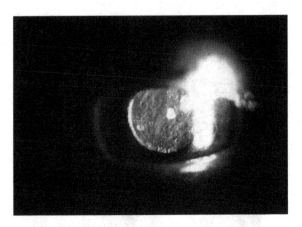

FIGURE 24.2 Posterior capsular opacity.

Pagani, JM. Digital Image of Posterior Capsular Opacity. 10 May, 2022. Author's personal collection taken at The Eye Institute, Philadelphia, PA

24.2.2 ASSOCIATED VISUAL CHANGES

A general blurring of vision at distance and near and sensitivity to bright lights are the characteristic symptoms reported by individuals with cataracts. As individuals with cataracts may experience progressively worsening myopia (nearsightedness), frequent prescription changes for spectacles may be required. Unfortunately, as cataracts progress, prescription changes can no longer correct the worsening vision. Cataracts can cause other visual disturbances; for example, as color vibrancy diminishes, an individual's vision may take on a yellow tint. Reading may become difficult due to cataract-induced reduction in contrast between letters and their background. Additional information regarding the rehabilitation of these individuals is provided later on in the chapter.

24.3 DIABETIC RETINOPATHY

Diabetic retinopathy is the most serious sight-threatening complication of diabetes mellitus and accounts for about 3.9 million cases of vision impairment globally.[1] It is the leading cause of blindness among US employed adults aged 20–74 years. It is projected that by the year 2050 the prevalence of diabetic retinopathy will increase to 13.2 million in the United States.[1] Diabetic retinopathy is caused by retinal changes occurring in patients with diabetes mellitus, Type 1 or Type 2, marked by blood vessel changes. There are two major forms of diabetic retinopathy: nonproliferative (NPDR) and proliferative (PDR). Given the increasing prevalence of an aging population, diabetic retinopathy will affect an increasing number of older adults.[2]

- NPDR is characterized by blood vessel abnormalities resulting in leakage of blood and fluid containing lipids and proteins (exudates) into the retina (Figure 24.3). In individuals with NPDR, progression to the more severe proliferative

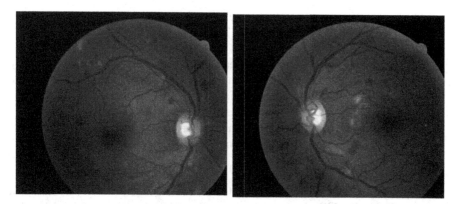

FIGURE 24.3   Right Eye and Left Eye with Nonproliferative Diabetic Retinopathy (NPDR).

Pagani, JM. Digital Image of NPDR OU. 10 May, 2022. Author's personal collection taken at The Eye Institute, Philadelphia, PA

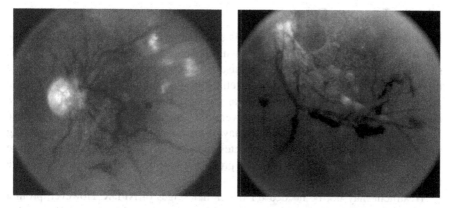

FIGURE 24.4   Right Eye and Left Eye with Proliferative Diabetic Retinopathy (PDR).

Pagani, JM. Digital Image of PDR OU. 10 May, 2022. Author's personal collection taken at The Eye Institute, Philadelphia, PA

form may occur. Sometimes retinal blood vessel damage leads to a buildup of fluid (edema) in the macula known as diabetic macular edema. The most common cause of vision loss in NPDR is macular edema. If macular edema reduces vision, treatment is required to prevent permanent vision loss.

- PDR is characterized by ischemia (lack of oxygen) in the retina that causes the growth of new, abnormal blood vessels on the surface of the retina and fibrous tissue proliferation (Figure 24.4). These new blood vessels are fragile and when they leak blood into the vitreous resulting in vitreous hemorrhages, vision may be severely affected. Fibrovascular tissue emanating from the growth of new blood vessels may also cause a retinal detachment. There is significant risk of vision loss in patients with proliferative retinopathy.[6]

In addition to the increased risk for severe vision loss, diabetes mellitus is associated with other disorders of the visual system. These include the following:

1. **Refractive error fluctuations** (nearsightedness or farsightedness) caused by changes in blood glucose levels may occur. As a result of these fluctuations, corrective lenses may not be initially prescribed until a repeat refraction is performed on a following visit to ensure the stability of the prescription.
2. **Neurological consequences** of diabetes may include eye movement disorders as well as changes in eye alignment resulting in reduced binocular depth perception and/or double vision.
3. **Blood vessel abnormalities** can also occur in the brain, causing hemorrhagic or ischemic strokes associated with visual field deficits as well as in an individual's ability to process visual information. Diabetes mellitus may also result in optic nerve disease affecting visual acuity, visual field, and contrast sensitivity. These changes increase the risk for mobility-related tripping and falling and associated injuries.

24.3.1 TREATMENT OPTIONS

Early detection of diabetic retinopathy is essential as early treatment can reduce the risk of severe vision loss. All patients with diabetes mellitus, whether Type 1 or 2, should have yearly dilated ocular health examinations screening for complications of this disease. Individuals with diabetic retinopathy are followed closely by eye care practitioners who determine the stage and progression of the diabetic retinopathy. Diet control and pharmacological interventions aimed at controlling blood glucose levels should be combined as the mainstay of treatment to prevent vision threatening retinopathy.[7,8] For individuals with diabetes, management to normalize blood glucose levels can delay the onset and slow the progression of diabetic retinopathy by more than 70%.[9]

Treatment may not be needed for the earlier stages of NPDR. However, prompt treatment of PDR and/or diabetic macular edema is essential to prevent irreversible vision loss. Treatment options might include the following:

1. Injections of vascular endothelial growth factor (VEGF) inhibitor drugs into the eye stop the growth of abnormal new blood vessels, reduce bleeding (hemorrhages) within the retina and vitreous, and decrease fluid buildup.
2. Panretinal photocoagulation treatment for PDR that uses laser generated light energy to stop the growth of abnormal new blood vessels throughout the retina that may lead to hemorrhages and retinal detachment.

In addition to ocular disease management, good glycemic, lipid, and hypertensive control will reduce the development and progression of diabetic retinopathy.[2] Individuals in conjunction with their health care providers and caregivers should work together for optimal management of their diabetes and reduction of risk factors associated with their disease.

24.3.2 Associated Visual Changes

Visual symptoms may not be present in the earliest stages of NPDR. They often occur as the disease progresses toward advanced or PDR. Symptoms can include distorted, blurry, or hazy central vision as well as impaired color vision.[10] Affected individuals may also experience floaters, a web-like phenomenon in their vision or even perceive a reddish discoloration of their visual environment. They may also experience distorted or non-seeing areas within their visual field and their vision may fluctuate throughout the day.

While devastating vision loss can be averted as a result of panretinal photocoagulation treatment, the treatment may in turn cause difficulty adapting to dimly illuminated environments[11] as well as result in a constriction of peripheral vision.[12] This is because the peripheral retina is populated by photoreceptor cells called rods which enable individuals to visually adapt to dimly illuminated environments and also transmit visual information about objects that are located in peripheral visual regions. The ability to safely travel in crowded environments, safely negotiate curbs and other mobility related drop-offs in the surrounding environment, as well as to see when moving from a sunny area to a shaded area may be negatively affected when these photoreceptor cells are compromised.

Magnifying devices, discussed later in this chapter, may also be useful for individuals with diabetes who have experienced a decrease in their central vision due to macular edema or optic nerve disease. These devices provide sufficient print enlargement to enable individuals to read, perform activities of daily living, and watch favorite programs on television. Diabetes mellitus, however, also results in physical changes that may affect a person's ability to use magnification devices. Loss of sensation in one's fingers as a result of peripheral neuropathy may make it difficult to hold a magnifier or focus a low vision telescope.[13] Individuals with severe visual impairment who also have peripheral neuropathy may not feel raised dot markings on stove dials or the dots on a braille publication. For individuals who find visual reading to be difficult, talking books and text-to-speech scanning software can provide alternatives to visual reading strategies. Nonoptical health care devices such as talking glucometers for measuring blood glucose levels, large print pill dispensers, large print syringes, talking scales, and blood pressure measuring devices enable individuals with diabetes and vision loss to keep up with their essential health care needs.

24.4 STROKE

Strokes occur when there is a disruption of blood supply to the brain. This is most commonly caused by blockage of a blood vessel (ischemic stroke) or, less commonly, by leakage of blood into surrounding brain tissue from a blood vessel in the brain (hemorrhagic strokes).[14] Strokes are more likely to occur in the older population, especially to those with high blood pressure, diabetes, or heart disease.[15] People who have experienced a stroke may develop a weakness of the right or left side of their bodies. They may also develop difficulty with speech or processing of information. Visual disorders are also associated with stroke. These can include full or partial peripheral visual field loss or a reduction in central vision.

24.4.1 TREATMENT

Treatment of strokes can be broken down based on the etiology of the stroke. Ischemic stroke can be treated by thrombolytic ("clot busting") drugs to break up blood clots causing the damage. Hemorrhagic strokes can be treated by different vascular surgeries.[16]

24.4.2 ASSOCIATED VISUAL CHANGES

A possible visual consequence of stroke is loss of peripheral vision. This may include the onset of right-sided or left-sided blindness (homonymous hemianopsia) in corresponding halves or quadrants of each eye's visual area or field. When there is accompanying one-sided weakness of the body, the area of the field loss corresponds to that side of the weakness. For example, a person who has experienced a weakness of the right arm and leg as a result of stroke may also experience a right-sided hemianopsia in the visual area of both the right and left eye. People with a right homonymous hemianopsia will typically not be able to detect objects that are located in their right visual hemisphere without turning their eyes and head into that visual region. They may bump into people who are on their right side or miss words that are on the right side of the page. A typical adaptation of individuals who have homonymous hemianopsias is to have a constant head turn toward their blind visual region. Individuals with a right homonymous hemianopsia will turn their head to the right in order to shift their remaining left visual field to the right of their postural midline. By doing so, they are better able to detect objects located on their right side. This head turn, however, may cause the individual to miss potential obstacles that are located in the unaffected left visual region as it creates an artificial peripheral visual field loss on the left side.

A complicating factor that may be associated with a stroke-related hemianopsia is that of visual spatial neglect. In addition to a loss of vision on the affected side, the affected individual will also exhibit a neglect of that side of the body and environment. In these cases, instead of adapting to the hemianopsia by turning their head toward the defective visual region, individuals will actually turn away from that visual region. For example, people with a left-sided neglect and a left hemianopsia will turn their head to the right, further reducing their ability to detect objects in the left visual region. It is essential that individuals with both visual spatial neglect and homonymous hemianopic visual field deficits work with professionals from the fields of occupational therapy and orientation and mobility in order to overcome both the visual and cognitive issues that are interfering with safe mobility.

24.5 CHRONIC OPEN ANGLE GLAUCOMA

Chronic open-angle glaucoma is an internal eye (intraocular) pressure-related progressive optic nerve disease with accompanying characteristic changes in the optic nerve and visual field (Figure 24.5). The Blue Mountains Eye Study found a 3% prevalence of open-angle glaucoma in people aged > 49 years. An exponential rise in prevalence was observed with increasing age.[17] Damage to the optic nerve from

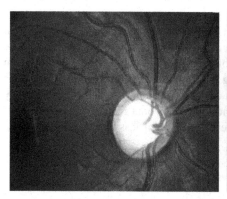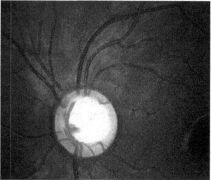

FIGURE 24.5 Right and Left Optic Nerve with Glaucoma.

Pagani, JM. Digital Image of Optic Nerve OU. 10 May, 2022. Author's personal collection taken at The Eye Institute, Philadelphia, PA

glaucoma can cause irreversible visual field loss. When untreated or uncontrolled, the visual field and visual acuity loss may progress to complete blindness. The aim of treatment is to prevent visual disability during a person's lifetime, or prevent progression of visual disability if it has occurred.[2]

Since age is the greatest risk factor for the development of glaucoma, prevalence is four to ten times higher in people over 40 years of age.[18] Approximately 7.7 million people worldwide are affected by blindness due to glaucoma (World Health Organization 2021 estimate). Of those people who are legally blind (visual acuity of 20/200 or worse in the better seeing eye with best correction and/or 20 degrees diameter or less of remaining visual field) because of glaucoma due to visual acuity or visual field loss, 75% are older than 65 years of age according to the American Academy of Family Physicians. Other factors that are associated with glaucoma include: race, positive family history, high blood pressure, diabetes mellitus, high myopia, and elevated intraocular pressure (IOP).[8]

24.5.1 TREATMENT OPTIONS

Current treatment for open-angle glaucoma is lowering the IOP. This is most commonly achieved through the use of pharmacologic agents such as topical eye drops. When topical treatment is prescribed for older adults, consideration should be given to the most appropriate agent. Multiple classes of drugs are used by doctors, either alone or in combination, to achieve and maintain a lower IOP. However, some of these medications may have adverse systemic side effects when combined with other medications or used by individuals with severe cardiovascular disease.

Individuals who cannot be adequately and safely treated with eye drops or are unable to physically instill eye drops, or are not compliant with treatment because of cognitive impairment may benefit from laser treatment or surgical intervention to reduce IOP.[2]

Glaucoma is a chronic, progressive disease with no known cure, and many patients may require a combination of treatment options. Individuals must be aware that treatment will be lifelong and requires their full compliance with medications, office visits, and testing, even in the absence of visual symptoms. Good communication between care providers is key in ensuring that treatment and management decisions are appropriate and include consideration of the individual's age and life expectancy as well as comorbidities. Individuals who have glaucoma require lifelong follow-up care considering the potential for irreversible visual loss and functional visual field loss.

24.5.2 ASSOCIATED VISUAL CHANGES

The rate of progression of vision loss from glaucoma depends on the level of IOP, the susceptibility of the optic nerve to damage, and the severity of the disease. The mild and moderate stages of open angle glaucoma are normally asymptomatic. However, as the disease progresses, the resulting vision loss is progressive and occurs over a period of years, primarily affecting peripheral vision. Typically, high detail central vision is spared until late in the course of the disease when the affected individual is left with only a small area of central vision. Individuals may report that they are aware that parts of objects appear to be missing in their peripheral vision. With disease progression, other visual changes may be reported. Individuals with glaucoma may complain of blurred vision or misty vision or haloes which may be associated with elevated IOP. Poor color perception and reduced night vision may also suggest glaucoma. Progression of glaucoma is also associated with a reduction in contrast sensitivity function or the inability to see low contrast objects in the environment. They may experience difficulty reading low contrast newspapers, detecting low contrast curbs, or identifying people's faces.[19] Glare sensitivity also increases with progression of glaucoma. Older adults with glaucoma may experience increased glare-related discomfort when they encounter glare in both outdoor and indoor environments.[20]

24.6 AGE-RELATED MACULAR DEGENERATION (AMD)

AMD is the most common cause of severe vision loss in people over 50 years of age in the Western world with over 25 million people suffering from this disease.[21] The single greatest risk factor for AMD is increasing age, although a number of other risk factors have also been suggested. Studies have found that the prevalence of AMD is associated with gender (women are more likely to have AMD than men), race (lower prevalence in Afro-Caribbean American and Hispanic people than Caucasian people), and cigarette smoking.[22] A positive family history of AMD also suggests that genetic factors may be involved. Other possible risk factors that have been identified include cardiovascular disease, elevated serum cholesterol levels, and hypertension.[23]

AMD is a degenerative disorder of the macula, which is an area near the center of the retina, which results in a gradual loss of central vision. The condition is classified into two types, according to the type of degeneration: non-neovascular macular

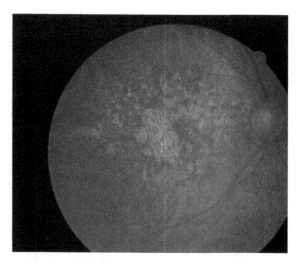

FIGURE 24.6   Dry Macular Degeneration.

Pagani, JM. Digital Image of Macula OD. 10 May, 2022. Author's personal collection taken at The Eye Institute, Philadelphia, PA

degeneration (also known as dry or geographic AMD) and neovascular macular degeneration (also known as wet AMD).

- Non-neovascular AMD accounts for approximately 80% of all AMD and is associated with the formation of drusen and retinal pigment epithelial changes (RPE) in the macula (Figure 24.6).
- Neovascular AMD is characterized by choroidal neovascularization (CNV), consisting of the formation of new, fragile blood vessels growing from the eye's choroid layer into the retina's macular region (Figure 24.7). These vessels leak blood (hemorrhages) and fluid, resulting in scarring in the macula that can severely affect visual acuity.

24.6.1 TREATMENT OPTIONS

Although gene therapy clinical trials are ongoing, currently there is no available effective therapeutic treatment to improve vision in dry AMD. However, there are some preventative measures that may reduce the risk of disease progression. Smokers should stop smoking as research has shown that smoking increases the risk of progression in dry AMD. There is also evidence that nutritional supplements (nutraceuticals) contribute to a reduction in risk of developing the advanced stages of AMD. Antioxidant supplements may prevent cellular damage and retard progression.

For most people with wet AMD, current treatment involves injection of anti-vascular endothelial growth factor (VEGF) agents into the eye. These agents have revolutionized the management and visual prognosis of wet AMD over the last decade, and frequently vision is stabilized or improved with treatment.

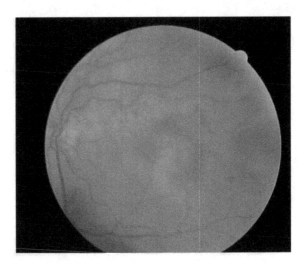

FIGURE 24.7 Wet Macular Degeneration.

Pagani, JM. Digital Image of Macula OS. 10 May, 2022. Author's personal collection taken at The Eye Institute, Philadelphia, PA

It has been proposed that in the future, older adults could undergo an AMD "risk factor assessment."[3] This could include family history, genetic testing for AMD susceptible genetic mutations, environmental and lifestyle assessment, smoking history, and dilated retinal evaluation. Older adults considered to be at risk could be alerted to possible preventive strategies, new and effective nutritional supplements, and developing new treatments which could reduce the incidence of advanced forms of AMD in older adults and could reduce the burden on society as a whole.[24]

24.6.2 ASSOCIATED VISUAL CHANGES

The macular/foveal region of the retina provides individuals with the ability to see fine visual detail, adapt to high levels of illumination and recover from glare, as well as to discern low contrast images and colors. Damage to this region from conditions such as AMD can affect all of these visual abilities. AMD is associated with reduced central vision and central areas of visual distortion or, in more advanced forms of the disease, central "blind spots." In the early stages of the disease, vision may only be mildly reduced (20/25–20/40). In advanced stages of the disease, vision may be at the 20/200–20/400 level representing severe visual impairment. Vision changes may be gradual, as in the case of "dry" AMD or sudden as in the case of the "wet" or exudative form of AMD. Despite the profound loss of central visual acuity that may result from AMD, the condition does not progress to the peripheral retina, therefore sparing the affected individual's peripheral visual region.

Magnification of visual detail (which will be discussed later in this chapter) may be beneficial for individuals with early stage as well as advanced AMD. In addition to magnification, individuals with AMD may benefit from an "**eccentric**

viewing" strategy or the use of healthier undamaged retinal areas that are adjacent to the damaged central macular visual region (preferred retinal locus)[25] while engaged in activities such as reading or watching television. This viewing strategy enhances vision both with and without the use of magnification devices by minimizing interference from central blind spots that are associated with AMD. Societal misperceptions sometimes discourage individuals from using eccentric viewing in public. This is typically due to the widely held belief that looking away from someone during conversations indicates a lack of honesty or interest. Educating family members as well as acquaintances about the reason for the eccentric viewing will eliminate those misconceptions.

AMD is associated with increased glare sensitivity. Individuals with AMD typically find that it takes them much longer to recover visually from glare in their environment.[26] A common complaint of individuals with AMD is that after a prolonged reading session, they experience temporary "blindness" as they look away from their reading materials. Reading at reduced distances, which is common with AMD in order to achieve necessary image magnification, may result in a significant amount of glare created by indirect light that is reflected from the white background of the reading material as well as direct light from bright overhead illumination, onto the foveal region of the reader. When the reader looks away to a less brightly illuminated environment, that area appears to be dimmer, and the reader may even feel "blinded," for a period of time. This is the same effect that is created in individuals without AMD when the intense burst of light emitted by a camera's flash results in the photographer's subject experiencing a central blind spot that obscures central detail for a brief period of time.

AMD is also associated with a reduced ability to identify low contrast visual detail. Any strategy that will enhance the contrast of print or other visual targets is of significant benefit to individuals with advanced AMD. Contrast-enhancing nonoptical devices such as felt tip pens and bold-lined paper enable individuals with AMD to keep up with their bills and correspondence. Another strategy that significantly reduces glare when viewing text on a computer screen, tablet, or iPhone is to reduce the brightness of the screen or to use reverse polarity presentation of text (white letters on a black background).

24.7 MANAGEMENT OF VISUAL DEFICITS WITH LOW VISION REHABILITATION

A low vision evaluation is a specialized type of intervention that is designed to address the functional needs of individuals whose reduced vision is not addressed by standard eye exams. It consists of modified exam elements in comparison to a primary care visual examination. It should be noted that this type of evaluation may be exhausting for some of the older adult population because of the time and energy exerted for exam elements.

The low vision evaluation should be a goal oriented with an emphasis on increasing independence and improving quality of life. Multiple studies and reviews have shown an improvement in quality of life and visual functioning through this intervention.[5] Goals may be modified or expanded based on individual needs that could change due to a host of factors including a decrease in vision or change in living situation.

24.7.1 HISTORY

A comprehensive history is a vital starting point for the low vision evaluation and provides an opportunity for an individual to share his or her story of the impact of visual impairment on all aspects of life. Although a systemic, social, family, and ocular history should always be obtained, a functional visual history can provide insights into how the individual is managing in day-to-day life. This would include an exploration of areas such as reading, activities of daily living, mobility, leisure, and vocation. Issues involving safety can be revealed at this time, including difficulty seeing the height of the flame on the stove, missing curbs, and steps due to visual field deficits and incorrectly identifying medication bottles due to the inability to see small print. In regard to reading, the following areas should be explored: reading needs and preferences, the current setup used for reading, type of reading activity (e.g., mail, books and magazines, medication bottles, price tags, computer, iPad, or smartphone), lighting, reading distance, and preferred print size. Activities of daily living are another important area to discuss. Common activities that should be explored include food preparation, money management, shopping, personal grooming, cleaning, medication management, and communication with others (e.g., phone calls, letter writing, social media, and video chats). An additional area to investigate is the current living situation of the individual which may include the type of housing, composition of household members, and support required to manage important activities of daily living. Mobility concerns should be explored by asking questions such as how the person feels navigating in familiar or new environments; if they have had any recent falls; if they are currently driving or using public transportation; or if they feel they are dependent on others to travel. Answers to these and other questions can often lead to a referral for orientation and mobility services which will be discussed in the following chapter. Leisure activities such as watching TV, playing cards, handcrafts, and socializing with friends can also be affected by a visual impairment. The loss of leisure activities due to vision can affect an individual's cognitive abilities, and morale and may lead to a feeling of isolation and depression.[27] Finally, if the older adult is still working, vocational needs should be explored. Tasks may include the completion of paperwork, money management, accessing information from a computer, and identifying signs in the workplace. In summary, the functional history provides a snapshot of how the individual is managing in everyday life and sets the tone and focus of the examination.

In addition to the functional history, psychosocial concerns should be explored. Questions about coping with the vision loss, currently available support systems, and anxiety experienced regarding loss of vision are asked. In a recent survey, 47.4% of Americans rated vision loss as the worst possible health outcome.[28] Factors such as the type of vision loss, family reaction, life stage, significant life events, expectations from the low vision evaluation, self-concept, and personality all factor into how an individual manages living with vision loss.

24.7.2 EXAM ELEMENTS

After the history, the low vision examination typically begins with visual acuity testing as the starting point for assessing visual status. Individuals with vision loss

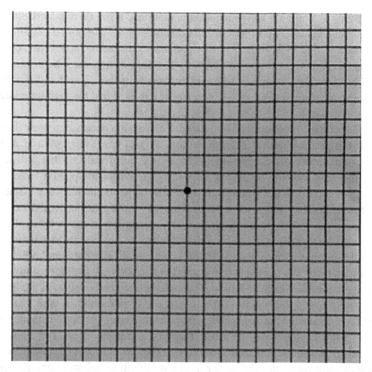

FIGURE 24.8 Amsler Grid: This portable grid can help detect early signs of retinal disease and monitor changes in vision.

often struggle with standard eye chart testing due to impaired central vision, peripheral vision, contrast sensitivity or brightness sensitivity.

Distance visual acuity is important for determining the degree of visual impairment as well as for recommendations for assistive devices to address distance visual goals. Functional near visual acuity is equally important for the determination of optical and nonoptical approaches that address reading goals. Reading is a complex visual task, which can be affected by visual field deficits, reduced contrast sensitivity and eye movements disorders.

In addition, peripheral and central visual field loss in individuals with low vision can impact multiple aspects of their life, including mobility, reading, and several activities of daily living. To help monitor the presence or progression of a central blindspot, an Amsler grid is a useful tool. It is a handheld graphic on a sheet of paper that contains a square grid pattern with a dot in the center. This portable and inexpensive grid is easy to use and can be taken home by the individual and viewed on a regular basis to detect changes in central vision (Figure 24.8).

24.7.3   REFRACTION AND LENS SELECTION

A low vision refraction is a pivotal point in the low vision evaluation and allows the clinician to determine a person's corrective spectacle prescription that will ensure optimal focus. A common misconception is that once a person has an ocular disease that results in a visual impairment, refraction and glasses will not be helpful. A study in 2010 showed that 11% of the 739 new patients evaluated at a low vision clinic over three years experienced improved vision by two lines or more on the distance acuity chart with refraction alone.[29] Just like other aspects of the exam that differ from the standard eye examination, a low vision refraction is modified for each individual with a visual impairment.

Once a spectacle prescription is finalized, a discussion occurs as to how the glasses should be designed to best meet the individual's visual goals. For example, a progressive lens, sometimes called a "no-line" bifocal, is a multifocal lens that provides clear focus at multiple viewing distances, depending on where the person looks through the lens. The top portion of the progressive lens contains the person's distance prescription, and, as they lower their gaze, they pass through an intermediate distance prescription zone before reaching their near prescription at the bottom of the lens. This type of lens, however, can create difficulty for those with visual impairment specifically due to the presence of central scotomas (blind spots) and other types of visual field deficits. If a person is trying to eccentrically view in order to enhance their remaining vision, they may be looking through the incorrect portion of the lenses and therefore not receive the benefit of the correction. If an individual with low vision is looking for the convenience of having one pair of glasses for both distance and near activities, a lined bifocal may be a better option as it provides a larger viewing area for accessing the required near prescription. Separate distance and near prescriptions are often the best recommendation for individuals with visual impairment, especially those with visual field loss.

The type of lens used to make up a pair of glasses may also be different for individuals with visual impairment. If someone is monocular, or has a better vision out of one eye, polycarbonate or Trivex safety lens material may be recommended. Polycarbonate and Trivex are types of plastic lens material that have more impact resistance in comparison to glass or other standard lenses. Polycarbonate and Trivex lenses are also beneficial for the protection of eyes in case of a fall or contact with an unanticipated object such as a tree branch or the corner of a cabinet.

As mentioned previously, individuals with low vision often have contrast sensitivity impairment as well as color vision discrimination issues. These can result in difficulty with activities such as recognizing faces, differentiating colors on traffic signals, and discriminating colors of clothing. Certain testing results can provide the clinician with information to determine appropriate contrast enhancing recommendations, which may include a lighting evaluation or the use of contrast enhancing devices.

24.7.4 MAGNIFICATION

When a section of the central macular area has been damaged by diseases such as macular degeneration, the section of the image being viewed that falls within the damaged retinal area is not processed, resulting in a "blind spot" within the image. See example here (Figure 24.9).

Obs..ve

Obs●rve

FIGURE 24.9   Central scotoma or "blind spot": A representation of how magnification may benefit an individual with a central blind spot reading the word 'observe.'

In order to minimize the effect of the damaged macular region or blind spot, magnification of images over a greater area of the retina enables a greater part of the image to be transmitted by the surrounding undamaged retinal areas to the visual center of the brain.

As vision loss increases in severity, greater image magnification is required. Image magnification can be achieved in four general ways: moving materials closer, enlarging print, using optical devices, or using digital technology to project larger images.

Magnification may also be beneficial for individuals with central vision loss or central blind spot. Other strategies may also be useful to achieve necessary magnification. An individual may achieve increased magnification by sitting closer to a television. Sitting twice as close to the television screen provides double the magnification of visual detail. Another strategy for increasing magnification is to obtain a larger television screen. A 52-inch screen will provide an image size that is approximately twice that available from a 26-inch screen. High-definition televisions may reduce the need for magnification as they often provide enhanced contrast and image definition. Large print materials may also enable people with central vision loss to read and resume favorite activities such as playing poker with large print playing cards. Fonts with uniform stroke widths such as Arial are easier for individuals with decreased central vision to see as they reduce variations in visual complexity within letters, numbers and words. Use of fonts such as Times New Roman which has less uniform stroke widths may create visual confusion as well as reduce reading accuracy and efficiency for the reader with AMD or other disorders with reduced central vision. Owing to their variability in stroke widths, Times New Roman fonts also require greater print magnification than fonts such as Arial to ensure that the narrowest stroke width within each letter is appropriate for the reader's visual acuity level.

24.7.5   Near and Intermediate Devices

Common near goals in the older population include reading, playing cards, sewing/ crocheting, and writing. In a study done by Owsley, approximately 86% of low vision adults in the United States have problems reading (see Figures 24.10 to 24.13).[30]

Here are examples of near device options:

- Spectacle microscopes: These are magnifying eyeglasses used for near visual activities. Magnification is achieved by holding the reading material closer and by the microscope's power correcting for that working distance. The higher the power of the microscope, the closer the individual will have to hold the reading material in order to focus the print clearly.
- Handheld magnifiers: This is a magnifying lens that is held by a handle and can be held at different distances from text, pictures, or other near visual detail. A common misconception about handheld magnifiers is that a larger magnifying lens is synonymous with greater magnification. The reality is the opposite. As the power of a magnifier increases, there is an accompanying increase in peripheral distortion of the images seen through the magnifier thus requiring the use of a smaller diameter lens in order to reduce the area of peripheral distortion.
- Stand magnifiers: This is a magnifying lens that is mounted at a fixed or adjustable distance from text, pictures, or other near visual detail and can be useful

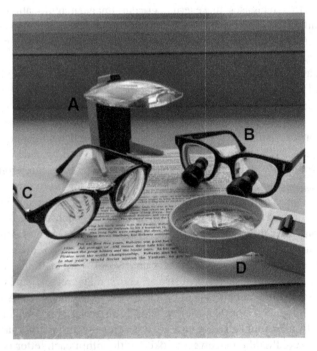

FIGURE 24.10   (A) Illuminated stand magnifier, (B) Telemicroscope, (C) Microscope, and (D) Illuminated handheld magnifier. Kenny, Erin. Photograph of Near Devices. 5 May 2022. Author's personal collection.

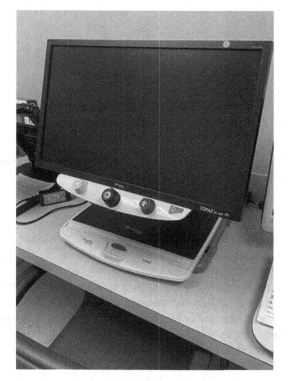

FIGURE 24.11   Desktop video magnifier. Kenny, Erin. Photograph of Desktop Video Magnifier. 5 May 2022. Author's personal collection.

FIGURE 24.12   Handheld telescope. Kenny, Erin. Photograph of Near Devices. 5 May 2022. Author's personal collection.

for individuals who have fine motor deficits or are trying to complete tasks requiring the use of both hands (e.g., sewing or writing).

- Telemicroscopes: These are devices that provide higher degrees of magnification at an increased reading distance. They are also known as loupes.
- Video/digital magnification: This type of device provides magnification on a high-definition screen.

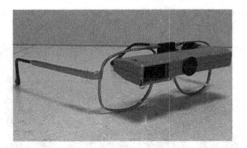

FIGURE 24.13 Spectacle mounted, bioptic telescope. Kenny, Erin. Photograph of Near Devices. 5 May 2022. Author's personal collection.

24.7.6 DISTANCE DEVICES

Common distance goals for individuals with low vision include identifying faces from a distance, reading signs, watching TV, and driving. In the same study done by Owsley, 50% of American adults who have low vision have a desire to drive.[30] While there are a range of devices that address near and intermediate activities, there are fewer magnification options for distance visual goals. Distance magnification can either be spectacle mounted or handheld. Handheld telescopes enable individuals with visual impairments to achieve "distance spotting" goals such as reading street signs, reading information on a bulletin board, and recognizing the faces of people who are waving to them from across the street.

24.7.7 NONOPTICAL OPTIONS AND ASSISTIVE TECHNOLOGY

In addition to the optical devices described, low vision clinicians may also recommend nonoptical strategies to address goals through the evaluation of task lighting, tints that address glare issues, enlarged printed material, and assistive technology. Common examples of how these options work in conjunction with magnification devices include focused lighting while reading with a microscope, high contrast pens for writing when using a stand magnifier and tinted lenses with telescopes to reduce glare. When combined with the appropriate optical device, low vision clinicians can further meet the goals of the individual with low vision (Table 24.1).

24.7.8 PERIPHERAL FIELD LOSS

Orientation and mobility specialists (who will be discussed in the next chapter) will provide individuals who have significant peripheral visual field loss with instruction in the use of remaining vision while walking. They will also evaluate the need for adaptive equipment such as mobility long canes or electronic mobility devices that will increase safety while traveling. Protective eyewear should be prescribed for individuals with severe peripheral vision loss in order to prevent eye injuries from obstacles, such as the corner of a kitchen cabinet, that are located in an individual's non-seeing visual region. Contrast enhancing strategies such as felt tip pens for writing, task lighting for reading, and the reduction of both indoor and outdoor glare

TABLE 24.1
Advantages and Disadvantages of Low Vision Devices Used for Near and Intermediate

Devices	Advantages	Disadvantages
Spectacle Microscopes	Familiar to individual Cosmetically appealing Largest field of view Hands free Good for individuals with hand tremors Good for prolonged reading Ability to write	As lens power increases, working distance decreases Lighting is critical Lens distortions Field size decreases May cause fatigue of neck, arm, shoulder May cause nausea, dizziness, eye fatigue, headache
Handheld Magnifier	Portable Illumination Relatively inexpensive Familiarity Good for spot reading Allows for extended working distance Large selection	Requires steady hands and coordination Decreased field of view as working distance increases Increased distortion in lens periphery as working distance increases Required to change battery/bulbs
Stand Magnifier	Extended working distance Portable (but not as much as handheld magnifier) Some can be written under Built in illumination source Good for hand tremors, poor motor control Wide selection available	Additional glasses are needed for most fixed focus stand magnifiers Decreased field of view compared to microscopes Bulky, cumbersome Need to change battery/bulbs May create posture/fatigue problems
Telemicroscopes (Loupes)	Increased working distance Can be portable	Critical focusing distance Need for addition lighting Small field of view
Video/Digital Magnification	Wide range of magnification Natural viewing posture Potential binocularity Maximum reading rates Contrast enhancement Color or black and white Reverse polarity Ability to write	Cost Portability (some models) Battery power in portable units

Source: Matchinski, Brilliant, & Bednarski [31].

with filters and sunglasses are typically recommended. Reading and writing guides help individuals with severe peripheral field loss to keep their place while filling out checks and reading mail. Nonoptical devices, such as guides for insertion of eye drops, will help those with significant vision loss to self-administer the drops that are essential in keeping the eye's internal fluid pressure under control and reduce the likelihood of further vision loss.

24.7.9 REFERRAL TO REHABILITATION

At the completion of the low vision evaluation, a clinician may also make recommendations for additional referrals to other low vision rehabilitation specialists. These specialists will focus on specific vision rehabilitation areas to increase independence in activities of daily living and overall quality of life. Chapter 25 will focus on these low vision rehabilitation specialists.

24.8 CONCLUSION

As the proportion of older adults in the United States grows, eye disorders and visual impairment are becoming an increasingly important public health care concern. The consequences of visual impairment are many and may be associated with social and functional decline, lower rates of workforce participation, use of community support services, higher rates of anxiety and depression, difficulty walking and falls, entry into nursing or assisted living homes, decreased quality of life and increased mortality.[32] In the United States, visual disability ranks among the top ten disabilities. This chapter has highlighted the need for effective care for older adults to minimize vision loss from age-related eye diseases such as cataract, diabetic retinopathy, strokes, open angle glaucoma, and AMD. Early diagnosis and prompt referral of older adults with common age-related eye diseases is important to maximize treatment benefits. When visual impairment is irreversible, referral for a low vision rehabilitation evaluation, vision rehabilitation services, and assistive devices will enhance the use of remaining vision in older adults with vision impairment. It is vital that primary care physicians, primary health care professionals and caregivers are aware of both the clinical signs and symptoms of these diseases as well as available ocular treatment, vision rehabilitative services, and resources. Armed with that knowledge, they can facilitate access to the necessary evaluation, treatment, and rehabilitation resources that will improve the quality of life and independence of older adults with visual impairment.

REFERENCES

1. Wittenborn J, Rein D. The future of vision: forecasting the prevalence and costs of vision problems, June 11, 2014, Presented at "Prevent Blindness," 1–76.
2. Green C, Goodfellow J, Kubie J. Eye care in the elderly. *Aust Fam Physic*. 2014 July;43(7):447–450.
3. Gibson J. Editorial: screening for eye disease in the elderly: is it worth It? *Br J Hos Med*. 2009 Oct;70(10):554–555.

4. Owsley C, McGwin G Jr, Scilley K, Meek GC, Seker D, Dyer A. Impact of cataract surgery on health-related quality of life in nursing home residents. *Br J Ophthalmol.* 2007;91(10):1359–1363.

5. Lee C, Gibbons L, Lee A, Yanagihara R, et al. Association between cataract extraction and development of dementia. *JAMA Intern Med.* 2022;182(2):134–141.

6. American Optometric Association Consensus Panel on Diabetes. Optometric clinical practice guideline care of the patient with diabetes mellitus. 1993. Developed by the AOA Evidence-Based Optometry Guideline Development Group © 2019 American Optometric Association. Available at: www.aoa.org/AOA/Documents/Practice Management/Clinical Guidelines/EBO Guidelines/Eye Care of the Patient with Diabetes Mellitus, Second Edition.pdf

7. Cooppan R. Diabetes in the elderly: implications of the diabetes control and complications trial. *Compr Ther.* 1996;22:286–290.

8. Harvey, P. Common eye diseases of elderly people: identifying and treating causes of vision loss. *Gerontol.* 2003;49:1–11

9. The Diabetes Control and Complications Trial Research Group. The effect of intensive treatment of diabetes on the development and progression of long-term complications in insulindependent diabetes mellitus. *N Engl J Med.* 1993;329:977–986.

10. Tan NC, Yip WF, Kallakuri S, et al. Factors associated with impaired color vision without retinopathy amongst people with type 2 diabetes mellitus: a cross-sectional study. *BMC Endocr Disord.* 2017;17:29.

11. Bavinger JC, Dunbar ET, Stem MS, et al. The effect of diabetic retinopathy and pan-retinal photocoagulation on photoreceptor cell function as assessed by dark adaptometry. *Invest Ophthalmol & Vis Sci.* 2017;57: 208–217.

12. Maguire MG, Liu D, Glassman AR, Jampol LM, et al. DRCR Retina Network. Visual field changes over 5 years in patients treated with panretinal photocoagulation or ranibizumab for proliferative diabetic retinopathy. *JAMA Ophthalmol.* 2020 Mar 1;138(3):285–293.

13. Lima KCA, Borges LDS, Hatanaka E, Rolim LC, de Freitas PB. Grip force control and hand dexterity are impaired in individuals with diabetic peripheral neuropathy. *Neurosci Lett.* 2017 Oct 17;659:54–59.

14. Lau LH, Lew J, Borschmann K, Thijs V, Ekinci E.I. Prevalence of diabetes and its effects on stroke outcomes: a meta-analysis and literature review. *J Diabetes Investig.* 2019 May;10(3):780–792.

15. American Heart Association. Heart Disease and Stroke Statistics – 2010 Update. Accessed February 23, 2022.

16. National Institute of Neurological Disorders and Stroke rt-PA Stroke Study Group. Tissue plasminogen activator for acute ischemic stroke. *N Engl J Med.* 1995;333(24):1581–1587.

17. Mitchell P, Smith W, Attebo K, Healey PR. Prevalence of open-angle glaucoma in Australia. The Blue Mountains Eye Study. *Ophthalmol.* 1996;103:1661–1669.

18. Leibowitz HM, Krueger DE, Maunder LR, Milton RC, Kini MM, Kahn HA, Nickerson RJ, Pool J, Colton TL, Ganley JP, Loewenstein JI, Dawber TR: The Framingham Eye Study monograph: an ophthalmological and epidemiological study of cataract, glaucoma, diabetic retinopathy, macular degeneration, and visual acuity in a general population of 2631 adults, 1973–1975. *Surv Ophthalmol.* 1980;24:335–610.

19. Jammal AA, Ferreira BG, Zangalli CS, et al. Evaluation of contrast sensitivity in patients with advanced glaucoma: comparison of two tests. *Br J Ophthalmol.* 2020 Oct;104(10):1418–1422.

20. Bertaud S, Zenouda A, Lombardi M, Authié C, et al. Glare and mobility performance in glaucoma: a pilot study. *J Glaucoma*. 2021 Nov 1;30(11):963–970.

21. Papadopoulos Z. Recent developments in the treatment of wet age-related macular degeneration. *Curr Med Sci*. 2020 Oct;40(5):851–857.

22. Velilla S, García-Medina J, García-Layana A, et al. Smoking and age-related macular degeneration: review and update. *J Ophthalmol*. 2013;895147.

23. Pieramici DJ, Bressler SB. Age-related macular degeneration and risk factors for the development of choroidal neovascularization in the fellow eye. *Curr Opin Ophthalmol*. 1998;9:38–46.

24. Camelo S, Latil M, Veillet S, et al. Beyond AREDS formulations, what is next for intermediate age-related macular degeneration (iAMD) treatment? Potential benefits of antioxidant and anti-inflammatory apocarotenoids as neuroprotectors. *Oxidative Medicine and Cellular Longevity*. 2020. doi: 10.1155/2020/4984927. PMID: 33520083; PMCID: PMC7803142.

25. Schuchard RA. Preferred retinal loci and macular scotoma Characteristics in patients with age related macular degeneration. Can J Ophthlmol. 2005 June;40(3):303-12

26. Wolffsohn JS, Anderson SJ, Mitchell J, et al. Effect of age related macular degeneration on the Eger macular stressometer photostress recovery time. *Br J Ophthalmol*. 2006 Apr;90(4):432–434.

27. Brunes A, Heir T. Visual impairment and depression: age-specific prevalence, associations with vision loss, and relation to life satisfaction. *World J Psychiatry*. 2020 June;10(6):139–149. doi:10.5498/wjp.v10.i6.13z

28. Scott AW, Bressler NM, Ffolkes S, Wittenborn JS, Jorkasky J. Public attitudes about eye and vision health. *JAMA Ophthalmol*. 2016;134(10):1111–1118. doi:10.1001/jamaophthalmol.2016.2627

29. Sunness JS, El Annan J. Improvement of visual acuity by refraction in a low-vision population. *Ophthalmol*. 2010 July;117(7):1442–1446. doi: 10.1016/j.ophtha.2009.11.017. Epub 2010 Mar 15. PMID: 20231036.

30. Owsley C, McGwin G, Jr, Lee PP, Wasserman N, Searcey K. Characteristics of low-vision rehabilitation services in the United States. *Archives of ophthalmology* (Chicago, IL: 1960). 2009;127(5):681–689. https://doi.org/10.1001/archophthalmol.2009.55

31. Matchinski, T., Brilliant, R., & Bednarski, M. (1999). Low Vision Near Systems I: Microscopes and Magnifiers. In *Essentials of Low Vision Practice* (pp. 201–233), essay, Butterworth-Heinemann.

32. Assi L, Rosman L, Chamseddine F, et al. Eye health and quality of life: an umbrella review protocol. *BMJ Open*. 2020;10:e037648. doi:10.1136/ bmjopen-2020-037648.

25 Blindness and Low Vision Rehabilitation Services

Lachelle Smith, MS, CVRT, Fabiana Perla, EdD, COMS, CLVT, Jamie Maffit, MS, COMS, CLVT, RYT, Kerry Lueders, MS, COMS, TVI, CLVT and Emily Vasile, MAT, TVI, MS, CLVT

CONTENTS

Older adults with visual impairments may benefit from Blindness and Low Vision (BLV) rehabilitation services, such as Vision Rehabilitation Therapy, Orientation and Mobility, Low Vision Therapy, and Assistive Technology. Each of these disciplines has its own focus, scope of practice, and certification, but they share the common goal of supporting individuals with vision loss realize their full potential. BLV services can be provided in the consumer's home or at rehabilitation facilities. Center-based rehabilitation may provide a comprehensive program of instruction as well as group instruction, while community-based services can tailor interventions to the senior's environment, fostering immediate application of the learned skills.

BLV professionals offer interventions that take into account the effect of visual impairment (VI) on independent function. They work collaboratively with clients to identify meaningful and relevant goals and serve on an interdisciplinary team with other vision, rehabilitation, medical, and community support professionals to provide holistic care to individuals across the lifespan. BLV services can be accessed through state vocational rehabilitation agencies, nonprofit agencies serving individuals who are blind/visually impaired, the Veterans Administration, or searching the ACVREP directory of certificants. These professionals are trained at institutions of higher education and are certified by the Academy for the Certification of Vision Rehabilitation and Education Professionals (ACVREP).[1] Certification status indicates the practitioner has met all professional standards, passed a national certification exam, and kept current with the best practices in their field through a combination of direct service and professional development. ACVREP maintains a directory of certified professionals on their website searchable by discipline, state, and/or name. Here is a summary of each discipline's scope of practice.

25.1   VISION REHABILITATION THERAPISTS

Vision rehabilitation therapists (VRT) provide adaptive independent living skills instruction and resources to individuals, who are blind and visually impaired, and their families for the performance of activities of daily living (ADLs). Areas of daily living addressed include home management (e.g., food preparation and eating, cleaning, organization skills), personal management (e.g., grooming, clothing, financial management and medication management), communication (e.g., handwriting, braille, reading), basic home maintenance (e.g., use of basic tools, managing heating/cooling, water, and electrical systems), recreation and leisure (e.g., hobbies, adaptive games and cards, adaptive entertainment), assistive/access technology (e.g., use of screen reader/magnification programs, mobile phone/tablet instruction), personal adjustment to vision loss (e.g., referral to support, mentoring, and consumer groups), indoor movement and orientation (e.g., human guide, seating in chair/car, body protective techniques, fall prevention), and employment (e.g., interview skills, job site assessment, job coaching). VRT professionals certified by ACVREP carry the "CVRT" credential.

25.2   ORIENTATION AND MOBILITY SPECIALISTS

Orientation and mobility (O&M) specialists support clients with VI in reaching their goals of safe and independent movement in their home, neighborhood, and community. O&M assessments determine current level of mobility in both familiar and unfamiliar environments, as well as effective use of sensory information (e.g., visual, auditory, tactual) for travel purposes. Following the assessment, O&M specialists work collaboratively with seniors to develop individualized goals and a plan for instruction. Strategies and tools to support O&M instruction may include mobility systems (e.g., long cane, guide dog), optical devices (e.g., telescopes, magnifiers), assistive technology (e.g., mobile apps), and nonoptical devices (e.g., glare remediation devices, tactile maps). O&M specialists certified by ACVREP carry the "COMS" credential and NOMA-certified specialists carry the "NOMC" credential.

25.3   LOW VISION THERAPISTS

Low vision therapists (LVT) serve an important role within the context of the clinical low vision team. They conduct assessments to determine how the individual is using vision for accomplishing their functional goals and work with the individual to apply into functional tasks the recommendations made by the low vision eye care specialist – Optometrist or Ophthalmologist – during the clinical low vision exam. Examples of functional tasks include using prescribed lenses and devices to read fine print, watch television, and recognize people's faces, teaching visual efficiency skills to locate objects in the environment, track people and vehicles, or use visual cues to anticipate the presence of stairs and drop-offs. LVTs provide services in settings such as optometric practices, low vision clinics, and Veterans Affairs Medical Centers. LVTs certified by ACVREP carry the CLVT credential.

25.4   ASSISTIVE TECHNOLOGY INSTRUCTIONAL SPECIALISTS

Assistive technology (AT) specialists provide instruction in the use of mainstream and assistive/access technology to enhance the clients' personal goals of employment, independence, hobbies, and education. Installation of hardware and software, along with assessment to determine appropriate instructional needs, are a few of the main tasks of the AT professional. These specialists work in a variety of settings, such as educational systems, rehabilitation centers, hospitals, senior living homes, individual homes, and nonprofit agencies. AT instructional specialists with specialization in VI and certified by ACVREP carry the "CATIS" credential.

25.5   BLV REHABILITATION ASSESSMENT

The importance of providing appropriate, individualized rehabilitation services cannot be overemphasized.

> The needs of older patients with low vision will depend on their circumstances: the region they come from, their economic status, their literacy levels, their family responsibilities, their attitude towards aging and disability, their general health, their motivation, and so on.
>
> *Simon, 2008, p. 30*

Through comprehensive assessments, BLV rehabilitation specialists identify and document the present skill levels of the assessed individual as well as their goals, forming the foundation for relevant and meaningful treatment plans.

Active participation on the part of the senior should be fostered as much as possible during the assessment process, goal setting, and instructional design. Careful consideration must be given to the senior's physical, cognitive, and emotional conditions during all facets of service provision. Strategies include scheduling appointments when the senior is most alert, providing information in their preferred communication medium, keeping sessions short if necessary, and allowing time for mental processing, among others (Ottowitz, 2020). Exhibiting interest and respect for the senior's cultural background is essential for a client-centered approach and to ensure goals and interventions are appropriate and relevant to each individual.

25.6 UNDERSTANDING EYE CONDITIONS

Older adults with VI may be unclear about their eye condition or its functional implications or have unrealistic expectations regarding optical interventions to address their visual goals. For example, they may believe that standard eyeglasses will meet all their visual needs, when in fact more specialized interventions may be required. BLV professionals are trained to explore the senior's understanding of their eye condition and provide appropriate feedback to develop greater understanding and self-advocacy skills when explaining the functional implications of their condition to family, friends, and other support professionals.

25.7 FUNCTIONAL VISION ASSESSMENT

Low vision eye care specialists assess clinical visual acuities and fields using standardized testing protocols. However, functional activities such as reading the mail, watching television, or reading street signs outdoors are anything but standardized. What is the font size in their utility bill? How far away do they sit from their television? What types of lighting are available in their home? How far away can they read a street sign? These and other areas are explored during a Functional Vision Assessment (FVA) performed by BLV professionals to better understand how the client is using their vision in real-world environments and to guide the design of a meaningful intervention plan.

Services by a BLV professional begin with a thorough review of recent eye examination reports and conducting interviews of the senior and other key individuals as warranted. Next, a series of acuity, field, and visual performance assessments are conducted. A *Functional Visual Acuity* assessment identifies the various distances at which the senior (1) is aware that a visual target (e.g., street sign, car, mailbox) is present; (2) first identifies what the visual target might be; and (3) is certain about what the visual target is. These tests are conducted multiple times for the tasks the senior prioritizes (e.g., reading grocery store signage, medicine bottles, viewing computer monitors, etc.) and take into account environmental conditions such as lighting, foreground and background colors, and the size of the visual target. A *Static Visual Field* assessment maps out the area that the senior sees while viewing a stationary target, keeping their head and eyes still, at various distances to demonstrate the effect of distance on field of view. An *Early Warning Visual Field* assessment identifies the first point the senior is aware of people or objects on their right and left sides and any functional blindspots. This assessment is valuable for understanding the functional implications of peripheral field loss due to conditions such as stroke or advanced glaucoma. A *Preferred Visual Field* assessment identifies the areas of the visual field that the senior uses when performing a task like walking a route and reveals where the senior naturally directs (or does not direct) their gaze. A *Functional Visual Performance* assessment identifies *critical incidents* or behaviors that are

FIGURE 25.1   Assorted wearable filters of various tints and frame styles.

demonstrated due to visual impairment, such as using one's foot to feel for the edge of a staircase, closing one's eyes due to glare, reaching inaccurately for objects due to poor depth perception, and so on. These critical incidents are tallied, clustered, and analyzed for qualitative themes in the evaluation data.

25.8   WEARABLE FILTER EVALUATIONS

Glare is one of the main reported challenges in completing ADLs for individuals with VI (Smith & Geruschat, 2010; Watson, 2001). Sources of glare can be internal (within the structures of the eye) or external (coming from the environment). While internal glare is best addressed by the individual's eye care specialist, external sources of glare are frequently managed with a wearable filter (WF). WFs are prescribed after completion of an individualized evaluation with a low vision rehabilitation specialist to address the specific glare challenges (e.g., bright sunny days, overhead fluorescent lights) in the actual or best simulated environment in which the challenges are experienced (Maffit et al., 2022). After completion of the evaluation, a filter will be recommended that adequately controls for glare while maximizing visual contrast. Frame selection is also an important part of this process and is included in the recommendations.

25.9   ENVIRONMENTAL ASSESSMENT

The incidence of VI increases with age (Dillon, Gu, Hoffman, & Ko, 2010), placing older adults with VI in a particularly vulnerable group for falls and constituting a grave threat to this population's quality of life and ability to live independently. Various features of one's place of residence have the potential to foster or hinder a senior's independence and safety. Therefore, an environmental assessment performed by a BLV professional is recommended to ensure that issues related to vision loss are taken into consideration during a home assessment. Using the senior's daily routine, the professional will explore how vision loss impacts everyday activities and safety, with special attention to maximizing independence and reducing fall risks, resulting in recommendations that support improved function and safety. The single most important factor for individuals with low vision is having adequate lighting while reducing or eliminating sources of glare (Watson & Echt, 2010). There are numerous types and possible sources of light, so it is important to explore a variety of options before making a recommendation. As seniors with low vision may experience fluctuations in vision, their needs may vary throughout the day. In addition, they may share their place of residence with others, resulting in conflicting needs. To address these issues, strategies for customizing lighting, such as dimmers and window blinds, are highly recommended. Other typical recommendations include using color/contrast to increase visibility, improving organization, eliminating clutter, and labeling or marking objects for easier identification. A thorough assessment will also reveal activities for which the use of other senses may prove more efficient, such as using talking clocks and timers and raised markings on the stove (Duffy, 2016).

Hazards need to be identified and eliminated as much as possible. Common areas of concern include loose rugs, electrical cords, and clutter. Observing the senior carry out everyday tasks and engaging them in the assessment process are critical to avoid modifications that could prove detrimental, and even increase the risk of falls. For example, while an assessor may recommend clearing a path by removing furniture, a senior with vision loss may have come to rely on the removed object for orientation or support which might have detrimental effects for the senior (Pynoos et al., 2010).

25.10 BLV REHABILITATION INSTRUCTION AND INTERVENTIONS

25.10.1 VISUAL EFFICIENCY SKILLS

Seniors with VI may benefit from learning strategies to be more efficient, systematic, and purposeful when using their vision. Following are sample visual efficiency skills addressed by BLV professionals:

Eccentric viewing is a method by which a senior with central visual field loss or a central blind spot (*scotoma*) directs their vision off of center (Watson & Echt, 2010; Whittaker et al., 2016). When the senior with a central scotoma looks straight ahead at a visual target, the light rays entering the eye go directly to the macular area of the retina, which is responsible for detailed and color vision. In individuals with age-related macular degeneration, the macula is affected, so when looking straight ahead, they may see a distorted image or a blackened area altogether. However, directing their gaze off to the side allows the light rays to contact a nonaffected area of the retina (assuming no other pathology), allowing for better viewing of the visual target. This alternate location on the retina is called *preferred retinal locus (PRL)* or *trained retinal locus (TRL)* (Whittaker et al., 2016). A BLV professional can work with the senior to identify their PRL/TRL, incorporating a combination of head and eye movements and change in positioning and magnification of the visual targets.

Visual tracing is a method by which an individual visually follows a stationary line in their environment (Smith & Geruschat, 2010), such as the line between the wall and the baseboard, or the border between a sidewalk and the grassline. Seniors using magnification through optical devices and video magnifiers may find visual tracing helpful for keeping themselves oriented to what they are viewing.

Visual tracking is a method by which an individual visually follows a moving object in their environment (Smith & Geruschat, 2010). Examples include tracking people in a grocery store, cars at an intersection, and a mouse pointer on the computer monitor.

Visual scanning is a method by which an individual visually searches their environment systematically (Smith & Geruschat, 2010). Visual scanning can be used to locate a specific object in the environment, or to preview the environment for unknown cues, hazards, and other visual targets. Examples of visual scanning include locating a traffic light on a corner, identifying pots and pans on a stovetop, and locating an office on a building directory sign.

Blur interpretation is a method by which an individual uses context clues to interpret information that is visually blurry due to reduced visual acuity (Fazzi & Naimy, 2010). For example, using color and shapes to identify smartphone apps without having to rely on the fine print label or certain products at a grocery store based on shape, color, size, and other visual cues.

25.10.2 ORIENTATION AND MOBILITY SKILLS

Vision loss may have a significant impact on a senior's ability to move safely and independently. O&M instruction is tailored to individuals' goals, to maximize use of available sensory information, and combined with specialized techniques and skills to support travel in their desired environments. Seniors may benefit from learning how to safely *travel with a guide*. This involves learning to interpret the guide's body movements, alerting them to turns, narrow spaces, as well as navigating through doors and up and down the stairs safely. Specific indoor, independent travel techniques can also be taught to support seniors' independent movement in their home or place of residence. These techniques include the *upper and lower protective techniques*, which provide advanced warning of head and body level obstacles (e.g., cupboard doors, chairs, tables, corners). Seniors may learn how to use *hand trailing* to follow a surface from one location to another, such as walking from the bed to the bathroom at night, or from their apartment door to the elevator. Other techniques can assist with crossing open spaces or hallways and maintaining orientation through the *use of landmarks and cues* when traveling independently. O&M instructors are also equipped to teach *problem-solving* strategies to regain orientation and to support continued, intentional travel. For seniors with advanced independent travel goals, instruction may include the use of *mobility devices* (e.g., long cane, guide dog, electronic travel aids), travel in *residential and small business* districts, *street crossing strategies,* and the use of *public transportation.* Finally, O&M instruction incorporates use of *optical devices* into lessons. Whenever seniors have been prescribed a magnifier, telescope, or field enhancement device, instruction in real-world environments (e.g., grocery store, coffee shop, bus travel) is provided to support integration of device use.

FIGURE 25.2 At a street corner, a woman is using a long white cane to position herself before crossing the intersection.

25.10.3 INDEPENDENT LIVING SKILLS

Independent living skills (ILS) are the visual and nonvisual techniques and strategies to perform everyday ADLs. There are two levels of ADLs: basic activities of daily living (BADLs) and instrumental activities of daily living (IADLs). BADLs are those related to personal care skills (e.g., eating, grooming, dressing, indoor orientation and movement, and access to information), while IADLs represent advanced, higher level skills (e.g., food preparation, cleaning, medical management, financial management, and personal and interpersonal communication) (Connors & Abbott, 2020).

Strategies for providing adaptive instruction to seniors impacted by vision loss may differ depending on their level of visual functioning and onset of vision loss. Those with acquired vision loss will most likely have experience performing many of the ADLs, and simple adjustments to the task, environment, or item may be sufficient for safe performance. Seniors with congenital blindness or low vision may have limited experience or lack understanding of the concepts associated with performing such skills (particularly if these skills have been traditionally performed for them by others). The VRT will provide comprehensive adaptive instruction in all needed areas and/or assist with training the support team of the senior. As mentioned in the *Environmental Assessment* section, the use of appropriate lighting, color/contrast, sensory development skills, and adaptive equipment are all effective strategies for ensuring the safe performance of all ILS. Examples of recommendations for the safe performance of a variety of ILS include the following.

Grooming and *Clothing Management*: use of high contrast wall colors in the bedroom or bathroom, appropriate height commode with grab bars (if needed), and safety bars with nonslip material in the bathtub are effective strategies for increasing safety. Organization techniques such as having designated storage places for items such as caddies, adjustable shelving, closet/drawer organizers, shoe organizers, and use of adaptive labeling techniques for the identification of all items (Fugate & Smith, 2020).

FIGURE 25.3 Photo of a woman using an adaptive technique for applying mascara by using her index finger as a guide. The woman rests her left index finger to the right of the nasal bone with the pad of the finger facing right. She uses her right hand to bring the tip of the mascara rod to the pad of the finger in a horizontal position so that when blinking the eye, the liquid makeup brushes the eye lashes.

Medication Management: the VRT can work with the senior on identifying their medication using low vision and nonvisual techniques (shape, size, color), cutting pills (if needed) and organizing and storing their medication in appropriate pill boxes. As diabetic retinopathy is a leading cause of VI (Swenor & Ehrich, 2021), seniors with this condition will likely need to learn adaptive strategies for monitoring glucose levels with auditory glucometers and/or through tactual feedback and use adaptive insulin measurement devices.

FIGURE 25.4 Various diabetic medical supplies including a talking glucose meter, testing strips, and insulin pen.

Food Preparation: strategies include organizing, storing, and retrieving kitchen items in ways that are efficient and memorable for the senior. A particular challenge may be using kitchen appliances such as microwaves, ovens, dishwashers, and stovetops. Working collaboratively, the most appropriate marking techniques (e.g., high contrast, tactile label) can be identified. In addition, a VRT can educate and train the senior on how to use their senses and specialized adaptive equipment to safely cut/slice food items, pour liquids, center pots/pans on the stove burner, turn and stir food while cooking (Smith & Fugate, 2020). Strategies for supporting seniors with diabetes and VI can include nonvisual or visual techniques for meal planning and preparation (e.g., accessing nutrition labels, measuring foods for portion control, etc.) to support optimal health.

Cleaning and Basic Home Maintenance: techniques include identification, use, and storage of cleaning products and tools using sensory cues and adaptive techniques. This can include strategies to access key information on cleaning product labels, techniques for measuring cleaning products, safe use of spray bottles, and systematic strategies for cleaning surfaces. In addition, operation of heating and cooling systems (adjusting their thermostat), management of in-home electrical system (e.g., circuit breaker, changing light bulbs, plugging in electrical devices), use of main water valve, and using basic tools such as a hammer, screwdriver, and tape measure are also addressed (Smith, 2020).

FIGURE 25.5   On a cold electric stove burner, an individual is centering a frying pan on the burner by forming both hands into the shape of a "C" and placing them at the three and nine o'clock positions around the burner and pan. This is a safe technique for centering pans on cold burners prior to turning on the heat.

Financial Management: strategies include techniques and equipment to identify their money (coins and bill notes), use check writing guides, large print financial documents, and AT programs to access banking information, utility websites, and pay portals (online banking). A VRT can assist the senior and their support team to access special funding programs to assist with meal services, rental assistance, adaptive transportation services, appropriate social security benefits, and other community services impacting personal finances (Smith & Fugate, 2020).

Communication Techniques: instruction may include adaptive techniques for handwriting notes, letters, and signing one's name, audio or visual strategies for reading and learning braille. AT can support accessible communication and access to support services, including use of mobile devices and mobile applications, connection to the National Library Service (NLS) for acquiring free talking books, and on-demand assistants (e.g., Aira, Be My Eyes, etc.), and accessing reader and driver services.

FIGURE 25.6 A mobile phone illustrating an individual dialing the phone on a high contrast, black-on-white background with extra large font size.

25.10.4 LOW VISION ASSISTIVE TECHNOLOGY

The field of AT continues to grow at a rapid pace, with new technologies and devices being developed daily, requiring specialized assessment and instruction on its use. A common example of AT is video magnification, which allows the user to customize the size, contrast and presentation of print and digital materials. Video magnification is available in the form of mobile applications, portable, transportable, and stationary devices. Many seniors use this technology for reading and writing activities, maintaining finances (e.g., writing checks, reading mail), and assisting in other areas of independent living skills, mentioned in the above section.

To support computer use, there are available built-in accessibility features as well as specialized software (e.g., ZoomText Fusion, JAWS, and VoiceOver) that provide both visual and nonvisual access. Seniors can learn to customize the on-screen output to increase contrast, font size, mouse controls, or navigate using keyboard commands and gestures (Lee & Ottowitz, 2020). Mobile devices, such as smartphones, tablets, and iPads have accessibility settings that allow the individual to magnify the screen as well as audio output for on-screen information (i.e., VoiceOver, TalkBack, Google Assistant). Seniors may also find success in using in-home technology (e.g., Amazon Alexa, Google Home, iHome, etc.) for voice control operation of household functions (e.g., lights, thermostats, etc.) and access to news and weather, as well as for leisure activities such as listening to music or audio books.

FIGURE 25.7   A white envelope with a return address is placed on the X-Y table of a desktop video magnifier, and the monitor displays a magnified image of the envelope with inverted colors.

NOTE

1   Orientation and Mobility specialists may also be certified by the National O&M Association (NOMA).

REFERENCES

ACVREP (2016). *Certified assistive technology instructional specialist for people with visual impairments certification handbook.* Academy for Certification of Vision Rehabilitation and Education Professionals

ACVREP (2016). *Vision rehabilitation therapists certification handbook.* Academy for Certification of Vision Rehabilitation and Education Professionals

Connors, E., & Abbott, P. (2020). *Braille and other forms of tactile of communication in foundations of vision rehabilitation therapy* (2nd ed., pp. 257–280). APH Press.

Connors, E., & Abbott, P. (2020). *Skills for independent living in foundations of vision rehabilitation therapy* (2nd ed., pp. 393–436). APH Press.

Dillon, C. F., Gu, Q., Hoffman, H. J., & Ko, C. W. (2010). Vision, hearing, balance, and sensory impairment in Americans aged 70 years and over: United States, 1999–2006. NCHS Data Brief 2010; 1–8. *National Center for Health Statistics*, www.cdc.gov/nchs/data/databriefs/db31.htm.

Fazzi, D., & Naimy, B. (2010). Orientation and mobility services for children and youths with low vision. In A. L. Corn & J. N. Erin (Eds.), *Foundations of low vision: Clinical and functional perspectives* (2nd ed., pp. 655–728). AFB Press.

Fugate, L., & Smith, L. (2020). *Personal management skills in foundations of vision rehabilitation therapy.* American Printing House.

Lee, H., & Ottowitz, J. (Eds.) (2020). *Foundations of vision rehabilitation therapy.* American Printing House.

Maffit, J., Lueders, K. S., & Vasile, E. (2022). Individualized glare evaluation for wearable filters. *Journal of Visual Impairment and Blindness*, 116(4), 552–557.

Ponchillia, P. & McMahon, J. (2020). *Psychosocial considerations in foundations of vision rehabilitation therapy* (2nd ed., pp. 34–64). APH Press.

Pynoos, J., Steinman, B. A., & Nguyen, A. Q. D. (2010). Environmental assessment and modification as fall-prevention strategies for older adults. *Clinics in Geriatric Medicine*, 26, 633–644.

Simon, L. (2008). Low vision and rehabilitation for older people: Integrating services into the health care system. *Community Eye Health*, 21(66), 28–30.

Siu, Y. T., & Presley, I. (2020). *Access technology for blind and low vision accessibility.* APH Press, American Printing House for the Blind.

Smith, A. J., & Geruschat, D. R. (2010). Orientation and mobility for adults with low vision. In A. L. Corn & J. N. Erin (Eds.), *Foundations of low vision: Clinical and functional perspectives* (2nd ed., pp. 833–870). AFB Press.

Smith, L., & Fugate, L. (2020). *Adaptive food preparation in foundations of vision rehabilitation therapy* (2nd ed., pp. 438–511). APH Press.

Smith, P. (2020). *Home management in foundations of vision rehabilitation therapy.* American Printing House.

Swenor, B. K., & Ehrlich, J. R. (2021). Ageing and vision loss: Looking to the future. *The Lancet Global Health*, 9(4), e385–e386.

Watson, G. R. (2001). Low vision in the geriatric population: Rehabilitation and management. *Journal of the American Geriatrics Society*, 49(3), 317–330.

Watson, G. R., & Echt, K. V. (2010). Aging and loss of vision. In A. L. Corn & J. N. Erin (Eds.), *Foundations of low vision: Clinical and functional perspectives* (2nd ed., pp. 871–916). AFB Press.

Whittaker, S. G., Scheiman, M., & Sokol-McKay, D. A. (2016). *Low vision rehabilitation: A practical guide for occupational therapists.* SLACK Incorporated.

26 Safety Considerations in Older Adults with Cognitive Impairment Including Dementia

Rachel Chalmer, MD

CONTENTS

DOI: 10.1201/9781003197843-26

26.1 INTRODUCTION

Dementia, or Major Neurocognitive Disorder (as the disease is now officially designated in the Diagnostic and Statistical Manual of Mental Disorders-5), currently affects an estimated 11% of older adults, or over 6 million Americans aged 65 years and older. This number is rapidly growing, due to the rising number of Americans living into older age. It is estimated that by 2050, over 12 million older Americans will be living with dementia. Life expectancy is also improving worldwide, and predictions are for a dramatic rise in global numbers of people living with dementia. Expectations are that the current count of nearly 47 million people living with dementia will increase by 10 million annually in the coming years.

There are no current proven treatments for dementia, and it is one of the fastest rising causes of death among Americans. Between 2000 and 2019, deaths from heart disease have decreased 7% while deaths from dementia have increased 145%.

Too often, dementia remains undiagnosed in affected individuals or is diagnosed late in the disease course. There are estimates that more than 60% of older adults with dementia have not received a formal diagnosis. Additionally, studies have shown that less than half of those with suspected dementia have received a comprehensive diagnostic evaluation. The problem of underdiagnosis costs precious opportunities to protect individuals living with dementia from the multiple safety complications that can arise from their disease.

Unfortunately, there are significant shortages predicted for the supply of health professionals trained to care for people with dementia, including Geriatricians, Psychiatrists, and Neurologists. Additionally, major shortages are projected for caregivers – both formal (trained aides) and informal (friends, family members, neighbors) – as well as for appropriate residential/institutional settings, nationally and globally. Healthcare leaders throughout the world are working to address this rising challenge.

All health professionals play a critical role in ensuring that safety issues which may arise in older adults with dementia are adequately recognized and addressed. Together, we can help protect vulnerable individuals from the preventable illness, injury, abuse and neglect, and other safety complications which reduce the quality of life and longevity of older adults living with dementia.

26.2 BACKGROUND ON DEMENTIA

26.2.1 WHAT IS DEMENTIA?

Dementia can generally be understood as a syndrome of brain failure, similar to how older adults might experience heart failure or kidney failure. As in the other conditions, the fact of having the organ failure does not imply only one specific disease process. Dementia is caused by one, or sometimes multiple, groups of disorders that lead to neurodegeneration (brain cell death) via various mechanisms. The most common causes for dementia include Alzheimer's disease, Cerebrovascular Disease and/or Stroke (causing "Vascular Dementia"), Lewy Body Dementia, Parkinson's disease Dementia, Frontotemporal Dementia, and Mixed Dementias in which there may be multiple diseases leading to dementia. In each of these cases, a mix of different genetic, molecular, cellular, and environmental factors combine to cause the neurodegeneration.

26.2.2 WHAT ARE THE SYMPTOMS OF DEMENTIA?

Neurodegeneration causes loss of cognitive (brain) functions, including short- and long-term memory, visuospatial skills, attention and planning abilities (also called executive function), language abilities, and social-emotional skills. These losses can lead to a wide array of common symptoms such as forgetting recent conversations, misplacing important items, having difficulty learning new information or learning to use a new device, changes in speech clarity or patterns, and displaying more emotional lability. Depending on the underlying cause of dementia, the losses in each of the cognitive domains may progress simultaneously, or certain cognitive domains may be more severely affected earlier than others. Whatever the cause of dementia, its overarching clinical manifestation is a progressive illness characterized by increasing loss of ability to manage one's own affairs. Generally, affected individuals first have trouble managing more complex tasks – the Instrumental Activities of Daily Living (IADLs) – such as paying bills, preparing meals, and managing medications. Eventually, the functional decline progresses to affect even simple tasks including the Activities of Daily Living (ADLs), such as getting dressed, eating, walking, and speaking.

Dementia is a disease; it reflects accelerated brain degeneration, and the symptoms are problematic and progressive. In contrast, in normal aging, there is a predictable relative decrement in cognitive functioning and the changes are usually minor, limited, and fairly stable. For example, an individual with normal aging may notice that it takes them a little longer to remember an unfamiliar person's name than prior, or to complete a task such as filing their taxes, reading a book, or cooking a favorite recipe. In normal aging, in contrast to dementia, older adults retain the ability to complete ADLs and IADLs on their own despite any mild cognitive symptoms. Between normal aging and dementia, there is an additional "stage" of cognitive aging called Mild Cognitive Impairment (MCI) in which the individual has noticeable symptoms of cognitive decline as well as objective cognitive loss (i.e., they score abnormally on clinical cognitive screening tests) but still retains the ability to compensate for their cognitive decline through adaptations. For example, an older adult with MCI may start to have difficulty memorizing their grocery list, and now need to write it down; they are thus able to compensate and complete grocery shopping on their own. For reasons that are not well understood, about 25%–50% of people with MCI never progress to having dementia. However, the majority of people with MCI do develop dementia and are therefore an additional high-risk group for which healthcare providers need to stay closely attuned to their cognitive and activity function over time.

In addition to the cognitive symptoms, *behavioral and psychological symptoms of dementia* (BPSD) are common, especially in the middle and late stages of dementia. Examples of BPSD include mood lability (fluctuations), anxiety, depressed mood, agitation, wandering, delusions, and hallucinations. BPSD are thought to be caused by a complex interplay of environmental factors (such as a caregiver with an aggressive communication style), cognitive factors (such as difficulty processing emotions and expressing needs), and prior experiences (such as a history of preexisting psychiatric disorder). Hence, the degree and type of BPSD vary greatly among patients with dementia, with the exception of some common trends in different types of dementia, such as hallucinations in Lewy Body Dementia, and disinhibited (socially inappropriate) behaviors in Frontotemporal Dementia.

26.2.3   How Is Dementia Diagnosed?

When a clinician evaluates an individual for memory concerns, they first elicit the history of the cognitive and functional decline, including what changes the person is experiencing and the timeline of those changes. They perform cognitive screening tests (such as a Mini-Cog, Mini Mental Status Exam, Montreal Cognitive Assessment, St. Louis University Mental Status test, or others) to document objective evidence of cognitive decline. If found, they then look for underlying medical conditions that could contribute to the cognitive decline and could be reversible, such as sleep disorders, alcohol use disorders, uncontrolled hypothyroidism, untreated vitamin B12 deficiency, untreated depression, and erratically controlled diabetes. If a patient is found to have a relevant history of cognitive and functional decline, objective findings with impairment on cognitive screening tests, and no other medical causes that better explain the symptoms, dementia is diagnosed.

A significant proportion of patients with dementia experience missed or delayed diagnosis. There are a multitude of reasons for this. Many people experiencing cognitive decline may not disclose symptoms to healthcare providers because of a lack of understanding of what is normal in aging and/or because of fear and stigma around a dementia diagnosis. For similar reasons, caregivers may also not share these signs with a patient's providers. Commonly, patients and their caregivers may think of cognitive and functional decline as "just part of getting older." Healthcare providers may also contribute to delays in diagnosis because of failure to ask older adults about cognitive and functional changes and may not routinely reassess patients as they age. Additionally, providers may not complete appropriate cognitive testing when symptoms do arise due to a lack of adequate training or education in in this area. Finally, health professionals may exhibit nihilism about the benefits of early diagnosis given the lack of curative treatment options for dementia.

The United States Preventive Services Task Force (USPSTF), widely considered the most important authority in preventive healthcare and screening, does not recommend universal screening of asymptomatic older adults for dementia. However, the USPSTF recommends early detection for older adults with symptoms of dementia. Early detection is beneficial, because it allows for patient-centered care in advance care planning, adjusting management of medical comorbidities, facilitated connections to support caregivers of people living with dementia, and prevention of safety complications that may arise. Because the symptoms of dementia can sometimes be subtle, normalized, or hidden by patients and families, it is critical that all healthcare providers who treat older adults be alert for symptoms and encourage affected individuals to speak with their Primary Care Provider (PCP), or a Geriatrician, Neurologist, or Psychiatrist, to confirm the diagnosis.

26.2.4   What Are the Stages of Dementia?

Mild dementia is often characterized by the person having lost the ability to manage IADLs such as shopping, cooking, and managing finances and medications, but retaining the ability to independently complete ADLs. Patients with dementia of moderate severity will be dependent on others for IADLs as well as dependent on others for assistance with some of the more complex ADLs such as bathing and dressing. Patients with severe or advanced dementia require assistance with all the IADLs and ADLs described above as well as with toileting, continence, transferring, and eating.

In patients with simple Alzheimer's disease dementia, their functional and cognitive loss generally precedes in a predictable pattern over ten years depicted by the Functional Assessment Staging scale (FAST) scale. However, in patients with other dementias or mixed dementias, the pattern of loss may precede in a nonlinear fashion and the progression may happen more quickly. For example, Lewy Body Dementia tends to progress to end stage over 3–8 years, and patients often develop more serious complications, such as falls, earlier in the course of the disease. In Vascular Dementia, the disease can have periods of more rapid progression of cognitive and functional loss followed by periods of relative stability (a so-called step-wise decline), and it can be more difficult to predict exactly how and when dementia will progress.

26.2.5   What Treatments Are Available for People Living with Dementia?

Dementia remains, unfortunately, a progressive disease without a cure. There are several medications, discussed below, which are approved by the United States Food and Drug Administration; these medications at best slow or temporarily pause the cognitive decline. The mainstays of treatment for dementia are the management of symptoms along with robust supportive care. Specific components of this type of care include determination of prognosis and goals of care; tailored management of medical comorbidities in light of prognosis and goals; medical and psychosocial management of cognitive and behavioral symptoms; optimization of physical function with reduction of risk of falls and injury; care coordination with legal and financial planning; and caregiver support.

Multidisciplinary, team-based models of care can most effectively and efficiently achieve care goals for people living with dementia. Disciplines involved include Geriatrics, Neurology, Neuropsychology, Geriatric Psychiatry, Rehabilitation Medicine, Social Work, and Palliative Care. These types of care models exist but are rare. A good option, for the majority of older adults with dementia, who do not have access to a team-based model of care, is a combination of follow-up with a

PCP for ongoing management of chronic medical conditions, along with intermittent consultative care from a specialist trained in dementia. Specifically, individuals with dementia who have atypical symptoms or characteristics – including early age of onset (before age 65), early or rapid language loss, rapid progression of dementia (normal to significant impairment in less than one year), or additional prominent motor symptoms (such as falls or tremor) – should be referred for Neurology consultation. A consultation with a Geriatrician may benefit older adults with dementia who have additional complications including multiple medical comorbidities, the use of greater than five medications, ongoing weight loss or nutritional issues, or caregivers with high levels of burden. A consultation with a Geriatric Psychiatrist may be helpful for people living with dementia who have severe neuropsychiatric symptoms.

For all older adults with dementia, and their caregivers, connection with community resources is critical to help manage day to day challenges. Although resources may vary by location, there are several that are widely available. Each state has a Department for the Aging, funded by the Older Americans Act, which serves as a nexus of local aging-related resources for housing and economic assistance, prevention of elder abuse/neglect, and social day programs. In addition, the Alzheimer's Association, with local chapters across the country and the world, offers education and training, advocacy, and support for people living with dementia and their caregivers. For example, they host regular support groups for different types of caregivers of people living with dementia as well support groups for patients themselves.

Pharmacologic treatment currently has a limited role in dementia. There are two classes of medications (N-Methyl D-Aspartate [NMDA] receptor antagonists and acetylcholinesterase inhibitors) available to help treat the cognitive symptoms of dementia. These medications, such as donepezil and memantine, may temporarily slow or pause the cognitive decline for about six months. Occasionally, families and patients find these pharmacologic interventions have meaningful, if limited, benefit. However, they also may have significant side effects with weight loss being among the most common and concerning. The most important and useful strategies for dementia management are non-pharmacologic interventions such as patient-centered care strategies and caregiver support interventions, as discussed above.

26.2.6 Dementia as a Fatal Disease; Advance Care Planning; and the Role of Palliative and Hospice Care

As dementia progresses to end stage and patients have trouble with the most basic activities, they are usually bedbound, frail, and vulnerable to life-threatening complications. These can include dysphagia (difficulty swallowing) often with aspiration (inhaling oral contents such as food or saliva into the lungs); susceptibility to infections including pneumonia (sometimes from aspiration) or urinary tract infections; pressure injury (skin breakdown over bony prominences) that can develop into painful or infected ulcerations; ongoing weight loss with protein-calorie malnutrition; and repeated hospitalizations as a result of the above. Affected individuals can die from any of these complications. At the end stage of dementia, patients' prognosis is generally limited to between six and eighteen months, regardless of any aggressive medical interventions that may be attempted when they develop the above complications.

For many individuals with dementia and for their families, understanding the likely progression of dementia can be scary and painful, but also compassionate and supporting. Given that the progression to end stage dementia removes an individual's ability to understand and engage in discussions around their values in life and their goals for their medical care, it is critical to give people living with dementia and their families the opportunity to discuss these topics in advance. In this way, an individual's values and preferences can be respected and families can feel supported in choices they may make for their loved one at the end of life. This issue is another reason that timely diagnosis of dementia is so important. Advance care planning can be done by patients and their family members, often in consultation with their physician. Preferences can be documented through completion of a Health Care Proxy (the legal document naming preferred persons to serve as an individual's decision makers when he or she is no longer able to make medical decisions) and MOLST (Medical Orders for Life-Sustaining Treatment) or POLST (Physician's Orders for Life-Sustaining Treatment) forms. These documents help ensure that an individual's preferences can be easily understood and followed by any health professional or decision maker.

Individuals in the end stages of dementia can benefit from palliative and hospice care. Palliative care refers to services and treatments geared toward ameliorating discomfort and suffering associated with any medical illness. Hospice is a set of services and treatments and an overall approach to medical care for individuals in their last six months of life which is covered under Medicare, Managed Medicare, and most other insurances. Hospice care incorporates palliative care but also provides more comprehensive wrap-around services including custodial care (assistance with ADLs), chaplain and social work services, medical equipment, and visiting nurse services. Hospice care can be provided in different settings including at home, in a nursing home, on a hospice unit within a hospital, or in a hospice residence. Hospice care provides an alternative to hospitalization (with its associated burdens and discomforts) and aggressive medical care and is a good choice when older adults or their families prefer to prioritize comfort and quality of life over longevity at the end of life. Individuals with end-stage dementia, especially when they begin to develop any of the complications mentioned in the prior paragraph, are often appropriate candidates for hospice care. Families and caregivers should be encouraged to explore this option, as many do not know that hospice care may be an option at the end of life.

26.3   SAFETY ISSUES IN PEOPLE LIVING WITH DEMENTIA

The ways that individuals are affected by dementia change with disease progression; therefore, the safety issues can also be different. Below, the various safety concerns are described according to early, middle, and late stages of dementia. However, it is important to note that the risks described in each stage can also present and persist at other stages of dementia, and ongoing surveillance is required.

26.3.1 SAFETY ISSUES IN EARLY DEMENTIA

In this stage, people with dementia are often still living independently. As such, this is perhaps the most vulnerable stage. There are four major areas in which they are vulnerable to safety risks: transportation, financial management, nutrition, and healthcare management.

26.3.1.1 Transportation

In early stages of dementia, independent travel can become unsafe. Whether individuals are traveling via public transportation (subway, bus, train) or driving themselves, they may have difficulty with navigation. They may get lost even in previously familiar surroundings. Those driving cars may also, due to attentional deficits caused by dementia, have a higher risk of accidents. Individuals using public transportation who are having difficulty navigating may, as a result, be more distracted and then more vulnerable to accidental injuries, or even to theft or assault.

Health professionals have a critical role in identifying transportation risks and counseling around cessation of driving or independent travel. Health professionals should routinely ask older adults about how they are managing transportation and ask specific questions around any safety incidents, since unsafe transportation habits can be an early sign of previously undiagnosed dementia. Also, because giving up travelling independently can be one of the most painful stages for older adults, caregivers may delay addressing it, despite witnessing risky behaviors. Often, the dementia-affected individual has reduced insight into their limitations and so may not understand why caregivers or clinicians move to restrict them. Caregivers may feel uncomfortable limiting another adult's autonomy, especially that of a spouse or parent. Caregivers also may not understand the way that dementia can affect insight and decision-making capability and often require education and reassurance around their role in preventing unsafe travel. Clinicians can collaborate with caregivers and patients in problem-solving to identify alternate safe means of transportation and new routines.

Legal requirements for healthcare providers to report potentially unsafe drivers due to dementia vary by state; providers should be familiar with local laws. Clinicians should also be familiar with available local programs that can formally evaluate an individual's driving safety; use of a formal assessment can help improve understanding of the need for driving cessation. More information on driving issues in older adults can be found in Chapters 13 and 14.

26.3.1.2 Financial Management

In early stages of dementia, affected individuals experience a decline in their ability to manage finances. They may have initially innocuous-appearing difficulties such as letting the mail pile up, misplacing or forgetting to pay a bill, or neglecting to balance their checkbook. When these difficulties are left unmanaged, they can increase in frequency and significance. Individuals can accumulate debt, exacerbated by late fees. They can experience power, gas, telephone, or water shutoffs, leaving them even more vulnerable. Furthermore, older adults, particularly those with dementia, are at some of the highest risk for identity theft and other financial scams (see Chapters 28 and 29 for more information on financial safety among older adults).

Health professionals play a key role in identifying financial management risks in people living with dementia. Studies have shown that a very early sign of cognitive impairment can be increased vulnerability to telephone or email financial scams. Health professionals should routinely ask older adults about how they are managing finances, including specific questions around any unusual occurrences. Caregivers and older adults may dismiss these changes as normal aspects of aging, and not report them to a clinician unless asked.

Health professionals can also counsel older adults and their caregivers around protective steps to maintain safe financial management. First, people living with dementia should be encouraged to create a Power of Attorney (POA); this document provides the named agent significant and broad ability to make financial decisions. POA documents can be created through online services, usually with in-person certification by a notary public, or they can be generated with an attorney. POAs can only be generated by the person granting the power; they cannot be retroactively created on behalf of someone who has lost the ability to make this decision for him or herself. A POA is critically important for an older adult with dementia, because future financial needs cannot always be fully anticipated. These questions may include how many years they will live with dementia and need personal care, and whether they will need nursing home placement. A POA document can help avoid cost and delay in meeting these needs in situations when the person can no longer engage in financial planning for him or herself.

In addition to completing a POA, older adults with dementia should be advised to add a trusted person as a comanager on existing financial accounts. This can facilitate ongoing monitoring for any evidence of financial abuse, theft, or mismanagement. This can also facilitate logistical ease in ongoing management of accounts, such as easily switching a utility account in the event of a move.

At times, when complex financial issues emerge, and an older adult with dementia has not completed a POA while they are still able, a court-appointed guardian is required. The lengthy and complex guardianship process is best avoided by both timely diagnosis of dementia and counseling on financial management.

26.3.1.3 Nutrition

In early stages of dementia, affected individuals generally retain the ability to independently feed themselves in the most basic sense: they can get food into their mouth, and safely chew and swallow. However, they may start to have difficulty with the many other tasks required to maintain nutrition.

First, they may have trouble procuring food. They may be unable to effectively create a meaningful shopping list, or have challenges travelling to the store or managing payment for goods. Second, they may have problems preparing food. They may attempt to cook, but forget to turn off the stove, causing dangerous kitchen fires. They may forget how to prepare even their most oft repeated recipes. They may develop unsafe food handling practices, such as leaving food out too long, and hence be more vulnerable to foodborne illness. Third, they may become disengaged from the acts of cooking and eating due to behavioral symptoms associated with dementia, such as depressed mood or apathy. Each of these changes in nutrition-related activities can lead to unintended weight loss and/or malnutrition.

Health professionals should routinely monitor older adults' weight changes, including asking specific questions around food intake and dietary habits when weight loss is noted. As with transportation and financial management risks, unexplained/unintended weight loss can often be a tell-tale sign of an undiagnosed dementia. Additionally, unintended weight loss in a person with dementia is particularly important to recognize because it portends an increased risk of death from any cause.

In general, nutrition-related safety risks can best be addressed by providing additional support around mealtimes. This can take various forms, including home-delivered prepared meals (such as from Meals on Wheels, a DFTA program); family members taking over shopping; and/or a home attendant who cooks and sits down to meals with the dementia-affected individual.

26.3.1.4   Healthcare Management

Older adults living with dementia generally have decreased ability to manage their own healthcare. First, dementia compounds the challenge of managing a complex medication regimen. Even older adults who previously had an organized system to help with medication adherence may begin to miss doses, develop confusion on what their regimen is supposed to be, forget to refill medications, or accidentally overdose themselves in their effort to make sure they take their medication. Helpful interventions may include caregivers pre-filling a weekly pill organizer; pharmacies providing pre-filled medications on a weekly or monthly basis ("blister packs"); alarms for reminders to take pills; and ultimately having a caregiver completely take over medication administration.

Older adults with dementia may have difficulty keeping track of their medical appointments; written reminders, a calendar, and alarms can help. Challenges with appointments often herald a stage in which the affected individual should also be accompanied by a caregiver or family member for all medical appointments. This person's role includes ensuring follow-through and adherence to changes in medical management.

26.3.2   SAFETY ISSUES IN MIDDLE STAGES OF DEMENTIA

Individuals in the middle stages of dementia are at a particularly dangerous juncture. They may retain basic physical abilities, yet often lack insight into their limitations. They may have profound difficulty carrying out household tasks on their own and may develop difficulty safely completing personal hygiene including bathing and dressing. Hence, accidents and injuries that sometimes end in tragedy are more common in these stages. Adequate education of caregivers and families about these potential risks and ongoing supervision and support of the person with middle-stage dementia can potentially prevent many of these disastrous outcomes.

26.3.2.1   Wandering

Wandering refers to the phenomenon of a person living with dementia traveling somewhere in an unplanned way, often having difficulty returning home on their own. Wandering episodes can vary from brief and local (knocking on other apartment doors on their hallway, trying to find their way back to their own apartment) to extended and distant (a car trip to visit a relative across town ending in a person being located by police two states over). Some people living with dementia have frequent wandering episodes; others have a one-time episode without recurrence. Interventions such as having the person with dementia carry a cell phone do not always mitigate the risk, because often the same cognitive dysfunction that leads to the wandering also leads the person to forget to turn on, charge, or bring the cell phone with them when they leave home. Additional technologies may be attempted including a "Wander Alert" or other tracking device, cameras, and door and bed alarms. However, the onset of wandering is usually an indicator that the person with dementia is best cared for in a more supportive living environment with a caregiver's presence around the clock.

26.3.2.2   Falls

Falls are common and problematic among older adults in general, but even more so in those with cognitive impairment. In certain types of dementia, such as Dementia with Lewy Bodies, falls can be particularly frequent and severe. However, across all types of dementia, falls risk increases in the middle stages of dementia when the individual often has profound cognitive impairments but retains basic physical abilities. For example, a person with impaired balance may forget to use their cane or walker and be at higher risk for falls. In the mid stages of dementia, disruptions in the sleep-wake cycle become more frequent, and the person may start walking around in the middle of the night, when it is dark, and there is often less supervision, both leading to increased risk of falls. Additionally, in the event of a fall, a person with mid-stage dementia may not remember the steps required to summon medical assistance and may thus experience worse fall complications. See Chapters 4 and 9 for additional information on falls risk in older adults.

26.3.2.3   Other Accidents/Injuries

If not already done during early stages of dementia, it becomes even more important during middle stages to take steps to mitigate potential risks in the home environment. This is because of the loss of insight and judgment, and sometimes motor control, along with the behavioral dysregulation (more in next section), that can happen during middle stages of dementia. Therefore, common household items that can cause significant harm if wielded unsafely – including knives, other sharps, and household chemicals with toxic potential – should all be stored securely and made inaccessible to the person with dementia. Further, more than one-third of older adults own a firearm or live in a household with someone who owns one. Therefore, it is critical that inquiry about the presence of firearms and counseling about safe firearm storage (unloaded, ammunition stored separately, and all stored securely) be a routine part of care for older adults with dementia.

26.3.2.4   Worsening Behavioral and Psychological Disturbances

In the middle stages of dementia, BPSD often peak in intensity and disruptiveness and can result in individuals incurring significant safety risks. Because of paranoia or delusions, they may be more prone to wandering episodes as they feel they are not at their home and need to leave to "get home." They may be more prone to agitation or aggressive behavior, which rarely is violent, presenting a safety risk to themselves and to caregivers. In these extreme cases, law enforcement is sometimes involved.

Severe behavioral disturbances leave an individual with dementia more vulnerable to abuse or neglect as their caregivers struggle under the stress caused by the behaviors. Individuals with severe behavioral disturbance are also more likely to be more heavily medicated for the symptoms, which can bring its own risk of drug-related side effects.

Addressing BPSD is a complex challenge. As these symptoms progress, caregivers and people living with dementia need ongoing support and education. This can be a critical time to connect caregivers with support organizations, and with respite (temporary external caregiver) services or a formal, long-term home attendant arrangement. Importantly, the availability of a long-term home attendant is dependent on a patient's financial and insurance status and informal caregivers (e.g., family and friends) deliver a large amount of unreimbursed supervision.

26.3.3   Safety Risks in Late Stage/Advanced Dementia

26.3.3.1   Caregiver Burnout

Dementia is generally a slowly progressive disease and caregivers can find themselves serving in this role for years on end. As dementia progresses to late stages, behavioral issues often persist including significant sleep disturbances, and caregivers may experience increasing levels of burden or burnout. Caregivers of people living with dementia are themselves at elevated risk of worse health outcomes, simply due to the burden of caregiving. They may neglect their own medical needs and they are at higher risk of depression and poor immune system function than non-caregivers. Older adults living with dementia, who are cared for by caregivers with high levels of burnout, are more likely to end up being institutionalized for ongoing care. Caregivers' burden and burnout can be addressed when their needs are assessed and then attended to through education, support, and concrete services such as respite. See Chapter 20 for more information about caregiver self-care.

26.3.3.2   Elder Abuse/Neglect

When high caregiver burden persists and caregivers are not able to effectively manage, they may be more likely to carry out elder abuse or neglect of their loved one living with dementia. Elder abuse/neglect is extremely common and comes in multiple forms and levels of severity. Health professionals should be alert to potential red flags for elder abuse including a person with dementia having significant weight loss or unexplained bruises. Additional signs of elder neglect include the affected individual not being brought to medical appointments, or given prescribed medications; not being provided adequate support for ADLs; or evidencing poor personal hygiene. In cases of elder abuse/neglect, an interdisciplinary team approach combining clinicians, social workers, and community service organizations can help protect and restore the safety of the affected individual through providing additional services and supports to the caregiver and patient. In rare severe cases, the state may move to name a new guardian for an elder abuse/neglect victim and forbid the original caregiver from being involved in their care. See Chapter 29 for additional information on elder abuse.

26.3.3.3   Burdensome or Unwanted Medical Interventions

As discussed earlier in this chapter, dementia is a fatal disease. Too often, people living with dementia and their families or caregivers do not have discussions around their preferences for end of life care (so-called advance care planning). This frequently leads to situations in which individuals receive aggressive, burdensome medical care at the end of life which is not in keeping with their wishes. Health professionals have a critical role in helping persons with dementia and their families, understand and address the decisions around their end of life care.

One specific potentially burdensome medical intervention is a feeding tube. When individuals with end-stage dementia lose the ability to be nourished adequately or safely by oral feeding (due to dysphagia and/or aspiration), families sometimes consider feeding tube placement to facilitate nutrition. Although it seems intuitive that feeding someone would help them live longer, in general this is not the result seen in patients with end-stage dementia. Large clinical trials have demonstrated that feeding tubes do not prevent aspiration pneumonia or other infectious complications of end-stage dementia. And feeding tubes can be burdensome, as they can become dislodged (accidentally or by the patient pulling it out, sometimes repeatedly) and require hospitalization or repeat procedures for reinsertion. Although it can be very difficult for families to grapple with the idea of allowing their loved one to enter a period when they have little or no oral intake, the support provided by a palliative care and hospice interdisciplinary team can provide the support needed at the end of life.

26.4   HEALTH DISPARITIES AS A SAFETY ISSUE IN DEMENTIA

Racial and ethnic health disparities exist in dementia prevalence and incidence. For example, in the United States, Black older adults are about twice, and Hispanic older adults about 1.5 times, as likely to have dementia as White older adults. At the same time, studies have demonstrated that Black and Hispanic older adults with dementia are even more likely than White older adults with dementia to experience a lack of or a delay in diagnosis. These disparities suggest that older adult members of racial and ethnic minority groups may be even more vulnerable to some of the safety complications arising from dementia, because they and their caregivers may not have received adequate and timely anticipatory guidance and education. Additional research is ongoing in this important area. See Chapter 31 of this book for further discussion on health equity in care for older adults.

26.5 INTERVENTIONS TO ADDRESS SAFETY ISSUES IN DEMENTIA

Successfully identifying and managing safety risks in people with dementia involves a combination of clinical and psychosocial assessment, education, connection to concrete services, and caregiver support. Interprofessional teams utilizing creative strategies, shared decision-making with the person with dementia and their caregivers, and ongoing problem-solving can optimize success.

For example, a PCP may identify a safety risk, such as incorrect use of blood pressure medications. A pharmacist member of the team can work with the PCP to simplify the medication regimen and then collaborate to devise a safer strategy, such as blister packs. The PCP can request a visiting nurse service to make a home visit to review all of the medications and make sure the blood pressure is at the desired goal on the updated medications. Finally, a social work consultation can facilitate accessing and establishing a formal caregiving schedule so that there is always someone in the home to assist with supervision of medications.

In addition to addressing the specific interventions needed to mitigate safety risks, attention must be paid to the ethical issues that may arise. Frequently, people living with dementia have reduced insight into their own cognitive and functional limitations. As such, they may strongly express desire for autonomy (e.g., continuing to drive) even when it is clear that their desired course of action would incur a high risk of harm to themselves or others. Caregivers and clinicians may face difficulty understanding when and how to override the autonomy of the person living with dementia. They may be unclear about who can, or should, be the decision maker; they may wonder whether the patient has capacity to decide for him or herself. They may wonder how to simultaneously maintain respect for the individual even when respect for their autonomy cannot be maintained. When conflicts arise between the desires of patients, the desires of their caregivers, and the responsibility of health professionals to ethical practice, all parties should engage in clarifying conversations. Health professionals should foster the dignity of the person living with dementia by involving them in shared decision-making as much as feasible, while also clearly explaining to them and to their caregivers where and why limits are necessary on the individual's decisions. In particularly complex situations, the advice of a clinician trained in decision-making capacity assessment, or a bioethicist, can be helpful. See Chapter 1 for additional discussion of the issue of considering risk in older adults.

26.6 SUMMARY

Older adults living with dementia are vulnerable to multiple and varied safety risks, from financial mismanagement, to wandering, to inappropriate feeding tube placement, to elder abuse and neglect, and many more. These dangers begin to present very early in the disease course and can persist and change as the disease progresses. Hence, attention to safety risks for older adults with dementia is not a "one-and-done" proposition. It is an ongoing process of assessment (clinical and psychosocial), intervention (with education, connection to concrete services, and caregiver support), and repetition. Interprofessional team approaches to care are often the most effective way to manage safety risks for older adults with dementia. All members of the interprofessional team can contribute to this complex and important effort.

BIBLIOGRAPHY

1. American Psychiatric Association. *Diagnostic and Statistical Manual of Mental Disorders, Abreviado (DSM-V)*. Vol 80; 2014.
2. Association A. 2021 Alzheimer's disease facts and figures. *Alzheimer's & Dementia: The Journal of the Alzheimer's Association.* 2021;17(3):327–406. doi:10.1002/alz.12328
3. Bessey LJ, Walaszek A. Management of behavioral and psychological symptoms of dementia. *Current Psychiatry Reports.* 2019;21(8). doi:10.1007/s11920-019-1049-5
4. Cordell CB, Borson S, Boustani M, et al. Alzheimer's Association recommendations for operationalizing the detection of cognitive impairment during the Medicare Annual Wellness Visit in a primary care setting. *Alzheimer's and Dementia.* 2013;9(2):141–150. doi:10.1016/j.jalz.2012.09.011
5. Jennings LA, Tan Z, Wenger NS, et al. Quality of care provided by a comprehensive dementia care comanagement program. *Journal of the American Geriatrics Society.* 2016;64(8):1724–1730. doi:10.1111/jgs.14251
6. Lennon JC. Review of dementia reimagined: building a life of joy and dignity from beginning to end by Tia Powell. *Journal of Alzheimer's Disease.* 2019;72(4). doi:10.3233/jad-190611
7. Livingston G, Huntley J, Sommerlad A, et al. Dementia prevention, intervention, and care: 2020 report of the Lancet Commission. *The Lancet.* 2020;396(10248):413–446. doi:10.1016/S0140-6736(20)30367-6
8. Lo B. *Resolving Ethical Dilemmas: A Guide for Clinicians*; 2013.
9. Manly JJ, Espino DV. Cultural influences on dementia recognition and management. *Clin Geriatr Med.* 2004 Feb;20(1):93-119. doi: 10.1016/j.cger.2003.10.004.
10. Oh ES, Rabins PV. Dementia. *Annals of Internal Medicine.* 2019;171(5):ITC33–ITC46. doi:10.7326/AITC201909030
11. Parker K, Horowitz J, Ingielnik R, Oliphant B, Brown A. America's complex relationship with guns. Pew Research Center. June 22, 2017. Accessed at www.pewsocialtre nds.org/2017/06/22/americas-complex-relationship-with-guns on April 7, 2022.
12. Patnode CD, Perdue LA, Rossom RC, et al. Screening for cognitive impairment in older adults. *JAMA.* 2020;323(8):764. doi:10.1001/jama.2019.22258
13. Reisberg B. Functional assessment staging (FAST). *Psychopharmacol Bull.* 1988;24(4):653–9. PMID: 3249767.
14. Tsoy E, Kiekhofer RE, Guterman EL, et al. Assessment of racial/ethnic disparities in timeliness and comprehensiveness of dementia diagnosis in California. *JAMA Neurology.* 2021;94158. doi:10.1001/jamaneurol.2021.0399.
15. Valcour VG, Masaki KH, Curb JD, Blanchette PL. The detection of dementia in the primary care setting. *Archives of Internal Medicine.* 2000;160(19):2964–2968. doi:10.1001/archinte.160.19.2964

27 Legal Safety Tools

David M. Godfrey and Charles P. Sabatino

CONTENTS

27.1 INTRODUCTION

Most people don't think of legal tools in terms of safety, but the legal planning tools discussed in this chapter have a great deal to do with protecting safety, specifically the safety and security of our ability to live our lives according to our own values, goals, and wishes, even in the face of injury or illness, and ensuring that we are not taken advantage of financially or personally. When no planning is done and one's capacity to manage his or her own affairs declines significantly, the default is the use of guardianship or conservatorship through the courts. That option results in significant, if not total, loss of control over one's own life. Consider what it means to have no say in where you live, with whom you can spend your time, or on what you can spend money. Thus, legal safety is just as important to aging well as one's physical safety.

Legal safety focuses on planning strategies that maximize the likelihood that future lifetime decisions will be made according to one's own values, priorities, directions, and preferences. We can't plan for everything, but we can help manage life's unknowns by talking openly about what matters to us and what we would want

DOI: 10.1201/9781003197843-27

most if we became seriously ill and then using the legal tools discussed in this chapter to provide the legal means to make that happen.

There are two broad areas to think about in legal planning – decisions affecting one's finances and decisions affecting one's health and personal care. The goal is to strengthen the security, safety, and integrity of individual decision-making while living. The chapter focuses on security, safety, and quality of life issues while one is living and not on decisions about making a will or other after-death estate plans. Accordingly, the chapter is divided into financial decision-making and health care decision-making, but with an initial discussion of a tool most adults use naturally but seldom think about – relying on formal and informal supporters in making decisions. We will explain why this is important.

Whether we ask a friend's opinion on what to wear or consult a professional mechanic about the funny sound we are hearing from our car's engine or seek our physician's advice on a health care decision, we all seek out and use input from others in countless ways nearly every day. In circumstances where our decision-making capacity is diminished, the role of those supporters and advisors becomes even more important. Without a meaningful network of supporters and advisors, we are left in isolation, and that is a sure formula for vulnerability and risk of exploitation. The reality is that older adults are at greater risk of social isolation compared to younger adults.[11] Therefore, more intentional efforts to identify trustworthy supporters become more important as we age.

27.2 SUPPORTED DECISION-MAKING

The law has adopted the terminology "supported decision-making" or SDM to capture these supportive relationships that assist in decision-making. It is increasingly discussed as an alternative to guardianship and conservatorship.[7] Several states have also given legal recognition to formal supported decision-making agreements.

There is no single definition or model of SDM, but a commonly used one describes it as

> a series of relationships, practices, arrangements, and agreements, of more or less formality and intensity, designed to assist an individual with a disability to make and communicate to others decisions about the individual's life.[2]

As this definition suggests, SDM is not a single tool or a single process, but really more of a paradigm for collaboration in decision-making with persons one trusts. A supporter helps you to understand the issues and choices in whatever kinds of matters you have identified as needing support. The supporter may ask questions about the topic and your wishes, provide explanations in language you understand, help explore the risks and benefits of options, and help communicate your decisions to others to the extent you want and need that help.

The role of the supporters or advisors is to support and advise and not to replace or supplant the wishes of the person. When these relationships are made formal by a written SDM agreement, the document helps define and communicate to others what decisions you want the supporter to help you with, how you want them to help you, and for how long you want their help.

Whether formal or informal, SDM can help in managing decisions with one's doctors, hospitals, banks, schools, service providers, governmental agencies, or others. Supporters need to understand that in their role, they do not make decisions for you but only help facilitate your decision-making in the way and to the extent that you want. Because the universe of daily decisions is vast, a network of supporters for different purposes is most effective and safer for preserving one's autonomy. The stronger one's social network of supporters, the less vulnerable one is to the undue influence of anyone. The challenge is to consciously develop that network of supporters, which may include family members, neighbors, caregivers, church members, or even the mailman. The National Resource Center for Supported Decision-Making provides useful guidance on the practice of SDM, state laws on SDM, and SDM agreements.[17]

The paradigm of SDM is applicable to any decision. But, there may come a time when, even with SDM, we are unable to make the decisions required to manage some or all of our personal affairs. The tools described below are grounded in the same person-centered philosophy of SDM in that they are always guided by your values, priorities, instructions, and preferences. The difference is that they provide a legally recognized agent or fiduciary to carry out your wishes or act as decision maker on your behalf when needed.

27.3 FINANCIAL DECISION-MAKING

Financial planning starts with assuring that income is received, and expenses are paid, using a variety of tools, including direct deposit, automatic payment and power of attorney (POA), with all of these ideally structured to provide oversight and account-ability. Planning for financial decision-making is especially important because older persons experience higher rates of abuse and exploitation than other adults. The National Center on Elder Abuse research summaries tell us that one in ten older adults 60 years of age living in the community experienced some form of abuse in the prior year, and a substantial portion of that abuse is financial exploitation.[20] While not a pleasant topic, the basics of identifying, preventing, and intervening in elder abuse is a topic in which all professionals need training.[6]

27.3.1 PUTTING MONEY ON AUTOMATIC PILOT

It has never been easier to automate our financial lives and doing so creates a safety net for times when we are distracted from managing our affairs and reduces the risk of unpaid bills or financial exploitation.

Direct deposit is a tool to assure that income is received and deposited. Nearly all income can be received by direct deposit. Direct deposit is mandatory for Social Security, other government pensions and benefits. Direct deposit eliminates delays in check deposits and loss or theft of checks. Direct deposit helps to assure that money is in the account when bills come due. Direct deposit is set up through the entity paying the income or benefit, by providing the bank routing number and account number and normally by providing a voided check. For defined contribution retirement plans, such as 401k and IRAs, periodic distributions can be set up with the financial institution holding the account. Some have an option for setting up distributions when the balance in a spending account reaches a predetermined low point. Income from rents or businesses can be set up to be deposited through financial institutions or management services.[1]

Automatic payment pays reoccurring bills directly and automatically from a bank account. This is set up either with the merchant, service provider, creditor, or with the person's financial institution. When the bill comes due, a notice is sent. The person has a few days to review and object to the invoice. Unless a question is raised, the invoice is paid by direct transfer. Nearly all bills can be set up this way. It eliminates issues with lost or late mail, lost or stolen checks, that can result in late payment or nonpayment when an adult is traveling, distracted, or unwell. This won't cover every expense, but it will cover most.

Both direct deposit and automatic payment should include the safety tools of oversight. Someone should verify that income due is received in the expected amounts, and someone should review invoices and bills to assure that the amounts are correct and that they are paid in a timely manner. The individual can normally do this easily. But, it is also advisable to have a third party (a supporter) able to check. Financial institutions are increasingly able to set up read-only access to financial accounts and statements. Read-only access does not allow the person to transact business, but only to look at the accounts and monitor. The notices or invoices for bills being paid by automatic payment are most often sent by email. Sharing access to the email account those messages come to with a trusted supporter allows the supporter to monitor the bills and highlight anything unexpected.

Another tool is *direct or online payment*. With direct or online payment, a notice or invoice is sent when the bill is due, and someone must approve payment. The approval is generally done online or on an app. This provides the oversight step up front, but at the risk that if the payment is not approved the invoice or bill is not paid. If the person is traveling, distracted, or unwell, this can result in missed payments. Delegating access to approve online payments can be done, in doing so you are authorizing a third party to transact business and that carries with a slightly higher risk, than automatic payment.[23]

Another tool for assuring payment of bills is *daily money management or bill paying services*. Originally these were hands on services that opened paper bills, reviewed, and approved them, and sent paper checks. Online banking and bill paying have increased the efficiency of these services with nearly everything being done electronically reducing costs in most cases. The actual cost will vary with the number of accounts that need to monitor and pay. These services provide oversight, accountability, and should be bonded against any misconduct. To find a reputable agency that provides daily money management, contact your local Area Agency on Aging, which you can locate through the Elder Care Locator.[4]

27.3.2 Use Joint Ownership Carefully

Joint ownership of property with a "right of survivorship" is a common ownership arrangement used both as a property management tool and a will substitute. Under this arrangement, all owners have access and control of the property, and a surviving joint owner becomes the sole owner (or owners if more than one) when one joint owner dies.

Bank accounts are commonly held this way, as well as title to automobiles, a house, or stocks and bonds. But there are substantial risks in joint ownership, especially if you are making someone a joint holder merely as a convenience for managing money.[21]

Consider these risks:

- Joint owners of financial accounts have complete access to the money in the account. If untrustworthy, they can take it all.
- The funds in the account may be available to pay the joint owner's debts.
- The owner of a car may be liable if the joint owner gets in an accident.
- If you have provided in your will that the funds in the account are to go to person other than the joint owner, the joint ownership may defeat your plan.

Use joint ownership only when the other parties are, in fact, co-owners of the account or property with you. If you are using joint ownership as a convenience for management of a bank account, make sure the account assets are limited to the amount necessary to the purpose of the account, for example, paying regular bills. Don't use it to provide control over major assets. Use a POA or trust, described below, instead.

If the intent is to limit joint ownership to proportional shares so that a survivor's share and control does not change, then that has to be made clear in the document that creates the ownership interests. This kind of joint ownership is called a "tenancy in common." For example, if there are three owners of a piece of property as tenants in common, and one owner dies, the one-third interest of the deceased passes according to that person's will or recognized will alternative; otherwise, it passes by intestate succession, that is, the state's default rules on who inherits property when there is no will or other applicable estate planning document.

27.3.3 Add a Trusted Contact to Accounts

A trusted contact is a person you authorize your investment firm or company to which you make essential monthly payments (e.g., gas or electric bills) to contact in limited circumstances, such as if there is a concern about activity in your account and they have been unable to get in touch with you. If cognitive difficulties develop in older age, the first signs of it often show up in poor financial decision-making.[19] By designating a trusted contact, you are authorizing the firm or company to contact someone you trust and disclose information about your account only in limited circumstances, such as in circumstances that suggest financial exploitation or loss of financial capacity.

A trusted contact may be a family member, attorney, accountant, or another third party. You will want to name someone who will respect your privacy and know how to handle the responsibility. The trusted contact does not have any authority to act on your behalf or make changes on your account. Their role is to help check on the status of the account holder, confirm current contact information, health status, or the identity of any legal guardian, executor, trustee, or holder of a POA who may now be helping to handle the affairs of the individual. FINRA, the agency that oversees broker-dealers, provides a helpful information about the use of trusted contacts for investment accounts.[9]

27.3.4 APPOINT AN AGENT IN A POWER OF ATTORNEY

An important tool to maintain control and safety is a POA. It is particularly helpful in managing titled property, like financial accounts, a house, or a car when an individual becomes incapacitated for any reason. When a person loses the ability to make financial decisions, no one has automatic legal authority to make decisions for the person. Without the right financial planning tools in place, family or other interested parties may have to turn to the courts to seek appointment of a guardian or conservator to act on the persons' behalf. The court may appoint a stranger with little or no knowledge of the wishes or values of the person and whose time and fees are paid out of the person's estate.

A POA is the most important planning tool to avoid this situation. It is a legal document that enables one person, called the principal or grantor, to authorize another person, called the agent or attorney in fact, to make some or all financial decisions for the principal. If intended to apply to all financial decisions, it is called a general power; if limited, it is called a special power. The document can also name a successor agent or agents if the primary agent becomes unavailable. State law specifies how they are to be executed and whether any special provisions are needed for certain transactions, such as selling real estate. The ABA Commission on Law and Aging publishes a general summary of selected issues in state POA law.[22]

While it is possible to find and download POA forms on the internet, do-it-yourself forms are not the best option. POAs are incredibly powerful documents and benefit from professional legal assistance to customize them to the individual's needs and circumstances.

Every state recognizes the concept of a POA being "durable," meaning that the authority remains in effect even if the grantor is unavailable or totally unable to act. In some states, the authority can be made to begin only upon incapacity of the principle, in which case it is called a "springing" POA. However, most POAs are written to be immediately effective in order to avoid legal disputes over whether the document has validly sprung into effect. Granting a POA does not mean that the principal can no longer make decisions; only that the agent can make them, too. The principal always retains the power to direct the agent or to revoke the agent's power while the principal has the capacity to do so.

A POA is a good way to authorize someone to transact business on bank accounts and other financial accounts, but there is a catch. Third parties are not necessarily compelled to recognize the authority of the agent. With banks, it is especially important to check ahead of time whether the bank recognizes the validity of the POA. Most will, but some banks ask for specific language in the POA or prefer that their own banks' POA form be used.

Making a POA Safe. A POA can be an incredibly powerful document, with the agent able to legally bind the grantor without consultation or permission. The first step in safety is careful selection of a trustworthy and responsible agent.[10] One needs to feel comfortable that the agent will make the kinds of decisions that the principal would make, and that they won't abuse the authority granted to them. The vast majority of agents under POAs do all of the right things for all of the right reasons, but when relationships go bad, it can be really bad. Even when well-intended, an agent may not understand that an agent is a fiduciary with legal obligations. The Consumer Financial Protection Bureau provides a useful guidebook to explain the agent's role.[12]

It is possible to draft protections into a POA, but you will not typically find these protections in form documents. In consultation with legal counsel, consider the following:

- Require an inventory of assets when your agent begins managing your affairs. This provides a baseline for future comparison. Require that it be shared with a trusted third person you name
- Require some form of annual accounting to a trusted third person. Powers of attorney are private agreements, normally with no monitoring or oversight after the principal loses capacity, unless you create the mechanism for monitoring and oversight. Having a second set of eyes on the money provides a higher level of transparency. Accountings should at least provide an inventory of assets and documentation of the dates, nature, and amount of all financial transactions.
- Appoint coagents or require a second signature for large transactions. Two agents can help spread the burden of responsibility and creates some checks and balances, but the agents must be able to work together. Alternatively, approval by a trusted third person for large transactions provides one extra step to ensure the appropriateness of large transactions, such as the sale of a home, or sale of a large investment.
- Clearly define and limit the power of the agent to make gifts. Many people want to continue their pattern of giving even after they lose capacity. Limiting gifting requires identifying the permitted recipients or class of recipients and limits on the amounts and frequency. There are times when gifting may be essential for tax or asset planning, but it is also the most easily abused authority. Defining the agent's authority for those purposes is complex and requires an experienced advisor.
- Limit any changes to beneficiary designations that would be contrary to your goals and values. This may include limits on changing rights of survivorship under bank accounts; or designations on payable-on-death or transferable-on-death accounts or property; or changes in beneficiary designations under wills, trusts, life insurance policies, annuities, investment portfolios, or other instruments. There are occasions when it may be appropriate to change some of these designations, but these kinds of changes can radically revise your estate and lead to financial abuse; they need to be carefully considered before signing a POA.[5] Additional POA safety resources can be found on the elder abuse web page of the ABA Commission on Law and Aging.[3]

27.3.5 Establish a Representative Payee

If one receives benefits from Social Security, Supplemental Security Income, civil service, railroad retirement, the Department of Veterans Affairs, or some state pension funds and needs help managing the income from those benefits, these agencies have their own processes for naming someone to manage their benefits paid on behalf of a beneficiary. Social Security and the federal Civil Services refer to this person as a "representative payee." The Department of Veterans Affairs calls this person a fiduciary.

Several things are important to think about when considering a representative payee or a fiduciary:

- Even if you have a POA, the agency paying your benefits may choose not to follow it, and require a representative payee or fiduciary be appointed.
- Someone must apply to the relevant administrative agency to be named as representative payee or fiduciary. The agency will then determine whether it should appoint the applicant as a representative payee or fiduciary, based on evidence of the beneficiary's incapacity. Medical and other evidence will be required.
- Social Security also offer the option to advance designate up to three individuals who could serve as payee if the need arises.
- Once a representative payee or fiduciary has been appointed, he or she has the authority and the duty to manage the relevant income for the benefit of the beneficiary. Under Social Security Administration rules, most representative payees must file fairly simple annual reports according to SSA instructions.
- The authority does not extend to any other income or assets.

Both the Social Security Administration[25] and the Department of Veterans Affairs[24] provide useful guides on this topic, as does the Consumer Financial Protection Bureau.[12]

27.3.6 Consider Establishing a Trust if Needed

A trust is a legal arrangement created by contract. Under a trust, a person or institution (the trustee) holds title to property for the benefit of another person or persons (the beneficiaries). The person who creates the trust is called the grantor or the settlor. There are many kinds of trusts, but here the focus is on a "living trusts" which people create during their lifetimes primarily to manage property if they become unable to do so. These trusts also provide for the disposition of the property at a future time, usually at the person's death.

A living trust can be most useful for the management of substantial assets during one's incapacity or where assets are located in more than one state. Trusts are especially important in planning for the care of one's minor children or disabled family members when the grantor is no longer available to provide care. However, beware of high-pressure sales tactics or sweeping generalizations that trusts are ideal for everyone or serve as a miracle tool to avoid probate.[18]

Several things are important to consider in the creation of a trust:

- Trust come in many variations and can serve wildly different purposes. As a result, trust language is complex and typically lengthy.
- A trust's terms can have undesired tax consequences, as well as consequences for future eligibility for public benefits such as Medicaid long-term care benefits.
- Trust drafting should be done by a lawyer experienced in estate planning and, if necessary, public benefits law.
- Trusts can be expensive. Costs include: the legal expense of having the trust drafted, the cost of transferring your property into the trust, and significant management fees if a professional trustee is used.
- You must make sure that property actually gets transferred into the trust. Because the trustee's authority is limited to only those assets in the trust, if property is not transferred, you will have wasted the money spent creating the trust.
- Financial institutions provide professional management of trusts but they are costly.
- If a family member or friend serves as a trustee, they should understand their duties as a fiduciary.

Family members or friends serving as trustee will benefit from the guide for lay trustees from the Consumer Financial Protection Bureau.[12]

27.4   HEALTH CARE DECISION-MAKING

Most people want a say in all their personal decisions and nothing is more personal than their health care. Most of us will experience times when we are unable to make health care decisions. These may be temporary, for example, after an accident that incapacitates us but from which we recover. Many of us will also experience some period of inability to make decisions as we near death. This is when our values, goals, cultural and personal preferences about our care should be guiding all care decisions. Yet, who will know those values, goals, and preferences? Unless one has spoken with loved ones and health caregivers and taken certain planning steps, there may be confusion and stress over what those decisions will be and who can make them for you. The planning process every adults need to engage in is called "advance care planning."

27.4.1   ENGAGE IN ADVANCE CARE PLANNING

Advance care planning has been described properly as "[a] process that supports adults at any age or state of health in understanding and sharing their personal values, life goals, and preference regarding future medical care."[14] When people think about this topic, they most often think about the legal documents that are available to use to express their wishes: Living Wills and Health Care Powers of Attorney. These documents are known more generally as Health Care Advance Directives. While important, these documents are not at the heart of advance care planning, but as a starting point, they need to be understood.

- A "living will" is a document that spells out any wishes, instructions, or preferences you have about your future health care, particularly end-of-life care. It may also be referred to as a "medical directive."
- A "health care power of attorney" is a document that names someone to make medical decisions for you, should you be unable to make them yourself. Depending on your state, it may have a different name. The person you appoint may be called your "health care agent," "health care proxy," "surrogate," "attorney-in-fact," or "health care representative." And the document itself may be called a "health care proxy," "medical power of attorney," and "durable power of attorney for health care," or another name.
- The two types of documents are commonly combined into one document referred to generally as a "health care advance directive." These documents can also include instructions about important decisions such as Organ Donation.

Documents that are recognized in one's state can usually be obtained from local hospitals or bar associations or offices on aging or from a variety of online sources. However, do not just pick up a form fill it out, and sign it in front of witnesses. Documentation is the last step after engaging in the process of exploring one's values, goals, priorities, cultural and personal preferences.

The most important of these documents is the health care POA because it ensures that the person you choose and trust will have the authority to make health care choices for you if you cannot make them yourself. If you don't appoint a health care agent, your relatives or health care providers may be called upon to make medical decisions for you in such cases. Although this might not be a problem for some people, the lack of direction to loved ones is an emotionally trying experience for those loved ones, and it increases the likelihood of disagreements arising or of your wishes being ignored. It also increases the likelihood of needing to resort to the courts to seek appointment of a guardian.

The most important message to remember is that these documents are only as useful and effective as the thought, discussion, and planning that goes into them. Advance care planning done well requires discussions with family members and health care providers about what the future may hold health-wise. It requires you to think about how you want your beliefs and preferences to guide decisions and how concerns like finances, family matters, and spiritual questions should be taken into account. These are not easy topics for contemplation or discussion, so guides and tools for understanding the issues and discussing them are extremely valuable. A compilation of guides and tool kits, and other resources are available from the ABA Commission on Law and Aging Health Resources web page.[8]

Advance care planning, if done right, accomplishes four things:

- It helps ensure that the person you want to speak for you has the legal authority to do so.

- It helps ensure your wishes about your health care are known and respected.
- It avoids unnecessary, intrusive, and costly medical treatment at the point you no longer want it.
- It reduces the suffering experienced by loved ones faced with making end-of-life medical decision for you, because they will have your guidance.

Keep in mind that you don't have to spell out detailed instructions about treatment and care. Many people have very mixed feelings about how they would cope with serious illness and just aren't sure what they would want. That's fine. Effective advance planning clarifies your most important values and priorities in living life. Health care providers possess the expertise to identify medical conditions and treatment options, but the patient or the patient's well-informed agent brings the necessary expertise to weigh those options in light of the patient's values, priorities, and life story.

Finally, a good rule of thumb is to reexamine your health care wishes whenever any of the "Six D's" occur:

1. Decade – when you start each new decade of your life.
2. Death – whenever you experience the death of a loved one.
3. Divorce – when you experience a divorce or other major family change.
4. Diagnosis – when you are diagnosed with a serious health condition.
5. Decline – when you experience a significant decline or deterioration of an existing health condition, especially when it diminishes your ability to live independently.
6. Domicile – when you move to an entirely new home or someone moves into your own home as a caregiver.

27.4.2 BE AWARE OF POLST

A shared decision-making process, known by many names, but most commonly POLST – short for Physicians (or Providers) Orders on Life-Sustaining Treatment, provides another dimension to advance care planning for persons with serious illness or frailty. It complements health care advance care planning by ensuring that care plans addressing high-probability critical care decisions have been discussed with the patient or the patient's surrogate and are now reflected in a highly visible and portable set of medical orders that follow the patient across care settings.

Virtually everyone knows of at least one instance in which a medical turn of events results in unwanted suffering for patients and family, because the patient's goals or wishes were not followed, even when their wishes had been documented in an advance directive. But in medical crisis where emergency medical services may be called and upon arrival, EMS personnel are required to provide all necessary interventions to resuscitate and stabilize patients and transport them to the emergency room. Only a recognized doctor's order can change that outcome.

This is where POLST comes in, for it is designed to enhance the process of determining the goals of care and creating a medical care plan for persons who are considered to be at risk for a life-threatening clinical event because they have a serious life-limiting medical condition, which may include advanced frailty.[16] In the early 1990s, Oregon experimented with a protocol for seriously chronically ill individuals, which they named Physicians Orders for Life-Sustaining Treatment, or POLST. Today, most states have some version of POLST, under varying names such as MOLST, POST, MOST, COLST, and TPOPP. In many states, advance practice nurses and certain other providers can assist with and sign these medical orders.

There are several ways to describe the POLST process, but four components should always be part of the process:

1. The use of POLST requires a discussion between the health care provider and patient or surrogate about high-probability end-of-life care treatment options. This is the point at which one's prior advance care planning impacts the clinical decision-making process most directly, especially if the patient is unable to speak for himself or herself at this stage of their illness.

2. The patient's goals of care and priorities are incorporated into doctor's orders and the orders are recorded on a unique, visible POLST form (and increasingly, in a highlighted part of the electronic health record). The orders address high probability medical events – CPR; the level of medical intervention desired in the event of emergency (comfort only/do not hospitalize; limited; or full treatment); and often the use of artificial nutrition and hydration.

3. POLST is reviewed and modified periodically as needed, especially when the patient's condition changes or the venue of care changes. As a person's medical condition declines, changes in goals of care and priorities of life may frequently change, and POLST should always reflect the patient's current goals of care.

4. Health care providers must ensure that the POLST form travels with the patient whenever the patient is transferred from one setting to another. This helps ensure continuity of care decision-making.

POLST is not an advance directive, but it is an advance care planning tool that reflects a seriously ill patient's here-and-now goals for medical decisions. It builds upon one's advance directive, but it also benefits patients who have no advance directive. And, if the capacity to make health care decisions is lost, the patient's health care agent or a default surrogate can engage in the POLST process on behalf of the patient. A final characteristic to understand about POLST is that it is outcome neutral. That is, it is not focused only on stopping medical interventions. It provides options for declining treatment, requesting aggressive treatment, and for gradations in between. National POLST provides extensive resources for both family and professionals on the use of POLST throughout the United States.[16]

27.5 CONCLUSION

Using legal tools to plan for safety increases the likelihood that the persons wishes will be honored at any time that a person is unable to speak up for himself or herself. These are times when the risk of a loss of rights or of abuse, neglect or exploitation is at its highest. No legal tool is a guarantee of safety, but the tools discussed in this chapter provide the mechanisms most effective in ensuring that your financial and health care goals, values, and priorities are known and followed. Without planning, the possibility of guardianship or conservatorship often arises. Benjamin Franklin put succinctly when he said, "if you fail to plan, you are planning to fail."

REFERENCES

[1]   Birken EG, Foreman D. Your complete guide to direct deposit. *Forbes.* www.forbes.com/advisor/banking/your-complete-guide-to-direct-deposit. October 2021.

[2]   Dinerstein RD. Implementing legal capacity under article 12 of the UN convention on the rights of persons with disabilities: the difficult road from guardianship to supported decision-making. *Hum. Rts. Brief.* 2012;19(2):8–12, at 8. https://digitalcommons.wcl.american.edu/cgi/viewcontent.cgi?article=1816&context=hrbrief

[3]   Elder Abuse. ABA Commission on Law and Aging. www.americanbar.org/groups/law_aging/resources/elder_abuse. June 15, 2021.

[4]   Eldercare Locator, a public service of the U.S. Administration for Community Living. https://eldercare.acl.gov

[5]   Five safeguards to consider adding to any power of attorney for finances, ABA Commission on Law and Aging, www.americanbar.org/content/dam/aba/administrative/law_aging/2017-poa-tip-sheet.pdf.

[6]   Godfrey D. Elder abuse basics: prevention, intervention, and remediation, national center on law and elder rights. National Center on Law and Elder Rights. https://ncler.acl.gov/getattachment/Legal-Training/Elder-Abuse-Basics-Ch-Summary.pdf.aspx?lang=en-US. January 2021.

[7]   Guardianship and Supported Decision-Making. ABA Commission on Law and Aging. www.americanbar.org/groups/law_aging/resources/guardianship_law_practice. August 16, 2021.

[8]   Health care decision-making. ABA Commission on Law and Aging. www.americanbar.org/groups/law_aging/resources/health_care_decision_making

[9]   Investing: establishing a trusted contact. FINRA. www.finra.org/investors/learn-to-invest/brokerage-accounts/establish-trusted-contact.

[10] Kirtland M. Choosing an agent for your power of attorney, Voice of Experience. www.americanbar.org/groups/senior_lawyers/publications/voice_of_experience/2020/may-2020/choosing-an-agent-for-your-power-of-attorney. May 2020.

[11] Livingston G. On average, older adults spend over half their waking hours alone. PEW Research Center. www.pewresearch.org/fact-tank/2019/07/03/on-average-older-adults-spend-over-half-their-waking-hours-alone. July 3, 2019.

[12] Managing someone else's money (series of guides). Consumer Financial Protection Bureau. www.consumerfinance.gov/consumer-tools/managing-someone-elses-money

[13] National Academies of Sciences, Engineering, and Medicine. 2021. *The challenges and opportunities of advance care planning: Proceedings of a workshop.* Washington, DC: The National Academies Press. https://doi.org/10.17226/26119

[14] National POLST. https://polst.org

[15] National Resource Center for Supported Decision-Making. www.supporteddecisionmaking.org/

[16] Palmer B. Should you set up a revocable living trust? Investopedia. www.investopedia.com/articles/pf/06/revocablelivingtrust.asp. June 2021.

[17]   Peres K, Helmer C, Amieva H, et al. Natural history of decline in instrumental activities of daily living performance over the 10 years preceding the clinical diagnosis of dementia: a prospective population-based study. *J Am Geriatr Soc. Jan*; 2008 56(1):37–44. [PubMed: 18028344]

[18]   Research, Statistics, and Data. National Center on Elder Abuse, https://ncea.acl.gov/About-Us/What-We-Do/Research/Statistics-and-Data.aspx

[19]   Robinson C. The trouble with joint bank accounts "Just in case," Kiplinger, www.kiplinger.com/article/retirement/t021-c032-s014-the-trouble-with-joint-bank-accounts-just-in-case.html. December 5, 2018.

[20]   Selected issues in power of attorney law. ABA Commission on Law and Aging. www.americanbar.org/content/dam/aba/administrative/law_aging/2020-chart-poa.pdf. April 2020.

[21]   Thayer A. Direct payments vs. online bill-paying: something for everyone, IU Credit Union, http://resourcecenter.cuna.org/16991/article/293/html. August 26, 2002.

[22]   U.S. Department of Veterans Affairs. Guide for Fiduciaries, www.benefits.va.gov/FIDUCIARY/docs/VA_Fiduciary_Guide_Apr2020.pdf. May 2020.

[23]   When people need help managing their money: representative payee. Social Security Administration. www.ssa.gov/payee

28 Older Adults and Scams

Ann Marie Cook and Daniel Lyon

CONTENTS

Tom, 85 years old, is a retired lawyer who is well known in the community. He serves on several nonprofit boards and has always been considered a community leader. He received a call from a scammer saying his grandson Jackson was in jail in New Mexico. (Jackson had been posting on social media about his trip to New Mexico.) The scammer explained that Jackson would be kept in jail unless he could pay his cash bail. Tom sent $15,000 to bail him out.

Tom became a victim of a scam along with millions of other people each year.

28.1 HOW MUCH DO WE LOSE FROM BEING SCAMMED?

Studies vary widely, but it is reported that scams and frauds cost between $3 billion and $36 billion per year!

28.2 OLDER ADULTS ARE TARGETED

Older adults are heavily targeted by scammers for several reasons.

- According to a 2016 AARP study, people 50 and older hold 83% of the wealth in America. Scammers target people who have money. (www.aarp.org/content/dam/aarp/home-and-family/personal-technology/2016/09/2016-Longevity-Economy-AARP.pdf)

DOI: 10.1201/9781003197843-28

- Researchers at Northwestern University have concluded that we become more trusting as we get older. (www.medicinenet.com/script/main/art.asp?article key=187687) This increase in trust puts older adults at higher risk for scams and fraud.
- Older adults are more accessible to scammers because they are home more often.
- Nearly one-fourth of adults 65+ are considered to be socially isolated (Nation Academies of Science, Engineering, and Medicine [NASEM]). Isolation and loneliness contributes to a higher risk of being scammed.

> Four out of five North American scam victims have a college degree; many intelligent, educated people fall prey to scams. Joseph Steinberg

- Some older adults may not be as technologically savvy, leaving them vulnerable to technology and computer scams.

28.3   HOW DO THEY SCAM US?

Scam artists use a **hook** to trick individuals into providing personal information and/ or sending money:

- **They prey on our love for family and our good will** – We all want to help those we care about.
- **They know that everyone needs/would like more money** – The promise of a great deal or the dream of sudden riches is often hard to pass up.
- **They play on our fear and desperation** – Being afraid can make us less cautious, more open to the promise of a quick fix or an unexpected cure.

28.4   SOME "POPULAR" SCAMS

Romance scams – Romance scams often start after a brief online or phone interaction. Scammers will often win the trust of the individual over time and spend time to get to know you and/or your family dynamics. This lengthy period is all about grooming you and making sure you get comfortable with the person.

That is exactly what happened to "Anna." Anna's story started innocently enough online with a gaming website. Anna received a direct message in the chat box from "Erik." Erik talked about liking online gaming and asked if she liked it as well. After several pleasant interactions, Erik encouraged Anna to connect with him on WhatsApp. After several months of texting, Erik said, *"I feel funny admitting this, but I'm in love with you. I want to meet you in person, but I just don't have the money to visit you."* Anna began sending Erik money to purchase a plane ticket. Then Erik told Anna he had purchased the ticket, but was having trouble taking off from work because he had overdue bills to pay. There was also the fee, he claimed, for changing his flight last minute. By the time Anna contacted us for help, she estimated that she lost over $100,000 to this scam.

Romance scams are up 50% from 2019. On average, a person over age 70 loses $10,000 per romance scam.

The IRS – Some individuals have fallen victim to "a call from the IRS." Preying on people's fear, they are told they will be jailed unless they pay "back taxes." Scammers sound official and threatening.

Remember: The IRS will NOT call you, email you, send you a text, or ask for your Social Security number. Government agencies, if they need to reach you for an urgent matter or something relative to your Social Security Number, they will contact you via mail.

Be mindful that government agencies will not be calling you on the phone. They will not threaten you with arrest. Hang up!

- *Grandparent scam – "Grandma, I am in jail and need money for bail. Please don't tell my parents. I know you would be the one person in the world I could call for help."*

So many older adults fall victim to this scam because they want to help their grandchild. They act out of love and have wired money before they can think it through. These calls come when you least expect them, such as in the middle of the night, to throw you off guard. The scam artist will in fact use your loved one's name. The "grandson" will claim they are calling from a police station, and they are under arrest for something like Driving While Intoxicated (DWI). The scammers use a variety of techniques to explain why their voice sounds differently or why they aren't contacting their parents for help. For example, they may say they sound oddly because they hit their face on the airbag and suffered a fat lip and or broken nose. The scam artist will have an answer to any and all questions to keep the victim they are preying upon focused on the issue. Often, they will hand the phone over to an accomplice who claims to be the police officer in charge of this investigation. Bail money in the form of cash, gift cards, or wire transfers will be requested with a time limit to pay before the grandchild goes in front of a judge for the arraignment. The scary new twist to this scam is when scammers send a local accomplice to the front door of your house to pick up the bail money. The victim hands over a box of cash for bail, believing they are helping their loved one.

Please know that the police will never come to your door to collect bail money. The scam may or may not be over after handing over a box of cash. The scammer may call you back and tell you the judge has reviewed this case and, because the DWI arrest involved a car crash with injuries, bail was revised and increased to $30,000.

We have seen scammers ask for money in the form of a wire transfer. Many people have never done a wire transfer, but scammers say, "No problem. We can walk you through it."

Bottom line is that grandparents are afraid of what could happen to their grandchild if they hang up the phone. Always verify through another party that their loved one is truly out of town and in trouble. Oftentimes, a simple call to the grandchild in question will prove that they are not, in fact, in any danger.

- *Tech scam* – This scam often begins with your screen flashing signs that you have a virus on your computer. More sophisticated versions of this scam will include bells and whistles to alarm you. The message on your screen will also provide you with a phone number claiming to be Microsoft, directing you to call for help to repair it.

The scam artist will also want to send you an email or a text with a link to click upon to download the software. The scam artist will need to convince you that this is a safe and easy way to fix your computer. In reality, once they have remote access to your computer, they have access to all of your personal information, photos, documents, Social Security Number, and any other information stored on the computer.

- Never allow remote access on your personal computer.
- Consider investing in a complete antivirus protection service through a reputable organization.
- If you do not know which program or service is best for you, consult with a local technology store or the store from which you purchased the computer.
- Ensure that antivirus and antimalware software is up-to-date and run regularly.
- Do not click on suspicious or unexpected links. Rather, go directly to the website in question.

Check scam – This scam starts in your mailbox. In a check scam, a person you don't know asks you to deposit a check and asks you to send some of the money to another person. You get to keep some of the money. The scammers always have a good story to explain why you can't keep all the money. They might say they need you to cover taxes or fees, you'll need to buy supplies, or something else.

Examples of check scams:

Claiming prizes. Sweepstakes "winners" are given checks and told to send money to cover taxes, shipping and handling charges, or processing fees. But that's not how legitimate sweepstakes work.

For example, you receive a check for $3000. The letter enclosed with the check will ask you to send $2,000.00 to another address. You believe you made an easy $1,000 when you are really losing over $2,000.00, plus bank fees. The bank will call you in 4–5 business days to advise you the check you deposited returned as fraudulent. You are out of money.

- *Gift card scams* – Only a scammer will ask you to go to a store and buy a gift card to pay for a fee, good, or service. This is a HUGE red flag.

Scammers know that buying any quantity of gift cards in the same location will also draw the attention of the clerk and store staff. Often, scammers will have done an internet search of your address and know what stores are closest to your home. They will direct you to purchase gift cards at a variety of locations to reduce suspicion and draw less attention to yourself. While most retail workers have been apprised of these scams, it does not look suspicious for a consumer to purchase a few gift cards, particularly around the holidays. It does, however, look very suspicious to buy ten gift cards from the same location for large amounts.

Please keep in mind no legitimate business will ever accept gift cards as a form of payment.

- *Lottery scams* – If you didn't play the lottery, you didn't win. There are a lot of lottery scams originating outside of the United States, and it is illegal to win any lottery which originates out of the United States, such as the Canadian Lottery. Scammers know that everyone wants to win! They inform victims that there are taxes and some fees the victim will need to pay to receive this prize. In reality, you never need to pay out of your pocket to receive any type of prize. Taxes and fees typically are removed from the winnings automatically. If you did not enter into a lottery or prize drawing, you didn't win.

  "If it sounds too good to be true, then it probably is."

- *Fraudulent GoFundMe sites* – Scammers either take someone else's story or make one up entirely designed to invoke an emotional response from victims. Money received by the scammers in these cases either lie about what the money was used for or take the money and disappear. While GoFundMe accounts can be a wonderful method of collecting money for needed causes, it's important to only give to folks who you personally know, have met, or have confirmed that their charitable cause is legitimate.

- *Charity scam* – A request for your money typically arrives in your mailbox with a letter requesting funds to support a cause. Unless you are familiar with the specific organization, you need to do some homework.

"Milton" was a victim of a charity scam. Wanting to help those in need, Milton gave $250 to a "charity." Milton's name was sold to other organizations, both legitimate and scam-related, as an active giver. Milton's mailbox was full every day from mail charity solicitations. He ended up sending thousands of dollars to organizations – leaving himself without enough money to pay his bills.

- Research companies – www.charitywatch.org.
- Consider giving to local organizations that you know.
- Throw out solicitations from organizations that you do not know.

28.5 HOW TO AVOID GETTING SCAMMED

- Read web addresses very carefully. Are any words misspelled? Poor grammar? Is it secure? Secure websites will either start with https:// or will show a closed lock icon to the left of the website address.
- DO NOT reply to any response such as "STOP" or "NO!" They are using fear tactics to elicit a quick response. Do not reply with a "yes" or hit numbers on your keypad such as 1 or 2 if prompted to do so. Clicking on numbers such as 1 or 2 to indicates to the scammer that your phone number is "live."

  By saying "yes," your voice can easily be recorded and later used. For example, the scammer may call your bank. The bank employee will ask if you give them permission to speak to someone on your behalf. The scammer will then play your recorded voice saying "yes." If an unsolicited call from an unknown source comes in, best to ignore it. Legitimate callers will leave a message for you and will not pressure you for a response.

- If you do answer the phone, it's fine to hang up. You're not being rude! You're being careful.
- It also may be helpful to call the main number of an organization. Look up the number. Do not redial the number that called you. Scammers can spoof the number to appear to be from legitimate source.

- Never click on a link in an email or text unless you're expecting it from someone you know. If you don't know the sender, it's best not to click on it at all. Delete it.
- Before responding to an email from a company, go to the business' website and access the account that way. You can also call the number directly *from the site* and speak to someone about the email.
- If you get a call asking for financial information, stop and think. Tell them you want to hang up and do more research. Call a trusted family member or friend to discuss the situation.
- If there has been a breach, sign up for fraud alerts on your credit cards and bank accounts.
- Invest in a quality cross-cut paper shredder that can also destroy ATM cards, credit cards, and other identifying paperwork and documents. Any mail you receive that has your name on it should be shredded before disposing of it. This includes not only credit card and banking statements, but also more innocuous items such as magazines and personalized advertisements. Scammers can glean a lot of information from your purchases!
- Cell phone users have the availability to add scam-blocking apps to their cell phones. The service provider at the store will be happy to assist you with adding it to your phone. Scam-blocking apps are not an end all cure but are another layer of protection.
- Isolation and loneliness are key factors making a person more likely to being scammed. Being homebound and/or lacking regular social interaction can increase vulnerability to scammers. Connecting with your local Office for the Aging or other aging service providers such as your local senior community center can be helpful. If able, one can also join other social or community-based groups such as church groups, Rotary or Kiwanis clubs, and others. Meeting up with friends and neighbors and talking helps keep you informed about local events and news, including the latest scams.
- Scammers will always use the tactics of telling their victims not to discuss with their family or friends. The scammers will often say this offer or prize you are eligible to receive will not be offered to anyone else so not to talk about it to anyone. Other scammers will go as far as to threaten you with physical violence, violence against your family or loved ones, or to "out" you in your community. Much like with other forms of abuse, scammers rely on anonymity and silence to accomplish their goals. They know all too well that if you start telling any of your friends or family members about your conversations with this scammer then they may be outed for their scam.

28.6 WHAT TO DO IF YOU HAVE BEEN SCAMMED?

- **File a police report:** Ask for a copy and keep it in a secure place for use with creditors and to get consumer transaction records.
- **Pull a free credit report for yourself** – (www.annualcreditreport.com) or call 1-877-322-8228. Request credit reports over the course of a year by pulling a report from one agency such as Equifax in January, wait four months (May) and pull one report from Trans Union, wait another four months (September) and pull your last available free credit report from Experian. If you cannot remember the last time, you reviewed your credit report then you are due. This is a simple and free way to really keep tabs on your credit, something that everyone should be doing regardless of being a victim of ID theft or not.
- **Create an Identity Theft Report:** Submit a complaint to the FTC 1-877-438-4338, or submit a complaint online (Identity Theft Affidavit) www.ftc.gov/complaint or https://identitytheft.gov
- **Notify:** All banks, creditors, utilities, insurance carriers, phone, internet service and cable television providers, libraries, and other memberships. Immediately close or put holds on your bank accounts.
- Notify all the credit reporting agencies – Equifax, Experian – and put a *freeze* on your credit.
- Change all your passwords, and make the passwords with multiple letters, lowercase, uppercase, and special characters at least eight digits long.
- Get on the DO NOT CALL REGISTRY 1-800-382-1222. Or simply register online (www.tfc.gov) which is quick and easy.

OTHER IMPORTANT RESOURCES

- www.charitywatch.org/charities
- www.directmail.com= Stop junk mail
- AARP (www.aarp.org/technology/privacy-security)
- The AARP website provides specifics on internet safety, how to protect your privacy, and the most up-to-date virus protections.
- FBI (www.fbi.gov/scams-safety/fraud/seniors/)
- U.S. Postal Inspection Service (USPIS) (www.uspis.gov/)

Here **Be vigilant.** If you've been scammed once, your name may well end up on a scammer's "sucker list." The best thing you can do is talk to a trusted person if you have any doubts.

BIBLIOGRAPHY

"How Does Financial Fraud Impact Financial Well-Being?" *Psychology Today.* www.psycholo gytoday.com/us/blog/the-fraud-crisis/202101/how-does-fraud-impact-emotional-well-being. January 3, 2021.

James, Bryan D., Boyle, Patricia A., and Bennett, David A. "Correlates of Susceptibility to Scams in Older Adults Without Dementia" (nih.gov). *J Elder Abuse Neglect*. 2014; 26(2): 107–122.

"Loneliness and Social Isolation Linked to Serious Health Conditions." Centers for Disease Control and Prevention. www.cdc.gov/aging/publications/features/lonely-older-adults. html. April 29, 2021.

"Memory, Forgetfulness, and Aging: What's Normal and What's Not?" National Institute of Health. www.nia.nih.gov/health/memory-forgetfulness-and-aging-whats-normal-and-whats-not. October 21, 2020.

"People May Grow More Trusting With Age." *Health Day News*. www.medicinenet.com/scr ipt/main/art.asp?articlekey=187687. March 27, 2015.

Shadel, Doug. "Confessions of a Con Artist." *AARP Magazine*. www.aarp.org/money/scams-fraud/info-09-2012/confessions-of-a-con-artist.html. 2021.

"Social Isolation is a risk factor for scam loss." Help Net Security. www.helpnetsecurity.com/ 2020/03/24/risk-scams/. 2021.

"Telemarketing Fraud and Victimization of Older Americans: An AARP Survey." AARP. Washington, DC: 1996.

"Trust Rises with Age." National Institute of Health. https://newsinhealth.nih.gov/2013/01/ trust-rises-age#. January 2013.

LIFESPAN OF GREATER ROCHESTER'S SCAM PREVENTION PROGRAM

Lifespan of Greater Rochester Inc. (Lifespan) is a not-for-profit in Rochester, NY, that helps older adults and their caregivers take on the challenges and opportunities of longer life. Lifespan operates over 30 programs. In 1995, Lifespan created the Scam Prevention Program to provide education to older adults about scams and to intervene in scam cases. This program is housed at Lifespan under Upstate Elder Abuse Center at Lifespan (UpEAC), providing statewide training and education as well as direct case management in a 12-county region in the Finger Lakes. When Lifespan intervenes to assist victims, it often leads to criminal prosecution and/or restitution. UpEAC serves hundreds of victims annually to stop elder abuse and financial exploitation and to remedy its effects.

29 Elder Abuse and Financial Exploitation

Daniel A. Reingold, Joy Solomon, and Tristan Sullivan-Wilson

CONTENTS

29.1 ELDER ABUSE

Ms. E, an 82-year-old woman, has been living alone in her longtime New York City apartment since her husband passed away 15 years ago. She loves to dance, has a dry sense of humor, and has a close relationship with her neighbors and the staff in her building.

Ms. E was recently diagnosed with dementia, requires assistance with many activities of daily living, and walks with a rolling walker. Living on her own with these increasing care needs created some difficulties for Ms. E, but she was committed to staying in her apartment and longtime community.

This year, Ms. E's niece moved in, saying she would help Ms. E at home. At the same time, Ms. E started to fall behind on her rent – something that had never happened in her decades living in that building. Concerned by this unexplained change, Ms. E's management company helped her contact the bank. They learned that Ms. E's niece had been consistently withdrawing money from Ms. E's account, not even leaving enough money to cover the rent. In total, her niece withdrew over $150,000 from Ms. E's account. Concerned for her safety, the management company contacted Adult Protective Services (APS). Once involved in the case, APS observed that Ms. E was struggling to manage

DOI: 10.1201/9781003197843-29

*her own finances. It appeared that her niece had been taking advantage of Ms.
E's need for assistance, financially exploiting her since arriving at Ms. E's
apartment. A guardianship proceeding was initiated and the court appointed an
independent guardian for Ms. E.*

*When the guardian tried to meet with Ms. E, her niece refused to allow them to
speak privately. Shortly after this strained first meeting, the niece removed Ms.
E from the apartment, moving her between short-term rentals in New York and
the surrounding states.*

*Finally, with the assistance of the local police department, the guardian gained
access to Ms. E. She had lost weight and did not look as strong as she had at
their initial meeting. The guardian immediately brought Ms. E to the hospital.
Afraid for Ms. E's safety in her own apartment with her niece, the guardian
referred her to a temporary elder abuse shelter.*

According to the Centers for Disease Control and Prevention, elder abuse is "an intentional act or failure to act, by a caregiver or another person in a relationship involving
an expectation of trust that causes or creates a risk of harm to an older adult." For
Ms. E, her niece exploited their seemingly trusting relationship and used *Ms. E*'s care
needs and dementia to financially abuse, displace, and harm her. Tragically, *Ms. E*'s
story echoes that of many older adults experiencing abuse.

29.1.1   PREVALENCE AND IMPACT

Every year, an estimated one in ten people over 60 in the United States experiences
some form of abuse. Worldwide, this number increases to one in six people. Even
these numbers likely fail to grasp the full picture. A study of older adults in New York
City found that, for every 24 older adults experiencing abuse, only one case was actually reported to law enforcement, a social service provider, or legal agency.[1]

There are a variety of reasons why an older adult may not report their experience of abuse, including fear of escalation, feelings of shame or guilt, lack of (or
perceived lack of) alternatives, impacted cognition, cultural barriers to disclosure,
and/or concern about consequences for the person causing harm. Concern for the perpetrator is central to the experience of many older adults experiencing abuse. In the
majority of elder abuse cases, the abuse is perpetrated by a family member. Notably,
many mothers experiencing mistreatment by their adult children do not self-identify
as a person experiencing abuse; rather, these mothers self-describe their concerns
and fears through the lens of motherhood and their experience mothering an adult
"difficult child."[2] For these women, the centering of motherhood places their experience outside the scope of elder abuse – including the reporting and service models
established to address elder abuse. This serves as just one example of how a failure to
acknowledge these complex relationships and competing interests creates a barrier to
effective identification of older adults experiencing abuse.

The impacts of elder abuse are wide ranging, impacting physical and mental
health, financial well-being, housing security, and social connectedness, resulting in
a higher risk of hospitalization and death. Awareness of and education about elder

abuse identification and resources is more vital than ever as the population of older adults increases worldwide: the proportion of Americans 65 and older has nearly quadrupled since 1900. These changing demographics demand a systemic approach to preventing and responding to cases of elder abuse and ensuring that existing infra-structure is equipped to meet the needs of our aging population.

Abuse takes many forms, including physical abuse, financial/economic exploit-ation, verbal and psychological abuse, sexual abuse, cultural/spiritual abuse, and neg-lect. Polyvictimization, the experience of more than one type of abuse, is frequent. Regardless of the specific form that abuse takes, all abuse centers around the exercise of power and control over another.

- *Physical abuse* is the intentional use of force that results in bodily injury, pain or impairment, including hitting, pushing, pinching, and physically restraining the older adult. Misusing medication to under or overmedicate an older adult is another form of physical abuse. For example, overmedicating an older adult as a chemical restraint or diverting pain medication for personal use, leaving the older adult without access to their prescription and in pain.
- *Financial/economic exploitation* is the improper use of another person's funds, property, or resources. Examples include theft, fraud, fraud, false pretenses, forgery, coerced or fraudulent property transfers, or prohibiting access to one's own assets.
- *Emotional or psychological abuse* is the infliction of mental or emotional anguish by threat, intimidation, harassment, or other verbal or nonverbal acts. Importantly, emotional or psychological abuse can arise from forced isolation or use of the "silent treatment" to harm an older adult.
- *Sexual abuse* is nonconsensual sexual contact of any kind, including forcing sexual contact, inappropriate touching, taking sexually explicit photographs, or forcing exposure to explicit sexual content. Sexual abuse of older adults is widely underreported. Ageist views of older adults as nonsexual obscures both healthy sexuality and sexual abuse, leading to lack of screening for sexual abuse of older adults or discussions of healthy sexuality and consent in the con-text of changing capacity.
- *Cultural/spiritual or identity abuse* is the use of spiritual, cultural, religious, or other identities to manipulate, coerce, or control an older adult. Examples include prohibiting use of culturally appropriate clothing, limiting access to culturally appropriate foods or cooking methods, destroying or preventing use of religious articles, denying access to religious services or rituals, and restricting or denying access to gender-affirming care.
- *Neglect* is the failure by a caregiver or other person in a position of responsi-bility to protect an elder from harm or the failure to meet needs for essential medical care, basic activities of daily living, or shelter, which results in a ser-ious risk of compromised health and/or safety. In order to rise to the level of abuse, the person causing harm must have a preexisting responsibility to pro-vide that care, for example, a home health aide or other person in a caregiving position.

Each of these types of abuse may present differently, but all cause significant harm to the older adult and must be taken seriously and addressed.

29.1.2   UNIQUE CONSIDERATIONS IN CASES OF ELDER ABUSE

Like other forms of abuse, perpetrators of elder abuse cause harm to exercise power and control over the target of abuse, attempts to leave the abusive situation can escalate the violence, and resources and support are vital to safely escaping the harm. However, the complex systems established to serve survivors of abuse frequently fall short for older adults. For example, elder abuse can be harder to identify or report due to isolation of the older adult. Further, traditional domestic violence shelters may only shelter women and are not equipped to meet the greater level of clinical care that some only adults may require due to complications arising from more complex medical needs or changing cognition.

Historically, capacity was seen as an all or nothing assessment. A more accurate, clinical understanding of capacity understands that capacity exists on a spectrum, variable between times and types of tasks. This nuanced understanding is key when engaging with older adults with diminished capacity: dementia impacts 8% of people older adults and people with dementia are at higher risk of experiencing abuse or neglect from a caregiver.

For people with dementia, abuse may present differently – the person causing harm can weaponize the diagnosis or symptoms to exert power and control over the person with dementia. A person causing harm may point to a dementia diagnosis to attempt to call allegations of abuse into question, utilize confusion to elicit feelings of shame or fear, intentionally disorganize a space to increase confusion, use dementia diagnosis to explain surveillance or confinement, or threaten institutional placement.

Changing cognition or a dementia diagnosis does not mean that an older adult lacks capacity for all tasks, and it is important to take allegations of abuse seriously regardless of a dementia diagnosis. Many people who carry a dementia diagnosis are able to credibly report experiences of abuse. A study of older adults across the capacity spectrum demonstrated consistent ability to indicate the cause of intentionally inflicted bruises when asked.[3]

Further, with the majority of elder abuse perpetrated by family member or caregiver, an older adult may have significant concerns about unmet care needs if they report the abuse. A lack of appropriate temporary shelters or other available caregivers presents a significant barrier to reporting and safety planning for many older adults experiencing abuse.

29.1.3   TRAUMA-INFORMED APPROACH

Trauma-informed work is essential when working with older adults who may have experienced abuse. Of course, the experience of abuse itself presents a traumatic event. For many older adults experiencing abuse, this may only be one in a series of traumas experienced over a lifetime. The experience of a national trauma, like war or other mass casualty events, can stack on top of individual trauma, like the loss of a loved one or experience of abuse. Trauma can also arise from the experience of cultural bias, like racism or homophobia. In a culture that places high value on youth, with ageism built into cultural values and systems, even aging itself can be a source of trauma.

Trauma can manifest in a variety of ways throughout one's lifetime, including through flashbacks, depression, outbursts of anger, headaches, more easily triggered startle response, and other physical, psychological, and cognitive effects.

The Substance Abuse and Mental Health Services Administration describes trauma-informed work as an organization or system that

> realizes the widespread impact of trauma and understands potential paths for recovery; recognizes the signs and symptoms of trauma in clients, families, staff and others; responds by fully integrating knowledge about trauma into policies, procedures and practices; and seeks to actively resist re-traumatization.

This requires all work to be infused with trauma-informed principles and practices.

GETTING STARTED: TIPS FOR PRACTITIONERS NEW TO TRAUMA-INFORMED CARE

BE CLIENT-CENTERED	COMMIT TO SAFETY	DON'T PATHOLOGIZE PEOPLE
Value the person's voice, perspectives and wishes in every aspect of care.	Create an environment and approach to providing care that feels safe to the individual you're serving.	Concentrate on the three e's—events, experiences and effects—as a way to focus on what has happened to the person as opposed to what is wrong with the person.
SEEK TO DISCOVER AND UNDERSTAND THE WHOLE PERSON	TALK LESS, LISTEN MORE AND DON'T RUSH	EMPOWER PEOPLE
Discuss traumatic events in the context of a person's entire life, with equal attention to positive experiences and personal strengths and achievements.	Make space for open-ended conversation and nonverbal expression (e.g., art, dance, yoga, etc.)	Look for any opportunity to give people a sense of control and mastery over their lives while avoiding actions that reconfirm the victim's feelings of helplessness.

Excerpt from *The Things They Carry: Advancing Trauma-Informed Responses to Elder Abuse*, Risa Breckman, Malya Levin, Leslie Mantrone, & Joy Solomon (2020).

29.1.4 IDENTIFICATION, INTERVENTION, AND PREVENTION

Elder abuse can occur anywhere and impacts older adults across race, gender, income, and location. For too many, abuse goes on for months or years before being reporting or identified.

29.1.4.1 Identification

Although there are some protective factors that are statistically correlated with lower risk of abuse, including social connectedness, professionals must remain aware of red flags of elder abuse and take reports of abuse seriously. Red flags of elder abuse can include:

- Bruising, lacerations
- Sudden changes in behavior
- Missing assistive devices
- Agitation
- Confusion
- Irregular withdrawal of money
- Abrupt changes to advance directives or a will
- Dehydration, malnutrition
- Change in affect
- Isolation

Despite the prevalence of elder abuse, many older adults never report their experience to professionals. Elder abuse and mistreatment screening tools assess experience of and/or risk of experiencing abuse and provide an opportunity for older adults to report abuse and for professionals to identify harms experienced that the older adult may not identify as abuse. The specific screening tools used should be chosen with the context and setting of use in mind. For example, the Weinberg Center Risk and Abuse Prevention Screen measures both risk of abuse and experience of abuse and must be administered by a professional. Additional tools have been developed for specific settings, including the Elder Assessment Instrument for use by nurses and the American Bar Association Section of Dispute Resolution Task Force Elder Abuse and Neglect Screening Tool developed for use by mediators.

These screening tools can be an important component of elder abuse identification. However, screening procedures must be paired with informed resource referrals. When considering implementing a screening procedure, an organization should first determine what appropriate referrals and resources are available in the event of a positive screen.

29.1.4.2 Interventions

The elder justice field is growing, continuing to expand the understanding of elder abuse and effective and accessible interventions. All effective interventions must center the individual client's needs, safety, and goals. Increasingly, elder justice professionals are attuned to diverse needs of older adults experiencing abuse, including the need for culturally appropriate services.

Older adults are an incredibly heterogeneous group, and focusing on age alone can obscure the complex and intersectional identities of older adults experiencing abuse. Professionals working with older adults must be open and understanding of the diverse experiences, values, and goals of this population. For example, for some older adults, this means supporting them through reporting to police and testifying in a criminal court. For others, this means understanding and valuing their decision not to engage with the criminal legal system and working with them to connect with alternate systems to reduce risk and meet their needs. While there are many emerging and established promising elder abuse interventions, there is still a long way to go in effectively meeting the needs of underserved populations and underfunded communities across the country.

For community-dwelling older adults, a coordinated community response is required to navigate different systems and meet the diverse and overlapping needs of older adults experiencing abuse. Enhanced Multidisciplinary Teams (EMDTs) are gaining traction as a hub for this coordinated response. EMDTs provide a venue for a diverse group of professionals to discuss and coordinate responses to complex cases of elder abuse and neglect. Each EMDT is comprised of representatives from adult protective services, law enforcement, District Attorney's Offices, local Offices for the Aging, attorneys, elder abuse prevention providers, medical professionals, and others working with older adults experiencing abuse. EMDTs are currently operating in many counties across the country.

Elder abuse shelter programs that can meet the skilled nursing needs of older adults experiencing abuse are a vital component of a complete coordinated community response. Without appropriate temporary housing solutions for older adults fleeing abuse, the range of other emergency and long-term services for older adults cannot adequately meet the needs of older adults experiencing abuse. Temporary elder abuse shelter programs provide safe shelter within a skilled nursing facility for older adults experiencing abuse that cannot access traditional shelters due to clinical or memory care needs. In addition to the clinical care provided at these shelter programs, residents work with a multidisciplinary team of case managers, social workers, and attorneys to address the emergent crises created by the abuse. Moreover, skilled nursing facilities have post-acute care capabilities, providing an opportunity to screen for and initially identify cases of elder abuse. The SPRiNG Alliance, a coalition of shelter programs across the country, is a hub for these programs and resources for those interested in starting and operating elder abuse shelters.

Similarly, emergency department-based elder abuse intervention programs, for example, the Vulnerable Elder Protection Team at Weill Cornell/New York-Presbyterian, are positioned to act as an important point of emergency intervention in cases of elder abuse. Older adults experiencing abuse often end up in an emergency department due to physical abuse or escalating health concerns due to abuse or neglect. An emergency department-based program specializing in elder abuse better equips the system to meet the complex needs of these patients through coordination with other providers and community partners.

For older adults experiencing mistreatment in residential care communities, including long-term care and assisted living, the state's Long-Term Care Ombudsman program serves as the first line of defense for an older adult experiencing abuse. Each state's federally funded program is required to respond to complaints, provide information about long-term services and supports, ensure access to services, represent resident interests to governmental agencies and seek appropriate remedies to protect residents, and provide input on changes to law and policy regarding residential care. The ombudsman can assist residents with a number of issues they may face, including abuse and neglect by staff or other residents.

29.1.4.3 Prevention

Social connection, access to resources, and education about elder abuse are at the heart of elder abuse prevention. Older adults must be able to connect with relevant providers and have meaningful access to systems of social connection and essential services. Professionals that work with older adults must be aware of red flags of abuse and appropriate resource referrals in their communities. Public education campaigns can be utilized to destigmatize self-reporting of elder abuse and lift up the importance of older adults in our communities.

Primary prevention of elder abuse addresses the root causes of elder abuse, including by taking affirmative steps to reduce ageism and affirm the value and importance of older adults in society more broadly. This requires systems designed with older adults in mind. The World Health Organization describes an "age-friendly society" as one that "adapts its structures and services to be accessible to and inclusive of older people with varying needs and capacities." The "Age-Friendly Health Systems Initiative" is a concrete example of a systematic approach to more reliable provision of high-quality, evidence-based health care for older adults.[4] Some communities have embraced this concept through creation of accessible public spaces, transportation that is responsive to the needs of older adults, and opportunities for older adults to engage with the community through cultural events and volunteering, but there is still significant work to be done across to ensure accessible and responsive systems for all older adults.

29.2 FINANCIAL EXPLOITATION: SPOTLIGHT ON SCAMS

In 2020, people over 60 lost approximately $1 billion to fraud. The number of older adults reporting financial exploitation has consistently increased since 2017, with a sharp increase in reports in 2020. Even this exorbitant number likely fails to grasp the scope of the problem due to underreporting.

29.2.1 PREVALENCE AND IMPACT

Unlike financial exploitation in the context of elder abuse, which occurs within trusting relationships, scams can be perpetrated by anyone who gains access to an older adult's money or assets by utilizing intimidation, threats, or relationship developed under false pretenses or government impersonation. In 2020, the most common scams targeting older adults were extortion, nondelivery of goods purchased, fake tech support, "grandparent scams," identity theft, phishing, spoofing, and "romance scams," and the average personal loss was over $9,000.

Scams are perpetrated over the phone, email, internet, or even in person and frequently target older adults regardless of cognition. While impaired cognition is certainly a risk factor for exploitation, new research demonstrates an increased risk of vulnerability to scams and financial exploitation for cognitively intact older adults without a dementia diagnosis, too. Older adults can be at higher risk of losing money to a scam due to typical changes to the brain and prefrontal cortex that are associated with a reduction in executive functioning and decreased ability to consider long-term consequences of financial decisions. "Age-associated financial vulnerability" describes recurring behavior that is inconsistent with past financial decision-making and results in an older adult being placed at substantial risk of loss of resources that would significantly change quality of life.[5]

Serious financial exploitation can have lasting impacts on an older adult's life. Financial exploitation causes feelings of fear, shame, and self-doubt, sometimes culminating in depression and hopelessness. For some older targets of scams, significant loss of assets – including a home, secure housing, and retirement savings – can destroy long-planned financial security and independence. For older, retired adults, there is little opportunity to rebuild personal financial security and independence. In fact, people over 50 who experience a sudden loss of wealth are at higher risk of death.

29.2.2 PREVENTION AND REPORTING

Scam perpetrators exploit their target's emotions and feelings of confusion, fear, sympathy, and love to fraudulently obtain money and other assets from their target. Effective scam prevention requires education to better instill confidence and savviness to consumers and other potential targets of scams. As more scams move to online spaces, with increasingly sophisticated tactics, internet users must be informed about internet safety, privacy, and red flags of scams.

Lack of awareness of financial scams and ever-changing tactics lead to older adults' increased vulnerability to exploitation and feelings of embarrassment and shame about falling prey to these scams are major contributors to underreporting. Increasing discussion of scams and public education and outreach campaigns that include peer contact (i.e., having an older adult who has experienced financial exploitation either colead sessions or appear on video) can increase older adult's ability to identify and avoid these scams in the first instance and reduce the stigma and embarrassment of reporting these scams after the fact.

Consistent reporting of scams is required to better understand scam prevalence, targets, and the ever-adapting tactics used by scammers. The federal government and state Attorneys General have established multiple avenues for reporting scams through the Federal Bureau of Investigation, Federal Trade Commission, Consumer Financial Protection Bureau, and state Offices of the Attorney General.

NOTES

1 Lifespan of Greater Rochester, Weill Cornell Medical Center, & NYC Dep't for the Aging, *Under the Radar: New York State Elder Abuse Prevalence Study* (May 2011).
2 Judy Smith, *Listening to Older Adult Parents of Adult Children with Mental Illness*, 15 *J Fam. Soc. Work* 126 (2012).
3 Laura Mosqueda, Kerry Burnight, & Solomon Liao, *The Life Cycle of Bruises in Older Adults*, 53 *J Am Geriatrics Soc.* 1339 (2005).
4 The John A. Hartford Foundation, *Age-Friendly Health Systems.*
5 Mark Lachs & S Duke Han, Age-Associated Financial Vulnerability: An Emerging Public Health Issue, 163 *Ann. of Internal Med* 877 (2015).

BIBLIOGRAPHY

ABA Dispute Res. Section Task Force on Elder Abuse and Neglect Screening Guidelines for Mediators, *Elder Abuse and Neglect Screening Tool* (June 2020), www.americanbar.org/content/dam/aba/administrative/law_aging/2020-elder-abuse-screening-tool-abadr-section.pdf

Admin for Comm. Living, *Projected Future Growth of Older Populations* (May 2021), https://acl.gov/aging-and-disability-in-america/data-and-research/projected-future-growth-older-population

Breckman, Risa, Levin, Malya, Mantrone, Leslie & Solomon, Joy, The Things They Carry: Advancing Trauma-Informed Responses to Elder Abuse (2020).

Cntrs. For Disease Control and Prev., *Violence Prevention: Elder Abuse* (June 2021), www.cdc.gov/violenceprevention/elderabuse/index.html

Dong, XinQi & Simon, Melissa, *Elder Abuse as a Risk Factor for Hospitalization in Older Persons*, 173 JAMA Intern. Med. 911 (2013).

Dong, XinQi et al., *Elder Self-neglect and Abuse and Mortality Risk in a Community-Dwelling Population*, 302 JAMA 517 (2009).

FBI, Elder Fraud Report (2020), www.ic3.gov/Media/PDF/AnnualReport/2020_IC3ElderFraudReport.pdf

Fulmer, Terry, *Screening for Mistreatment of Older Adults*, 108 Am. J Nursing 52 (2008).

Lachs, Mark & Han, Duke S, *Age-Associated Financial Vulnerability: An Emerging Public Health Issue*, 163 Ann. Internal Med 877 (2015).

Levin, Malya, Solomon, Joy & Lok, Deirdre, *Putting the "Sex" in Sexagenarian: Older Adults, Dementia and the Case of Henry Rayhons*, NY State Bar Assoc., Elder Law and Spec. Needs LJ (2016).

Mosqueda, Laura, Burnight, Kerry & Solomon Liao, *The Life Cycle of Bruises in Older Adults*, 53 J Am Geriatrics Soc. 1339 (2005).

Nat'l Cntr. Of Elder Abuse, *Tools Inventory*, https://ncea.acl.gov/Resources/Tools-Inventory.aspx (last viewed November 3, 2021).

Nerenberg, Lisa, Elder Abuse Prevention: Emerging Trends and Promising Strategies (2008).

Stanford Medicine, *Elder Abuse: Patient Barriers to Disclosure*, https://elderabuse.stanford.edu/screening/pt_barriers.html (last viewed October 31, 2021).

Teresi, Jeanne et al., *Methodological Approached to the Analysis of Elder Abuse Screening Measures*, 31 J Elder Abuse & Neglect 1 (2019).

The John A. Hartford Foundation, *Age-Friendly Health Systems*, www.johnahartford.org/grants-strategy/current-strategies/age-friendly/age-friendly-health-systems-initiative (last viewed October 31, 2021).

United State Dep't of Justice, *Elder Justice Network Locator Map*, www.justice.gov/elderjustice/elder-justice-network-locator-map (last viewed October 31, 2021).

US Dep't of Justice, *Older Adult Financial Exploitation* (2019), www.fbi.gov/file-repository/vsd-older-adult-financial-exploitation-brochure-2019.pdf/view

US Dep't of Justice, Red Flags of Elder Abuse, www.justice.gov/elderjustice/red-flags-elder-abuse-0 (last viewed October 31, 2021).

Wiglesworth, Aileen et al., *Screening for Abuse and Neglect of People with Dementia* 48 J Am. Geriatrics Soc. 493 (2010).

World Health Org., *Elder Abuse* (October 2021), www.who.int/news-room/fact-sheets/detail/elder-abuse.

30 Financing Care
How Clinicians Can Support and Prepare Their Patients

Nicole Braccio, Pharm D and Rebecca Kirch, JD

CONTENTS

30.1 FINANCIAL HEALTH LANDSCAPE OF OLDER AMERICANS

Preserving financial health and reducing risks of distressing medical debt for patients and their families are critically important factors for quality health care, patient safety, and improved health outcomes. Medical debt is the most common kind of debt in collections and the most common cause of bankruptcy in the United States.[1] The resulting financial stress has been shown to cause inflammation and can increase risk for a variety of health conditions from anxiety and depression to heart disease and memory impairment.[2] To avoid health risks, financial viability is particularly important to maintain because older adults on fixed incomes may have limited resources available to cover medical costs on top of their household expenses.

National level data demonstrate the extent of financial risk aging adults confront in the United States. Nearly one in three older adults aged 65 and up are living on the brink of poverty with less than $26,000 annual income available.[3] About half of households headed by an adult aged 55 and older have no retirement savings.[4] Notably, Black and Hispanic populations are twice as likely to experience poverty according to the 2020 Census data.[5] These factors interfere with older adults' ability to afford basic living needs and absorb financial shock of serious illness diagnosis or other unplanned circumstances.[6]

Looking beyond income and savings, food, energy, and housing security are also factors affecting health and safety for aging adults.[7] One in five food-insecure households include an older adult.[8] Concurrently, many seniors spend more than 30% of their income on housing[9] in the face of rising demand for affordable, accessible

DOI: 10.1201/9781003197843-30

housing along with expenses for home modifications, durable medical equipment, and other necessities to age in place.

Fortunately, safety net programs, direct assistance and personalized financial and social needs navigation programs and resources are available that provide skilled help to:

- Identify individuals' basic needs and primary concerns
- Demystify health insurance complexities and improve coverage literacy
- Decipher medical bills, disability eligibility, and other benefits
- Dispute coverage denials and harmful utilization review practices
- Negotiate payment plans and charitable program assistance opportunities
- Locate responsive safety net programs and community resources
- Minimize financial and other burdens related to social determinants of health

A diverse array of health and social service professionals that include patient and nurse navigators, community health workers, financial advocates, and social workers are often well-connected to local support networks and community resources and can help older adults coping with complex and chronic conditions work through their most pressing financial worries as part of treatment. While these services are not yet consistently accessible in all health care settings, advocacy efforts are underway to expand availability of high-quality needs navigation services so they can reach all patients and families who would benefit.[10]

Pro Tip: Never assume patients have what they need to afford treatments and services you recommend. It is important to ask whether costs are a worry so you can refer patients and caregivers to available assistance programs, resources, and/or skilled financial navigation services.

Here are prompts produced by VitalTalk (www.vitaltalk.org) for comfortably and effectively initiating cost of care conversations:

- "Many patients have financial concerns. How much are the costs of everything related to your medical care an issue for you?"
- "I am so glad you told me that finances are an issue. There are things we can do."
- "I have learned that finances can be the most stressful part of medical care."
- "I want to make sure that you receive all the resources that are available."
- "We have a specialist who can connect you to financial counseling and resources."
- "Please keep me updated on this and if you don't mind, I will check in with you later."

A patient- and caregiver-facing tip sheet available online from National Patient Advocate Foundation at www.npaf.org/resources/can-we-talk-about-my-costs-of-care/ can help them prepare for talking about total costs of care.

30.2   THE BASICS OF HEALTH INSURANCE FOR OLDER ADULTS

Obtaining health insurance coverage can be a complicated and stressful process for older adults and seniors aged 65 and older. Early discussions and planning can mitigate their distress. Older adults typically have three health insurance options to choose from:

1. *Original Medicare or a Medicare Advantage (MA) plan* when they turn 65 years old. Covered benefits, services, and associated costs are largely standardized in Original Medicare while MA plan coverage and costs vary considerably by the insurance issuer.
2. If an individual qualifies for Medicaid due to their income, they may be *dually eligible for a combination of Medicare and Medicaid*. This combination offers a range of covered benefits at an affordable cost to patients including long-term care that will be discussed later in this chapter.
3. Finally, an individual may choose to *extend their group health coverage from a former employer* (retiree coverage). This can also serve as a secondary insurance to supplement Original Medicare. Covered benefits and services of retiree coverage vary by plan which is administered by private insurance companies.

30.3   MEDICARE'S COVERED BENEFITS

Most seniors in the United States rely on Medicare for their insurance coverage.[11] Enrollment has continued to increase rapidly as the rest of the baby boomer generation ages into Medicare.[12] The chart below outlines the basic structure and covered benefits of Original Medicare (administered by the federal government) and MA plans (administered by private insurance companies). MA plans are intended to manage and coordinate beneficiaries' care.

Original Medicare		Medicare Advantage (Part C)
Part A – Care in a hospital, skilled nursing facility, hospice, some home health		Part A – Care in a hospital, skilled nursing facility, hospice, some home health
Part B – Doctor visits, medical equipment	– OR –	Part B – Doctor visits, medical equipment
Part D – Medications (add on separately)		Part D – Medications (usually included)
Supplemental plan you can add on to help pay for costs (Medigap or retiree plan)		Most plans include extra benefits like vision, hearing, and dental

Pro Tip: Remind older patients nearing 65 that *Social Security* will send a packet of information for Medicare three months before their 65th birthday.

Late enrollment penalties may apply for not signing up when it's first available (for both Part B and D).

Individuals consider a variety of factors when deciding between the two options such as limits on out-of-pocket spending, extra benefits, preapproval requirements, and whether their doctor is a participating provider in the local MA plan network. Survey results from the Centers for Medicare and Medicaid Services (CMS) suggest that MA and Original Medicare beneficiaries have comparable health care experiences and express similar satisfaction with the quality of care. Importantly, both groups of beneficiaries reported that their out-of-pocket costs do create barriers to needed health care.[13] Over the last decade, enrollment in MA has doubled and comprises a larger percentage of overall Medicare enrollment each year.[14]

Pro Tip: Refer your patients to a trained Medicare counselor through their State Health Insurance Assistance Program for free, personalized help with evaluating their options. They can find a counselor online at www.shiphelp. org or call 877-839-2675. Patients can prepare by reviewing National Patient Advocate Foundation's Step-By-Step Guide for Choosing a Medicare Plan found at www.npaf.org.

Medicare pays for much, but not all, of the costs for covered health care services and supplies. In fact, most Original Medicare beneficiaries have supplemental coverage, either through Medigap which offers ten different options, employer-sponsored retiree health coverage or Medicaid.[15] To prepare for future health expenses, it is worth discussing with your patients the importance of supplemental coverage if they are considering Original Medicare.

Older adults should also be aware that Original Medicare does not cover every item or service they may need such as:

- Long-Term Care also known as custodial care (i.e., institutional or home/community settings including personal care services that help with activities of daily living)
- Most dental care
- Eye exams related to prescribing glasses
- Dentures
- Cosmetic surgery
- Acupuncture
- Hearing aids and exams for fitting them
- Routine foot care[16]
- Health care outside of the United States with some exceptions[17]

MA plans may offer the above extra benefits, but actual benefits and coverage vary by plan.

Due to the phasic nature of Medicare's Part D benefit design, prescription drugs may expose patients to significant out-of-pocket costs, particularly for high-cost specialty drugs. As of January 1, 2020, the Part D benefit effectively has three different coverage phases each with their own cost-sharing obligations shown below:[18]

Initial Deductible Phase *Begins on January 1 each year*	Patient pays 100% of costs up to $480*
Initial Coverage Period + Former Coverage Gap "the donut hole" *This period begins once deductible is met.*	Patient pays 25% of costs
Catastrophic Benefit Period *When an enrollee's total out-of-pocket spending reaches $7,050**	Patient pays 5% of costs

*Thresholds for each phase are subject to change and may increase from year to year.

Prior to 2020, patients were exposed to 100% of the costs in the donut hole and shouldered even more financial responsibility for their prescription drugs. Even without the donut hole, the Part D benefit design often causes confusion at the pharmacy counter as patients experience vastly different cost-sharing obligations for the same medication at different points in the year.

Pro tip: Discuss the Part D coverage phases and associated costs with your patients, especially if they take specialty drugs that would expose them to high cost-sharing amounts before the catastrophic phase.

Your patients may worry about paying for health care costs in Medicare given the coverage gaps and excluded benefits which can be expensive as described above. Fortunately, many beneficiaries with income and resources below a certain limit may qualify for one or more of the financial assistance programs administered through Medicare and listed in the following chart.

For Help With:	The Following Programs May Be Applicable:
Prescription Drug Costs in Part D	**Extra Help** is a program for people with limited income and resources to assist in paying Medicare drug costs. If you qualify for programs 1, 2, or 3 below, you'll automatically qualify for Extra Help. This is also known as "**Part D Low-Income Subsidies.**" Beneficiaries can apply through Social Security.
Medical costs such as premiums and cost-sharing	Beneficiaries may qualify for one of the following four **Medicare Savings Programs**:

For Help With:	The Following Programs May Be Applicable:
Part A and/or B costs	1. Qualified Medicare Beneficiary (QMB) Program This program also ensures that patients won't be billed for deductibles, copays, or coinsurance
Part B premiums only	2. Specified Low-Income Medicare Beneficiary (SLMB) Program
	3. Qualifying Individual (QI) Program
Part A premiums only	4. Qualified Disabled and Working Individuals (QDWI) Program

Charitable assistance programs may also be an option available for those who need help with Medicare's costs. Nonprofit hospitals must offer financial assistance, or charity care, for emergency or medically necessary care. The eligibility requirements vary by State. Patient Advocate Foundation (www.patientadvoc ate.org) offers a consumer-friendly online financial resources directory as well as personalized needs navigation support services by phone and other assistance for referred patients and caregivers to get insurance, access, and affordability guidance they may need.

30.4   FINANCING LONG-TERM CARE

Most people will eventually need long-term care, services, and supports to help sustain their quality of life as they age with complex chronic or disabling conditions. This personal care assistance can be provided through paid professionals or unpaid caregivers and often lasts several weeks, months, or years when patients have trouble completing self-care tasks and activities of daily living.

Long-term care broadly refers to institutional care as well as in-home and community-based aid with activities of daily living such as bathing, dressing, housekeeping, and managing meals.[19] Services in long-term care may include:

- Nursing facility care
- Adult daycare programs
- Home health aide services
- Personal care services
- Transportation
- Supported employment
- Family caregiver assistance
- Care planning and coordination

Long-term care is often expensive and typically requires people to rely on personal funds, government programs, and private financing options to pay for it. Medicaid is the primary payer for a range of institutional and community-based long-term supports and services that is reserved for low-income patients and families with eligibility and covered services varying by State.[20] The Program of All-Inclusive Care for the Elderly, known as PACE, is another important Medicare and Medicaid program available in some states that covers community-based medical, social service, and long-term care costs for frail people ages 55 and older who otherwise would need care in a nursing home.[21]

Pro Tip: Inform patients as they near retirement age about the importance of planning financially for long-term care expenses. Refer patients to a needs navigator, social worker, or financial counselor to determine eligibility for Medicaid, other government programs, and financing options. Consulting with an attorney may also be helpful for patients to assess the legal options associated with qualifying for Medicaid through a "spend down" approach.

Financial navigators and counselors are best suited to assist patients with understanding private financing options including long-term care insurance, reverse mortgages, certain life insurance policies, annuities, and trusts. The long-term care insurance market has changed greatly in recent years with introduction of hybrid options that are tied to life insurance. Experts suggest that while everyone needs a plan for long-term care, not everyone needs long-term care insurance. Some considerations include:[22]

- Affordability of monthly premiums which increases based on age at the point of purchase.
- Waiting periods before benefits commence may span 30, 90, or 180 days.
- Patients are not denied coverage for preexisting conditions; however, the policy may not necessarily cover care related to that condition.

30.5 LOOKING BEYOND DIRECT HEALTH CARE COSTS

Older adults, especially those with limited resources, often worry about a constellation of financial and social concerns including household material hardships. Government-funded safety net supports can help many people afford basic living expenses on top of medical care. A few examples are listed below:

- **Social Security Disability Insurance**, a program available to adults and children with disabilities who have limited income and resources, though it largely serves older adults (50 years and older) with severe physical or mental impairments.
- **Supplemental Security Insurance (SSI)**, a program for low-income older adults and people with disabilities. SSI payments are also made to people aged 65 and older without disabilities who meet the financial qualifications.
- **Supplemental Nutrition Assistance Program** (formerly food stamps). National Council on Aging estimates that almost 60 % of older Americans eligible for the program are not enrolled.

Pro Tip: Refer patients to the National Council on Aging's free and comprehensive tool that locates programs to save money on health care, housing and utilities, nutrition and more at www.benefitscheckup.org.

Another valuable resource for older adults is the Aging Network which was established by the Older Americans Act in 1965. It is comprised of Area Agencies on Aging (AAA) which serve as local entities who, either directly or through contract with local service providers, oversee a comprehensive and coordinated service system for the delivery of social, nutrition, and long-term services and supports to older individuals.[23] AAA services and supports include:

- Outreach, information and assistance, care management, transportation
- Nutrition programs and education
- Home and community-based services like home care, adult day care, national family caregiver support program
- Disease prevention and health promotion
- Vulnerable elder rights protection
- Additional help may be available for Native American elders

Pro Tip: The Eldercare Locator is a public service that connects older adults and their families to their local AAA. Access the tool online at www.eldercare. acl.gov or by phone at 1-800-677-1116.

30.6 CONCLUSION: INCLUDE COSTS CONVERSATIONS EARLY AND OFTEN

Financial viability is a critically important goal of care for many patients, yet they may be apprehensive about proactively raising their cost concerns or financial situation with you. In today's health care environment where continuously rising costs are the norm, clinicians and health systems can no longer afford to overlook patients' financial worries and distress. The consequences for patients and families in terms of treatment adherence, quality of life, and other outcomes can otherwise be dire. Every patient encounter offers a prime opportunity to normalize these conversations by acknowledging that costs concerns are common and can be alleviated using available resources, specialized financial and social navigation services, and other direct assistance support. Using the pro-tips and resources highlighted in this chapter will help practitioners meet patients where they are by addressing the financial and social supports they typically need.

NOTES

1   Financial Health Network Blog. Reducing the Impact of Healthcare Costs on FinHealth. June 28, 2021.

2 Wellmark. The costly impact of financial stress: A problem that impacts the health and productivity of your employees. Updated April 27, 2020.

3 Juliette Cubanski, Wyatt Koma, Anthony Damico, and Tricia Neuman. Issue Brief: How Many Seniors Live in Poverty? Kaiser Family Foundation. November 2018.

4 Government Accountability Office. Report to the Ranking Member, Subcommittee on Primary Health and Retirement Security, Committee on Health, Education, Labor, and Pensions, U.S. Senate. Most Households Approaching Retirement Have Low Savings. GAO-15-419. May 2015.

5 Emily A. Shrider, Melissa Kollar, Frances Chen, and Jessica Semega. Income and Poverty in the United States: 2020. Report Number P60-273. Table B-5. Poverty Status of People by Age, Race, and Hispanic Origin: 1959–2020. September 14, 2021

6 Andrew Dunn, Thea Garon, Necati Celik, Jess McKay. Financial Health Pulse. 2021 U.S. Trends Report. The Financial Health Network.

7 Food Research and Action Center. Hunger Is a Health Issue for Older Adults: Food Security, Health, and the Federal Nutrition Programs. December 2019.

8 Food Research and Action Center. Hunger Is a Health Issue for Older Adults: Food Security, Health, and the Federal Nutrition Programs. December 2019.

9 Joint Center for Housing Studies of Harvard University. Projections & Implications for Housing a Growing Population: Older Households 2015–2035. December 2016.

10 National Patient Advocate Foundation "Health *Needs* Navigation" advocacy campaign at www.npaf.org/issues/needs-navigation/ (accessed November 21, 2021).

11 Centers for Medicare and Medicaid Services Fast Facts Mobile Site. CMS Program Data – Populations. July 2021 Edition.

12 Leslie Read and Claire B. Cruse. Deloitte Analysis. Understanding the next wave of Medicare enrollees. 2017.

13 Gretchen Jacobson et al., Medicare Advantage vs. Traditional Medicare: How Do Beneficiaries' Characteristics and Experiences Differ? Commonwealth Fund, October 2021.

14 Meredith Freed, Jeannie Fuglesten Biniek, Anthony Damico, and Tricia Neuman. Medicare Advantage in 2021: Enrollment Update and Key Trends. June 21, 2021.

15 Wyatt Koma, Juliette Cubanski, and Tricia Neuman. Kaiser Family Foundation. A Snapshot of Sources of Coverage Among Medicare Beneficiaries in 2018. Published March 23, 2021.

16 What's not covered by Part A and Part B? Medicare.gov.

17 Department of Health & Human Services. Medicare Coverage Outside the United States. CMS Product No. 11037. Revised January 2021.

18 National Council on Aging. Tools & Training for Professionals Donut Hole: Who Pays What in Part D. August 17, 2021

19 National Institutes of Health. National Institute on Aging. What Is Long-Term Care? Web page available at: www.nia.nih.gov/health/what-long-term-care

20 U.S. Department of Health and Human Services Assistant Secretary for Planning and Evaluation Office of Disability, Aging and Long-Term Care Policy. An Overview of Long-Term Services and Supports and Medicaid: Final Report. May 2018.

21 The Centers for Medicare and Medicaid Services. Medicare.gov Web page available at: www.medicare.gov/your-medicare-costs/get-help-paying-costs/pace

22 Kim Painter. Understanding Long-Term Care Insurance Coverage Basics, Premium Costs and Policy Options to Make an Informed Decision. AARP. October 13, 2021.

23 Kirsten J. Colello and Angela Napili. Congressional Research Service. Older Americans Act: Overview and Funding. Updated April 22, 2021.

31 Health Inequities
Closing the Disparities Gap in the Aging Population

Juliana M. Mosley-Williams and Melissa A. Vitek

CONTENTS

31.1   INTRODUCTION AND BACKGROUND

People are aging and living longer. The population in the United States is aging rapidly in large numbers due to the Baby Boomer generation at an estimated 73 million, who will all be at least age 65 by 2030.[1] More specifically, the number of Americans age 65 will nearly double to 98 million by 2060, and those age 85 and older, the group most often needing help with basic care, is projected to more than double from 2016 to 2040.[2] This is causing a demographic shift referred to as the "gray tsunami."[1]

DOI: 10.1201/9781003197843-31

Another shift focuses on racial/ethnic communities of color who constitute the fastest growing segment of the elderly population.[3] With these demographic changes, traditional health care sector-focused interventions have proven to be insufficient as a primary strategy to address population-level health disparities, including those identified in the aging population.[4]

31.1.1 AGING

The experience of aging within and among various populations can be vastly different. Geroscience is an emerging interdisciplinary field that focuses on the association between aging and chronic age-related diseases (ARDs) and geriatric syndromes (GSs).[5] For example, it has been determined that there are some molecular mechanisms known to cause aging that occur in several ARDs/GSs, such as frailty, sarcopenia, chronic obstructive pulmonary disease, cancer, and neurodegenerative diseases including Alzheimer's and Parkinson's diseases.[5] A result of the analysis conducted by members of this field is the shift in the focus of medicine to combating the aging process instead of focusing mainly on the diagnosis and management of specific ARD/GS.

31.1.2 AGEISM

How a person is viewed by their communities impacts their aging experience. Ageism is defined as stereotyping, prejudice, and discrimination based on a person's age.[6] Ageism is prevalent across the world, and the United States is no exception. Americans' tendencies to exclude and undervalue older people are well documented.[6] One 2009 Pew study found significant differences between younger people's negative perception of aging and the actual experiences of older people. In every area of the survey, including illness, memory loss, inability to drive, and sexual activity, younger adults predicted that they would have more negative experiences as they aged than older adults actually report. Ageism impacts different segments of the population in unique ways and its impact needs to be incorporated into strategies for improved quality of life for all members of our aging population.

31.1.3 HEALTH DISPARITIES

As people age, concern for health becomes a greater reality. Health is impacted by many dimensions of identity and social factors that affect people and the characteristics of their community. These factors are referenced as social determinants of health (SDOH) and defined as the conditions in the environments where people are born, live, learn, work, play, worship, and age that affect a wide range of health, functioning, and quality-of-life outcomes and risks.[7] Differential access to SDOH both creates and maintains unjust and avoidable health inequities.[8] Health disparities are differences in any health-related factor such as disease burden, diagnosis, response to treatment, quality of life, health behaviors, and access to care that exist among population groups.[7] Health disparities research related to aging is the study of biological, behavioral, sociocultural, and environmental factors that influence population-level health differences.[7] There are numerous SDOH (race/ethnicity, economics, culture, disability status, education, geographic location, biological, health care, and behaviors), which are commonly grouped in five categories listed below, and serve as the guide for how these factors lead to health disparities and inequities of our aging population.

- *Behavioral factors* and psychological processes can positively or negatively impact health, while chronic stress can increase vulnerability for poor health.
- *Social factors* such as individual and systemic racism, discrimination, and bias can shape the everyday experience of marginalized or vulnerable populations.
- *Cultural factors* related to norms, traditions, and both individual and collective responses to aging and health. They can influence how individuals manage stress, diet, physical activity, and other critical health/coping behaviors.
- *Environmental factors* related to education, income, occupation, retirement, and wealth impact health over the life course.
- *Biological factors* that are influenced by environmental and sociocultural factors may alter the course, severity, and acceleration of disease and disability.[9]

This chapter will (1) provide an overview of the factors that lead to health disparities in older patients and (2) identify strategies to reduce those disparities.

31.2 FUNDAMENTALS AND FRAMEWORKS

Population-level differences in health status and life expectancy are well-documented.[7] Older US racial and ethnic populations of color (Asian, Black/African American, Hispanic, and Native American) suffer premature morbidity over the life course. Other defining population features that play a role in population health differences have been identified as age, gender, disability status, and gender identity. The National Institute on Ageing Health Disparities Research Framework was developed to illuminate and categorize factors examined in health disparities research related to aging.[7] Their approach includes the identification of factors at multiple levels of analysis and provides the opportunity to more effectively differentiate causal pathways to health experiences, and disease manifestations linked to several patient important outcomes. Ultimately, considering multiple levels of analysis broadens the scope of interventions and benchmarks for reducing health disparities in the aging population. Furthermore, utilizing a life-course perspective allows for implementation and measurement of impact along the entire course of a patient's lifespan.

31.2.1 Behavioral Influences on Health

There are different resources individuals bring to challenging situations due to variations in psychological capacities such as problem-solving skills, emotional regulatory strategies, and propensity for seeking support.[7] Individual attitudes, including optimism, pessimism, and sense of control, can also serve as additional risk or resilience factors that shape encounters with social and environmental stressors.[7] These characteristics fall under the broad umbrella of individual determinants of health.[12]

Health behaviors represent a more linear relationship with how a person addresses stressful circumstances. Engaging in behaviors that are categorized as healthy may help regulate emotional responses to stressors and foster a person's sense of control over a particular situation. Exercise, meditation or prayer, and social engagement are all linked to more positive responses to stress. Conversely, tobacco use, alcohol abuse, illegal substance abuse, and poor nutritional habits are associated with diminished health status and when sustained over several years, compromise healthy aging, and even lead to a reduced life expectancy.[7]

31.2.2 SOCIAL INFLUENCES ON HEALTH

Social factors of racism and discrimination have a long and varied history with vulnerable populations in the United States. These factors influence sociocultural norms and values through individual-level self-concepts, perceptions, and cognition.[7] In addition, conscious and unconscious biases shape individual behavior and may impact interpersonal relationships and interactions with social institutions leading to the compromise of health at the individual level.[7] For example, the limited availability of COVID-19 testing within Black/African American, Hispanic, and Native American communities disproportionately affected treatment and morbidity in those populations, solidifying perceived restricted access to quality health care as institutional racism and discrimination. These perceptions may lead historically vulnerable populations of color to further distrust health care systems and negatively impact health status and life expectancy. Institutional racism can also harm health through stigma, stereotypes, and prejudice, which can be factors to stagnate socioeconomic mobility and reduce access to societal resources and opportunities required for health.[10]

31.2.3 CULTURAL INFLUENCES ON HEALTH

Culture is defined as a cumulative deposit of knowledge acquired by a group of people over the course of generations.[11] Culturally associated health beliefs may provide additional insight as to why disparities exist as well as provide information on how to provide culturally appropriate services to marginalized racial/ethnic older adults. Different cultural beliefs about health and health care may influence the type of treatment that is sought, the types of providers a patient more readily trusts, and how illness is addressed and managed.[3] Self-perceived health status is informed by cultural beliefs and is defined as an individual's own beliefs about their physical and mental health status and social well-being.[12] Cultural factors affect vigilance and confidence that individuals may summon when faced with limited access to health resources. For example, while Blacks/African Americans have endured chronically low socioeconomic status, structural bias and inadequate access to quality health care, this population group has historically created and participated in productive networks of exchange and support.[7]

31.2.4   ENVIRONMENTAL INFLUENCES ON HEALTH

This category encompasses a broad spectrum of components, including income, education, occupation, and wealth, that may affect exposure to harmful chemicals, air pollution, poverty, crime, and violence.[7] Vulnerable populations from lower income, education, or occupational status experience worse health and die earlier than those from higher socioeconomic backgrounds.[8] More specifically, aging populations from low socioeconomic status experience greater disability, more limitations in daily living activities, and more frequent and rapid cognitive decline.[13] Structural factors that are geographic and political play a role and are linked to residential segregation that limits opportunities for social mobility and constrains some population groups to risky physical environments.[7] Inadequate educational preparation associated with poorly funded public schools may weaken chances for college entrance and needed professional training to enter the workforce, which may be compounded and limiting over the lifecycle.[4] High levels of unemployment, underemployment, and employment dissatisfaction may result, restricting important resources for protecting health. One very important resource for protecting health is access to quality health care. While health care availability and quality are critical for accurate diagnosis and treatment of disease, this resource may be influenced by residential segregation that affects health insurance, quality nursing homes, and availability of hospice.[7]

31.2.5   BIOLOGICAL INFLUENCES ON HEALTH

The epidemiology of the leading causes of death in the United States by race/ethnic groups reveals striking differences that are not fully explained by the environmental, social, cultural, and behavioral factors.[7] There are biological processes that can help explain mechanisms of observed differences in disease incidence and outcomes that are most relevant for populations that suffer premature mortality as well as those who endure harsh environmental conditions with exceptional resilience. Specifically, there are many connections that have been made between stress levels and cellular stability.[7] Additionally, groups of individuals that experience harsh social environments that are perceived as stressful may also engage in coping behavior that sparks certain biological processes that accelerate aging and/or undermine health.[7]

31.3   STRATEGIES FOR REDUCING HEALTH INEQUITIES IN THE AGING POPULATION

31.3.1   ACTION PLANS ON AGING AND HEALTH

The World Health Organization Global Strategy and Action Plan on Aging and Health include five strategic objectives (see Table 31.1).[14]

31.3.2   ALLOCATION OF RESOURCES

Life expectancy increased throughout the second half of the twentieth century but in recent decades this increase has come with more years spent in poor health.[15]

TABLE 31.1
Strategic objectives for aging and health as defined by the World Health Organization

Commitment to action on healthy aging in every country.	• Establish national frameworks for action on healthy aging • Strengthen national capacity to formulate evidence-based policy • Combat ageism and transform understanding of aging and health
Developing age-friendly environments.	• Foster older people's autonomy • Enable older people's engagement • Promote multisectoral action
Aligning health systems to the needs of older populations.	• Orient health systems around intrinsic capacity and functional ability • Develop and ensure affordable access to quality older person-centered and integrated clinical care • Ensure a sustainable, appropriately trained, and managed workforce
Developing sustainable and equitable systems for long-term care.	• Establish and continually improve a sustainable and equitable long-term care system • Build workforce capacity and support caregivers • Ensure the quality of person-centered and integrated long-term care
Improving measurement, monitoring, and research for healthy aging.	• Agree on ways to measure, analyze, and monitor healthy aging • Strengthen research capacities and incentives for innovation • Research and synthesize evidence on healthy aging

People living with socioeconomic disadvantage are more likely to develop disease or die earlier than those living in more advantageous circumstances. This pattern has been described as the social gradient, where the risk of poor health tends to increase in-step with declines in socioeconomic position.[15] The social gradient demonstrates the need for policies and interventions that elevate the health of the worst off to the highest level achievable within society. One approach to responding to this need is "proportional universalism" and is described as policies that are universal and benefit everyone in society, but that are at a scale and intensity that are proportionate to the level of disadvantage.[15] Extensive evidence supports the conception that in order to reduce health inequalities at older ages, policies and interventions need to address SDOH early in life and across a person's life course.[15]

31.3.3 RESEARCH AND DATA COLLECTION: PROMOTING EQUITY AND LIFE-COURSE PERSPECTIVES

Current intervention strategies to reduce health disparities do not typically take a "life-course perspective" and tend to be disease-specific, often targeting individual and health systems factors without addressing social determinants.[7] There is a growing consensus that utilizing a lifepath perspective, meaning focusing on how experiences early in life can impact health over a lifetime and even across generations, is one of the most significant strategies to improve the nation's health and is critical to eliminating population-level health disparities.[7] Research efforts should also be placed on strengthening the evidence base through research aimed at more comprehensively understanding the economic impact of widespread implementation of social determinants-targeted interventions.[4] Additionally, new directions for health disparities research should include multiple levels of analysis including the integration of key SDOH by interdisciplinary teams.[7] Increased understanding of the complex relationships among these factors within and between population groups can lead to critical insights and effective implementation of interventions that can improve the lives and aging experience for all.[7]

31.3.4 UNDERSTAND RACIAL/ETHNIC AND SOCIOECONOMIC HEALTH DISPARITIES

Marginalized communities are often highlighted and discussed regarding their limitations of wealth, power, and privilege, which overshadow their vast knowledge of marginalization based on lived experience. It took a pandemic to magnify voices of marginalized people. The COVID-19 pandemic raised consciousness and positioned a national conversation and agenda on the impact of race and racial discrimination on health disparities for communities of color (specifically Blacks/African Americans, Native Americans, Hispanics, and Pacific Islanders) in the United States.

While the volume of this national conversation appears to have increased more recently, the research study, Trends in US Public Awareness of Racial and Ethnic Disparities in Health (1999–2010), conducted by the NORC at the University of Chicago (NORC), with the US Department of Health and Human Services Office of Minority Health, revealed the extent to which the US population is aware of racial and ethnic health disparities, and how these perceptions have changed over the last decade.[16] In general, the data showed that people are more aware of disparities in health care access (i.e., health insurance coverage, costs, and access to providers and quality care). The study also revealed that while Blacks/African Americans are more likely to be aware of disparities, including the disproportionality of serious diseases and conditions, awareness of disparities is relatively low for the overall population.[16]

While racial disparities contribute to the severity of diseases and health conditions, access, quality, and intensity of health care are also negatively impacted by lower socioeconomic levels, which are often correlated with communities of color. Racial/ethnic differences in SES are large and contribute to racial/ethnic differences in health.[17] While the median incomes of Asian, Black/African American, Hispanic, and White households all increased from their 2018 medians, income disparity remains for Black/African American and Hispanic households when compared to Whites. In 2019, the median Black/African American household earned only 61 cents for every dollar of its White counterpart, while the median Hispanic household earned 74 cents.[18] This disparate difference in income may affect other areas of SES to include education, occupational status, access to health insurance, and housing, which affect specific communities or individual's power and privilege. Research indicates that SES tends to be a stronger factor related to variation in health than race, and SES disparities in health are evident within each racial group.[17] These effects amass throughout a person's lifetime with devasting outcomes. In summary, there is compelling evidence that race/ethnicity and SES are linked with disparities in health outcomes. Policies that seek to weaken that link will lead to better health outcomes for all.

31.3.5 UNVEIL THE BIOLOGICAL AND HEALTH EFFECTS OF DISCRIMINATION

Racism and discrimination have a storied history in the United States dating back to colonization, affecting all areas of society, including health. Research indicates that Black people in the United States are more likely than their White counterparts, to have comorbidities, greater severity, and more rapid progression of diseases. Where data is available, these factors show similar patterns for other populations of color, to include Native Americans, Native Pacific Islanders, Hispanics (US-born or long-term residence), and Asians (from lower SES).[17] When levels of income and education are equal, a growing body of evidence suggests that bias (conscious and unconscious) leads to patterns of racial/ethnic discrimination in employment, housing, applying for loans, and even daily tasks of hailing a taxi and shopping, which are a source of psychosocial stress.[19] An overview of several studies revealed that self-reported measures of discrimination were adversely associated with several indicators of health (i.e., hypertension, incident asthma, incident breast cancer, poor mental health, and all-cause mortality), several indicators of clinical disease (i.e., inflammation, visceral fat, obesity, coronary artery calcification, and cortisol dysregulation), and individual health behaviors (i.e., poor sleep quantity and quality, cigarette smoking, and substance use).[17] In addition to experiencing discrimination, the threat of discrimination was also related to increased cardiovascular response, symptoms of poor mental health, and hypertension. The most pervasive form of discrimination for people of color is racism and an older population that has been victim to life-long personal instances or the effects of institutional racism are arguably more vulnerable and likely to experience increased disparities in health.

31.3.6 REDUCE THE IMPACT OF STRESS RELATED TO SOCIOCULTURAL INFLUENCES

Adverse exposure to sociocultural factors related to social marginalization (systematic racism, discrimination, and bias) and SES (income, education, and occupation) can tax individual capacity to maintain control and stability, leading to psychosocial stress.[7] This stress, continued long-term or layered from multiple factors, can become chronic, leading to physiological effects, including adverse metabolic, autonomic, and brain effects such as hippocampal atrophy.[13] While the systemic effects of racial disparities in health care warrant a comprehensive dismantling of its integrated structures, greater awareness, education, and training among health care providers related to sociocultural factors could be one means to reduce stress for vulnerable aging populations.

This awareness should begin with providers realizing the pervasiveness of disparities, the ways in which bias can influence provider decision-making and behavior, and a commitment to learning that minimizes those practices.[17] Health care institutions, schools, and organizations should work to increase diversity within the

TABLE 31.2

Interventions to reduce the impact of race-related stress as defined by the American Psychological Association

Understand the impact of racism	Racism, felt daily, also intersects with other forms of discrimination, including ageism, classism, sexism, ableism, and heterosexism. Aim to understand the lens through which individuals view their experiences.
Listen with empathy	Recognize and acknowledge past and present experiences. Provide support for the thoughts, feelings, and experiences shared.
Create safe spaces	Provide opportunities for dialogue around race, culture, gender, sexuality, and socioeconomic issues.
Support, strengthen and enhance resilience	Resilience refers to the ability to successfully adapt and overcome negative life events such as stress and trauma. Support spirituality, family, ties, and strong positive racial group identity.
Celebrate culture	Organize and encourage activities that celebrate life, history, culture, customs, and norms.
Be mindful of triggers	Pay attention to the impact of stressful events and incidents showcased in the media (e.g., television, radio, or newspaper). When possible, limit exposure.
Be aware	Consider your own cultural background and its influence on your values, beliefs, assumptions, and biases. Seek out training opportunities on culturally competent geriatric care and race-related stress.
Refer patients to mental health services and support	When needed, culturally competent mental and behavioral health services should be made available.
Advocate	Support and promote local and national efforts to increase access to community and national resources for African Americans across the lifespan.

health professions and work toward reducing discrimination and its adverse effects on health and health care.[17] Through its fact sheet, African American Older Adults and Race-Related Stress, the American Psychological Association provides the following suggestions for health care providers as a guide in the implementation of interventions to reduce the impact of said stress (see Table 31.2).[20]

31.3.7 CULTIVATE INCLUSIVENESS AND A SENSE OF BELONGING

The Frameworks Institute's research on how people think about aging, older people, and elder abuse found that many people swiftly and unconsciously draw a line between older people and the "rest of us."[6] This "us vs. them" mindset is not restricted to how we think about older people and the problems it generates it contributes to a range of social issues, including racial and economic discrimination.[6] Regardless of the circumstances, this mentality diminishes support for the inclusion of certain groups in public policies by creating a zero-sum mentality: more for "those" or "them" means less for "me" and "mine."[6] Some research also suggests that, during challenging times, people's bias for and positive feelings about individuals in their "in-group" (people like them) heighten while their positive feelings about "out-group" members decline.[6] Through quantitative experiments and focus-group research, it was found that messages were effective that centered on the idea that justice requires that we recognize all members of society as equal.[6] They increased people's support for policies designed to support older people of all cultural and racial backgrounds.[6] Advocates need to continue to shape messaging so that it appeals to justice, illustrates what ageism, racism, and discrimination look like, and be purposeful about utilizing inclusive language. Doing so will help policy makers, journalists, and the public recognize that all older people – like all people – can bring value to communities everywhere and deserve to be included equitably.[6]

31.4   CONCLUSION

America is aging and health care is advancing, yet inequities and disparities continue to plague the health care system and impact this already vulnerable population. It is important to understand how social determents of health (behavior, social, cultural, environmental, and biological factors) influence individual's health and health care. Continued research, understanding, and a broader commitment to dismantle structural and systemic inequities are needed to close the disparities gap. Strategies offered as more immediate measures to increase equity in health and health care include better allocation of resources, research and data collection that promotes equity and life-course perspectives, increased awareness and understanding of racial/ethnic and socioeconomic disparities, an unveiling of the biological and health effects of discrimination, reducing the impact of stress related to sociocultural influences, and the cultivation of inclusiveness and belonging. Ultimately, the elderly population, their health, and health care needs must be treated as priority, given their growing status in society.

REFERENCES

[1]   America Counts Staff. United States Census Bureau. www.census.gov/library/stories/2019/12/by-2030-all-baby-boomers-will-be-age-65-or-older.html. Published December 10, 2019. Accessed September 24, 2021.

[2]   Gabriel B. By 2040, One in Five Americans Will Be Over Age 65. AARP.org. www.aarp.org/politics-society/history/info-2018/older-population-increase-new-report.html. Published May 7, 2018. Accessed November 1, 2021.

[3]   Jimenez D, Bartels S, Cardenas V, Daliwal S, Alegría, M. Cultural beliefs and mental health treatment preferences of ethnically diverse older adult consumers in primary care. *American Journal of Geriatric Psychiatry*. 2012 June; 20(6): 533–542. doi:10.1097/JGP.0b013e318227f876

[4]   Thornton R, Glover C, Cené C, Glik D, Henderson J, Williams D. Evaluating strategies for reducing health disparities by addressing the social determinants of health. *Health Affairs (Millwood)*. 2016 August 1; 35(8): 1416–1423. doi:10.1377/hlthaff.2015.1357

[5]   Franceschi C, Garagnani P, Morsiani C et al. The continuum of aging and age-related diseases: common mechanisms but different rates. *Frontiers in Medicine*. 2018 March 12; 5:61. doi:10.3389/fmed.2018.00061

[6]   Kendall-Taylor N, Neumann A, Schoen J. Advocating for age in an age of uncertainty: how the COVID-19 crisis is amplifying ageism, and how advocates can push back. SSIR.org. https://ssir.org/articles/entry/advocating_for_age_in_an_age_of_uncertainty. Published May 28, 2020. Accessed October 21, 2021.

[7]   Hill C, Perez-Stable E, Anderson N, Bernard M. The National Institute on Aging Health disparities research framework. *Ethnicity and Disease*. 2015 Summer; 25(3): 245–254.

[8]   Stanley J, Harris R, Cormack D et al. The impact of racism on the future health of adults: protocol for a prospective cohort study. *BMC Public Health*. 2019 March 28; 19(1): 346. doi.org/10.1186/s12889-019-6664-x

[9]   Goal F. Understand health disparities related to aging and develop strategies to improve the health status of older adults in diverse populations. National Institute on Aging: Strategic directions for research, 2020–2025. www.nia.nih.gov/about/aging-strategic-directions-research/goal-health-disparities-adults. Published 2019. Accessed November 1, 2021.

[10] Williams DR, Priest N, Anderson NB. Understanding associations among race, socioeconomic status, and health: Patterns and prospects. *Health Psychology*. 2016; 35(4): 407–411. doi:10.1037/hea0000242

[11] Nair L, Adetayo OA. Cultural competence and ethnic diversity in healthcare. *Plastic and Reconstructive Surgery. Global Open*. 2019 May 16; 7(5): e2219. doi:10.1097/GOX.0000000000002219

[12] Dreachslin J, Gilbert JM, Malone B. *Diversity and cultural competence in health care: a systems approach*. San Francisco, CA: Jossey Bass, 2013.

[13] Fiscella K, Williams DR. Health disparities based on socioeconomic inequities: implications for urban health care. *Academic Medicine*: 2004 December; 79(12): 1139–1147.

[14] Global strategy and action plan on ageing and health. World Health Organization. www.who.int/publications/i/item/9789241513500. Published 2017. Accessed November 1, 2021.

[15] MacGuire F. Reducing health inequities in aging through policy frameworks and interventions. *Frontiers in Public Health*. 2020 July; 8: 315. doi:10.3389/fpubh.2020.00315

[16] Benz JK, Welsh VA, Espinosa et al. Trends in U.S. public awareness of racial and ethnic health disparities (1999–2010). NORC University of Chicago. www.norc.org/Research/Projects/Pages/trends-in-us--publics-awareness-of-racial-and-ethnic-disparities-in-health-1999-2010.aspx. Published September 30, 2010. Accessed October 21, 2021. doi.org/10.13016/ktq6-yzp9

[17] Williams DR, Wyatt, R. Racial bias in health care and health: challenges and opportunities. *JAMA*. 2015 August 11; 314(6): 555–556.

[18] Semega J, Kollar M, Shrider EA, Creamer JF. Income and poverty in the United States: 2019. United States Census Bureau. www.census.gov/library/publications/2020/demo/p60-270.html. Published September 15, 2020. Accessed October 21, 2021.

[19] Pager D, Shepard H. The sociology of discrimination: racial discrimination in employment, housing, credit and consumer markets. *Annual Review of Sociology.* 2008 January 1; 34: 181–209.

[20] Adomako F. African American older adults and race-related stress: how aging and health-care providers can help. American Psychological Association. www.apa.org/pi/aging/resources/african-american-stress.pdf. Published 2020. Accessed January 11, 2021.

32 The Value of Age-Friendly Public Health Systems to Older Adult Health and Well-Being

Megan M. Wolfe, JD

CONTENTS

32.1 THE AGING POPULATION IN THE UNITED STATES

Over the last ten years, the number of adults in the United States aged 65 and over increased by over 34%, largely due to the aging baby boomers and advances in public health, health promotion, and disease prevention (United States Census Bureau, 2020). Projections indicate 98 million, or 24% of the US population, will be over 65 by 2060 (Administration for Community Living, 2018). Those over 85 years will grow from 6 million to 20 million by 2060. Although, according to the US Centers for Disease Control and Prevention (CDC), the COVID-19 pandemic has reversed the positive longevity trajectory in the United States (Centers for Disease Control and Prevention [CDC], 2021c), the demographic of individuals living longer will continue to expand exponentially.

This rise in the number and proportion of people achieving longer life is important, but only part of the picture as the older adult population is also becoming more racially, ethnically, and culturally diverse. In 2014, 78% of older adults were non-Hispanic white, 9% were Black/African American, 8% were Latino, and 4% were Asian. By 2060, the percentage of non-Hispanic whites is expected to drop to 55%, while the proportion of other racial groups will increase, with 22% of the population

DOI: 10.1201/9781003197843-32

Latino, 12% Black/African American, and 9% Asian (Federal Interagency Forum on Aging-Related Statistics, 2016). These changes are important because of racial and ethnic inequities in health and access to resources, as well as cultural differences in expectations of informal and formal care. The impact of persistent inequalities, those social and economic conditions in life that can cause cumulative effects on health, may be contributing factors to the multiple chronic diseases and disabilities experienced by low-income adults (Carr, 2019). Older adults in communities of color face caregiving, housing, and social challenges as well as higher rates of healthcare-related harms, delays in care, bias in care, and discoordination due to a fragmented payment and delivery system.

32.2   THE NEED FOR PUBLIC HEALTH ENGAGEMENT IN HEALTHY AGING

Healthy aging has not historically been central to the public health agenda and most federal and state policies designed to support older adults' independence, such as Medicare, Medicaid, and the Older Americans Act, have not explicitly identified roles for public health, nor provided the necessary resources and funding for public health departments to expand programs and services (Kane, 1997). Indeed, prior to the COVID-19 pandemic, it was rare for public health agencies to have included older adults in planning and assessments and to have targeted funding or programs focused on people aged 65 and older. And yet, as more Americans live longer lives, their health needs may become more complex as well as costly. Chronic diseases account for two-thirds of all health care costs and 93% of Medicare spending (Gerteis et al., 2014). According to the CDC, 80% of Medicare beneficiaries have one chronic condition and nearly 70% have two or more (CDC, 2011). Evidence shows that public health prevention programs are effective at mitigating and reducing the effects of chronic diseases such as diabetes, and these are well-within the domain and expertise of public health departments (CDC, 2021a).

32.3   CREATING AN AGE-FRIENDLY PUBLIC HEALTH SYSTEM

Recognizing the important roles that the public health sector plays in promoting optimal health across the lifespan, and the public health contributions to longevity of American lives, Trust for America's Health (TFAH) began to see the need for a public health role in the lives of older adults. In 2017, TFAH brought together experts in public health, health care, aging services and academic leaders in aging to explore these roles and build a model for public health to use to consider policies and programs aimed at successful and healthy aging. This convening resulted in the *Framework for Creating Age-Friendly Public Health Systems (Framework)*, which outlines five functional areas in which public health could expand practice and programs to improve and support older adult health and well-being (Lehning et al., 2018). [MW: The framework was revised in 2022 to add another "C".]

The new "6C"s are:

1. Creating and leading policy, systems, and environmental changes to improve older adult health and well-being.
2. Connecting and convening multi-sector stakeholders to address the health and social needs of older adults through collective impact approaches focused on the social determinants of health.
3. Coordinating existing supports and services to help older adults, families, and caregivers navigate and access services and supports, avoid duplication, and promote an integrated system of care.
4. Collecting, analyzing, and translating relevant and robust data on older adults to identify the needs and assets of a community and inform the development of interventions through community-wide assessment.
5. Communicating important public health information to promote and support older adult health and well-being, including conducting and disseminating research findings, and emerging and best practices to support healthy aging.
6. Complementing existing health promoting programs to ensure they are adequately meeting the needs of older adults.

With generous funding from The John A. Hartford Foundation (JAHF), TFAH tested the Framework through a pilot in Florida, working directly with 37 of Florida's 67 county health departments. The 18-month pilot resulted in the creation of new data systems to help target resources and programming, establishment of new collaborations and partnerships to leverage existing programs and services, inclusion of older adult health priorities in community health assessments and planning, and expansion of existing programs (e.g., nutrition and health literacy) to include older adults. Significantly, almost all the public health practitioners who participated in the pilot noted that the increased awareness of the needs of older adults in their community has had a profound impact on their planning and assessment processes. The success of the Florida pilot has led to national momentum, expansion of AFPHS into additional states, and public health engagement in other age-friendly initiatives.

32.4 STATE MOMENTUM AND EXPANSION

TFAH's expansion work is also primarily funded by JAHF and includes significant new efforts in Mississippi and Washington, as well as some smaller programs in Colorado and Georgia, and further deepening age-friendly efforts in Florida. Based on the success of the Florida pilot, TFAH initiated a partnership with the Michigan Public Health Institute with funding from the Michigan Health Endowment Fund, to build an Age-Friendly Public Health System in Michigan. TFAH is also collaborating with the Institute for Healthcare Improvement and the Michigan Health and Hospital Association to build a model for a seamless older adult care continuum. This effort aims to identify gaps in care coverage and explore how best to bridge those gaps through alignment of the AFPHS Framework with the Age-Friendly Health Systems model. The initiative in Georgia is based on a partnership with the Association of State and Territorial Health Officials to identify synergies between the Georgia Department of Public Health and the Georgia Department of Aging

Services by crosswalking the State Health Improvement Plan and the State Plan on Aging. That crosswalk helped the two Georgia agencies to identify concrete collaborative actions, again based on the state's unique agency structure, existing relationships, and identified needs.

32.5   AGE-FRIENDLY PUBLIC HEALTH SYSTEMS RECOGNITION PROGRAM

In keeping with TFAH's prioritization of healthy aging, an Age-Friendly Public Health System Recognition Program was developed to incentivize and provide guidance for all state and local health departments to make the age-friendly transformation. The highest level of recognition by TFAH requires evidence of completion of ten action steps that, although not intended to be costly or burdensome, would facilitate meaningful changes in public health practice. The steps are:

1. Collect, analyze, and disseminate data and other information on the health and well-being of older adults.
2. Engage older adults to identify and address priority issues on their health and well-being (social determinants of health).
3. Meet and partner with organizations serving older adults such as the local Area Agencies on Aging and AARP.
4. Designate one or more staff persons to be the department's coordinators and lead and champion its healthy aging efforts.
5. Review existing public health programs to assess if and how they serve older adults.
6. Adapt work of existing programs to address the needs of older adults.
7. Ensure that public health emergency preparedness addresses the needs of older adults, particularly the most vulnerable, and their caregivers.
8. Participate in AFPHS trainings and educational programs offered by TFAH and others.
9. Engage with and support AARP/WHO Age-Friendly Communities efforts.
10. Engage and align with hospitals, health systems, emergency departments, or nursing homes participating in the Age-Friendly Health Systems movement.

32.6   COVID-19 AND OLDER ADULTS

The COVID-19 pandemic has taken a terrible toll on adults aged 65 and over and their families. Adults 65 and older account for 16% of the US population but CDC data show that 80% of COVID-19 deaths in the United States have been individuals 65 and over (Freed et al., 2020). In most states, the number of older adults who have died from COVID-19 is still higher than from all other causes (Freed et al, 2020). The highest percentage of older adult deaths have occurred in long-term care facilities (Freed et al, 2020). Furthermore, older adults of color are two to three times more likely to be hospitalized or die if infected with COVID-19 (CDC, COVID data tracker, 2021b). And despite the higher rate of vaccination of older adults, their vulnerability to the Delta variant remains high as vaccination rates in many states and communities lag behind overall vaccination rates in the United States.

The pandemic, which necessitated the implementation of mitigation strategies such as physical distancing, masks, and quarantine to protect vulnerable older adults from COVID-19 infection, significantly exacerbated the negative impacts of social isolation with the closure of elder care facilities to outside visitors, effectively separating residents from their loved ones (Sepúlveda-Loyola et al., 2020). In addition, the pandemic has caused fear, financial worry, depression, and other psychological problems (Lebrasseur et al., 2021). The impacts of physical distancing and face mask requirements on older adults living with dementia are far worse, as these individuals rely on routine and familiar faces to function.

Public health agencies at the federal, state, local, territorial, and tribal levels have played a vital role in the COVID-19 pandemic response, working tirelessly at the frontlines to protect the public's health. The pandemic has exposed significant gaps in the nation's public health system as well as the expertise needed to provide infection prevention education and support in the nation's elder care system. Despite these limitations, health departments that have embraced their roles in older adult health, and particularly those that have committed to or have become AFPHS, have been able to more effectively meet the needs of their older adult neighbors and experienced greater efficiencies in meeting their more complex health and social challenges during the pandemic. The collaborations built as a result of the Florida AFPHS pilot between county health departments and the aging services network contributed to a free flow of information about older adults who were isolated, needed home food deliveries, or needed vaccines at home. The development of the Aging in Florida profiles that resulted from the Florida pilot enabled county health departments to have a clearer picture of the needs of people in their communities and to track important health data.

32.7 COVID-19 VACCINE ACCESS FOR OLDER ADULTS WHO ARE HOMEBOUND

Early in 2020, it became clear that despite the prioritization of older adults for COVID-19 vaccination, those who were considered "homebound" – and those definitions varied greatly from state to state – received vaccines at lower rates than the general population. National and state vaccination plans focused first on mass vaccination sites to get as many individuals vaccinated as quickly as possible to protect the general population. However, considering the millions of people who are confined to their homes in the United States for a variety of reasons, this strategy left people who are homebound with limited or no access to the vaccine. The challenges for this population range from those related to the impact of their physical or behavioral diagnoses or conditions, to those related to physical and logistical barriers, to those related to distrust of the healthcare system (Trust for America's Health, 2019). Furthermore, for people of color and members of Tribal Nations, longstanding discriminatory practices and policies have resulted in additional barriers, such as lower rates of health insurance coverage, less access to quality health care, and fewer resources to pay for in-home care.

TFAH partnered with JAHF and the Cambia Health Foundation to explore the challenges and develop policy and practice recommendations to ensure access to the COVID-19 vaccine for this vulnerable population. In collaboration with public health leaders, home-based care providers, and aging services leaders, TFAH created a policy brief that includes recommendations such as

- Prioritizing the administration of the COVID-19 vaccine to people who are homebound, especially older adults and those with disabilities, and their caregivers
- Developing a standardized operational definition of people who are homebound
- Guaranteeing equitable vaccine access to homebound persons and ensure that none are underserved or overlooked due to race, ethnicity, age, socioeconomic status, urban or rural location.

TFAH learned that one of the most significant contributors to implementing successful homebound vaccination programs is collaboration across the public health, health care, aging, and home-based primary care networks.

32.8 THE AGE-FRIENDLY ECOSYSTEM: A SYSTEMS APPROACH TO HEALTHY AGING

When the World Health Organization launched the Age-Friendly Communities project, it sparked worldwide momentum among community leaders, health care providers, employers, public health officials, and other stakeholders to mobilize and develop programs and policies to better serve a rapidly growing demographic of older people. Everyone deserves to live in age-friendly communities to be sure, but age-friendly must include safe and effective health care, ensuring what matters most to these individuals regardless of care setting. The JAHF engaged the Institute for Healthcare Improvement to develop an Age-Friendly Health Systems framework that builds upon geriatrics care and models to help hospitals and hospital systems assess and act on the 4M's: What Matters, Mentation, Medication, and Mobility. As noted above, JAHF's partnership with TFAH is designed to elevate healthy aging as a core public health function.

Existing age-friendly initiatives remain separate, siloed frameworks, lacking shared vocabulary, measures, and goals and are just beginning to explore coordinated activities to achieve collective impact. Mutually beneficial partnerships among age-friendly leaders can lead to the creation of an age-friendly ecosystem, providing energy and coordination to strengthen the intended age-friendly impact. The leadership of JAHF, the Institute for Healthcare Improvement, AARP, and TFAH, in partnership with national and international experts, provides an unprecedented opportunity to identify effective strategies and best practices and rethink how to reorganize health care, social services, and public health, particularly for the most vulnerable populations.

REFERENCES

Administration on Aging. "Profile of Older Americans: 2017." April 2018. www.acl.gov/sites/default/files/Aging%20and%20Disability%20in%20America/2017OlderAmericansProfile.pdf

Carr, D. *The golden years? Social inequality in later life.* New York: Russell Sage Foundation, 2019.

Centers for Disease Control and Prevention. *Healthy aging at a glance, 2011: Helping people to live long and productive lives and enjoy a good quality of life.* Atlanta, GA: Centers for Disease Control and Prevention, US Dept of Health and Human Services, 2011. http://stacks.cdc.gov/view/cdc/22022

Centers for Disease Control and Prevention. About Chronic Diseases, 2021a. www.cdc.gov/chronicdisease/about/index.htm. Accessed September 7, 2021.

Centers for Disease Control and Prevention. COVID Data Tracker Weekly Review. February 16, 2021b. www.cdc.gov/coronavirus/2019-ncov/covid-data-covid-view/

Centers for Disease Control and Prevention. Vital Statistics Rapid Release, Number 015 (July 2021c). www.cdc.gov/nchs/data/vsrr/vsrr015-508.pdf

Federal Interagency Forum on Aging-Related Statistics. *Older Americans 2016: Key indicators of well-being.* Washington, DC: U.S. Government Printing Office, August 2016.

Freed, M., Cubanski, J., Neuman, T., Kates, J., and Michaud, J. What Share of People Who Have Died of COVID-19 Are 65 and Older – and How Does It Vary by State? Kaiser Family Foundation, July 24, 2020. www.kff.org/coronavirus-covid-19/issue-brief/what-share-of-people-who-have-died-of-COVID-19-are-65-and-older-and-how-does-it-vary-state-by-state/

Gerteis, J., David, I., Deborah, D., Lisa, Le., Richard, R., Therese, M., and Jayasree, B. *Multiple chronic conditions chartbook.* Rockville, MD: Agency for Healthcare Research and Quality (2014): 7–14.

Kane, R. L. "The public health paradigm." *Public health and aging. Baltimore: Johns Hopkins University* (1997): 3–16.

Lebrasseur, A., Fortin-Bédard, N., Lettre, J., Raymond, E., Bussières, E. L., Lapierre, N., Faieta, J., Vincent, C., Duchesne, L., Ouellet, M. C., Gagnon, E., Tourigny, A., Lamontagne, M. È., and Routhier, F. (2021). Impact of the COVID-19 Pandemic on Older Adults: Rapid Review. *JMIR Aging, 4*(2), e26474. https://doi.org/10.2196/26474

Lehning, A. J. and De Biasi, A. *Creating an age-friendly public health system: Challenges, opportunities and next steps.* Washington, DC: Trust for America's Health, March 2018. www.tfah.org/wp-content/uploads/2018/09/Age_Friendly_Public_Health_Convening_Report_FINAL__1___1_.pdf

Sepúlveda-Loyola, W., Rodríguez-Sánchez, I., Pérez-Rodríguez, P., Ganz, F., Torralba, R., Oliveira, D. V., and Rodríguez-Mañas, L. (2020). Impact of Social Isolation Due to COVID-19 on Health in Older People: Mental and Physical Effects and Recommendations. *The Journal of Nutrition, Health & Aging*, 1–10. https://doi.org/10.1007/s12603-020-1469-2

Trust for America's Health. Building Trust in and Access to a COVID-19 Vaccine Among People of Color and Tribal Nations: A Framework for Action Convening (2019). www.tfah.org/report-details/trust-and-access-to-covid-19-vaccine-within-communities-of-color/

United States Census Bureau. 65 and Older Population Grows Rapidly as Baby Boomers Age. July 25, 2020. (census.gov).

33 The Impact of Aging on Healthy Eyes

Elise B. Ciner, Zahava Nilly Brodt-Ciner, Marcy Graboyes, and Sarah D. Appel

CONTENTS

> ... *'it's not the years in your life that count. It's the life in your years'*
> *attributed to Abraham Lincoln*

33.1 INTRODUCTION AND BACKGROUND

There are many changes that occur as we age. In our work with older adults, it is especially important to understand those that relate to vision. These changes are common among all people, although there may be varying timelines from individual to individual. They must also be differentiated from eye disease and pathology that require medical intervention. While aging changes that may occur in healthy eyes do not necessarily create poor vision or vision impairment, they can have an impact on the performance of a number of activities of daily living that could create risks for safety and well-being. Examples of these activities may include: tasks requiring a near viewing distance (e.g. reading, working on an iPad or computer), mobility (e.g. walking on uneven surfaces or navigating down steps) or driving (e.g. a car or other vehicle). Being attuned to these changes is important when providing care consistent with an Age-Friendly Health System, as defined by the Institute for Healthcare Improvement and the John A. Hartford Foundation.[1] In an Age-Friendly Health System, care providers pay particular attention to an individual's mentation (cognitive health), mobility, medications and what matters to them most. In attending to these four evidence-based elements of high-quality care, known as the '4Ms', the interdisciplinary team can substantially reduce risk of adverse events and promote independence. Healthy vision is intrinsically linked to each of the 4Ms. It is,

DOI: 10.1201/9781003197843-33

therefore, important for older adults, their families as well as professionals in their care network to understand normal age-related changes in vision.

The normal visual changes that occur as we age can be optical, retinal and/or neural. These anatomical and physiologic structural changes along the visual pathway from the eye to the brain in turn impact our functional visual skills. Visual skills refer to how visual detail is seen (e.g. clarity of images), how quickly and accurately those images are processed and travel along the visual pathway and how a person then integrates this information with other senses to allow them to perform tasks and activities safely and effectively. This chapter will review some of the more common ocular or visual pathway anatomical and physiological changes that occur with age in the absence of disease, along with their functional implications with respect to safety and impact on daily living and quality of life. Strategies to compensate for issues created by these changes will also be discussed. It is not meant to be fully comprehensive as there is a large compendium of literature on vision and aging. There are also other visual skills which change with age that are beyond the scope of this chapter that we are beginning to understand and learn more about. These include visual attention/processing speed,[2] shape discrimination,[3] dual task effects on visual attention,[4,5] motion processing,[6] visual crowding[7] and other higher levels of cortical processing involving vision which change with age.[8,9] It is also not meant to be a complete compendium of the vast amount of informative and detailed research that has taken place over the last decades which together shed considerable light on our understanding of these normal aging changes. Some of these excellent resources are cited here and throughout this chapter and provide overviews and/or more in-depth understanding of the changes in vision that occur with age for those interested in further reading.[10–21]

33.2 ANATOMICAL AND PHYSIOLOGICAL CHANGES IN OUR EYES WITH AGE

Beginning with the front of the eye (Figure 33.1), our **tears**, **tear film** (produced by the lacrimal gland for each eye) and **eyelids** change with age. Increased flaccidity of the lids can create droopiness (ptosis), and a weaker and less efficient blink impedes the normal spread of tear film across the eye. There is also a decrease in tear production over time so that at 40 and 80 years of age, our tear production may be approximately one-half and one-quarter of childhood levels, respectively. The composition of our tears and tear stability also change so that the time it takes for the tear film to begin thinning or breaking apart between blinks can decline by approximately half between ages 8 and 80. Age-related alterations of the eye or ocular surface result in a continuum of change in adults. Even in the absence of active ocular surface disease, the decreased tear production, increased tear viscosity, structural changes in the lids and decrease in reflex tearing can cause increased discomfort including burning, stinging and excess tearing that can be uncomfortable and distracting for the older adult. These can also occur as side effects of medications or signs of an underlying or impending disease process. These tear film related issues may also affect the corneal surface resulting in fogging of vision which may lower visual acuity and reduce the ability to detect low contrast visual detail. More frequent digital screen use and

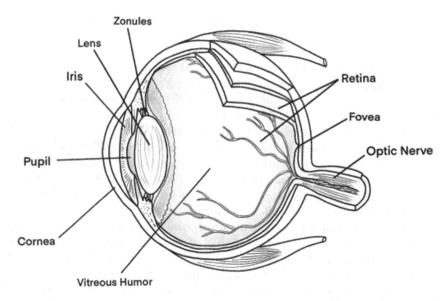

FIGURE 33.1 Schematic showing structures of the eye (illustration by Siva Meiyeppen OD).

associated alterations in blinking dynamics have also caused an increase in symptoms associated with aging changes.[22]

Clinical Pearls: 'The older eye tears less but cries more'.[23] These age-related changes may be helped with an increased blink rate, use of artificial tears or ocular lubricants and medications to increase tear production along with punctal occlusion (blocking of the small openings in the corner of the eyes to help retain tear film in the eye). Periodic breaks from extended screen time (20-20-20 rule: look 20 feet away for 20 seconds after 20 minutes of screen time[24]) may also help alleviate symptoms of normal changes related to tears that occur in older adults. In addition, the avoidance of ceiling fans, air drafts and extended or overnight contact lens wear and planned disposable contact lens use can be helpful.

The **cornea**, which is the clear, front layer of the eye, thickens with age and shows decreased sensitivity to touch throughout life and more significantly after the fifth decade.[25,26] The point or threshold at which an individual first notices discomfort doubles between age 10 and age 80[17,23,26] resulting in older adults being less sensitive to signs of corneal problems.

Clinical Pearls: Decreased corneal sensitivity is advantageous in adapting to contact lenses. The disadvantage is that the presence of corneal lesions including abrasions or ulcers may not be as noticeable or cause much discomfort. Left untreated, these issues could lead to permanent damage to the cornea and potentially obscure vision. It is therefore important to monitor the integrity of the cornea more frequently even in the absence of symptoms.

Behind the cornea is the **iris** or colored ring-shaped structure made of muscular and connective tissue with the **pupil** or circular opening in the center. The pupil controls the amount of light that enters the eye (similar to a camera's aperture). The pupil becomes miotic (smaller) at all light levels with age due to deterioration or atrophy of the iris dilator muscle fibers and increased rigidity of the iris blood vessels.[27] There is also an increased latency (time) for pupillary responses with an overall reduction in size fluctuations. The iris also becomes thinner and less pigmented.[17,28,29]

Clinical Pearls: Miotic pupils direct light through the thickest part of the lens resulting in an overall reduction in light reaching the retina. This may make it more difficult for an older adult to see in, or adjust to, poorly lit settings.

Light traveling through the iris next reaches the **lens**. The lens continues to thicken slightly throughout life with an increase of about 28% between 20 and 70 years of age.[17,29] Concurrently, the lens stiffens or becomes less malleable, resulting in a loss of focusing flexibility (accommodation).[29,30] There is a selective absorption of light as the lens thickens, its fibers become less organized and pigment is deposited, all of which contribute to increasing yellowing and opacification.

Clinical Pearls: Lens changes are a normal part of the aging process, occurring in over 90% of 75 to 85-year olds.[31,32] The need for reading glasses to compensate for the loss of lens flexibility and near focus typically occurs after age 40 in the fifth decade of life, with changes complete at approximately 55 to 60 years of age. As the lens becomes more opaque and visual acuity is compromised, it is diagnosed as a cataract which can be surgically removed and replaced with an artificial lens that improves vision.

The **vitreous body** or gel-like fluid filling the posterior part of the eye behind the lens undergoes synchysis (increase in liquid spaces or liquefaction), and syneresis (increased movement) resulting in an increase in the speed and amplitude of any vitreous debris (floaters) present. As a result of these processes, its fibers that are attached to the retina can begin to pull away after age 50 resulting in posterior vitreous detachment in approximately 50% of individuals.[33] A person who experiences this may see flashes of light and an increase in floaters. Posterior vitreous detachments also increase the risk for more serious complications including retinal tears, retinal detachments or hemorrhages within the vitreous.

Clinical Pearls: Older adults should be encouraged to report the presence of flashes and floaters to their eye care provider as these can be signs of either posterior vitreous or retinal detachments, retinal tears or intravitreal hemorrhages and need to be evaluated in order to prevent loss of vision. Mild floaters associated with normal aging changes of the vitreous may initially be visually and mentally distracting during activities such as reading and mobility. Individuals may, however, adapt to their presence as the brain can sometimes edit out this visual disturbance resulting in only intermittent awareness of floaters.

Aging changes in the **retina** in the absence of eye disease are not often readily observable during a clinical exam. There are, however, subclinical changes that occur in the retina that affect vision. With respect to the **photoreceptors** or light-receiving retinal cells (rods and cones) that convert light into electrical energy, there is a displacement (both rods and cones) and decrease in the number of photoreceptors (primarily rods) especially at the **fovea** or point of best vision.[34] This occurs throughout adult life, but at a faster rate after age 40.

Clinical Pearls: While each person's 'tipping point' is different, the vast number of photoreceptors (more than 100 million in each eye) and redundancy in how visual information is processed may result in no observable signs or symptoms of these changes by some individuals.

The **optic nerve** originates in the eye and carries visual information from the eye to the visual cortex of the brain by way of electrical impulses through long neurons or nerve fibers called retinal ganglion cells. There are reported age-related changes in the nerve fiber layer originating from the retina[35,36] and a reduction in these nerve fibers in the central retina (of which the fovea is at the center) by approximately one-fourth in 66–86 year-old donor eyes compared to young adults, along with a decrease in cell density.[35–37] Others report a loss of approximately 4900 nerve fiber axons per year[38] and a selective loss of those nerve fibers.[39]

Clinical Pearls: There are over one million retinal ganglion cells in each eye which allow for redundancy in the system. Although the exact effects of these changes are unknown, large quantity of cells provides protection from observable signs or symptoms of these normal aging changes in some individuals.

33.3   VISUAL SKILLS CHANGES WITH AGE

Each of the structural changes discussed above has varying levels of impact on the visual skills discussed in the next section. The changes that occur in our visual abilities with age can be due to optical, retinal and/or neural factors.

Visual acuity is a measure of the clarity of central vision determined by the smallest letters, numbers or symbols that an individual can read or identify on a standard eye chart. The average visual acuity with corrective lenses decreases slightly with age in older adults with increased variability among individuals. Many adults however, maintain visual acuity in the normal range (Figure 33.2) from approximately 20/20 to 20/40 with adults in their 80s averaging visual acuity of 20/25 (one line above 20/20).[12,40] Optical, retinal and/or neural factors in the normal aging eye can contribute to these changes. Average visual acuity (in the Beaver Dam Study) was found to be approximately 20/20 through age 64 years, dropping slightly to 20/25 for 65–74 year olds and averaging 20/40 for individuals older than 75 years.[41] Importantly, many older individuals continue to retain 20/20 visual acuity in the absence of disease. The Framingham Study showed the following percentage of individuals with 20/20 vision at increasing ages (includes those with and without eye diseases): 50% in 60 to 69 year olds; 25% in 70 to 79 year olds and 14% over the age of 80.[32,42] Of note, visual acuity can be further reduced when viewing moving targets, called dynamic acuity.[43] The decline in acuity with target velocity (speed of moving targets) increases with advancing age.[44,45] This may have a direct impact on driving and reaction time to other moving vehicles, people or objects. Alternative measures of visual performance to evaluate safety when driving include the 'useful field of view' (UFOV) which relies on higher order processing skills including selective attention and visual processing speed.[46–55] The UFOV is a computer-generated task requiring identification of a central task while simultaneously identifying a peripherally presented target.[46]

Practical Pearls: The potential for many older adults to attain and maintain high levels of visual acuity emphasizes the importance of regular vision care and updated optical prescriptions as needed. While reductions in dynamic acuity may not be detected during standard visual acuity chart measurements where the letters are

FIGURE 33.2   Range of 'normal' visual acuity (outlined in dashes) from 20/10 to 20/40 on a standard chart.

stationary, an individual may express safety concerns with mobility and driving, especially in fast moving environments such as highways and high-traffic areas. While vision is one component of safe driving,[51] measures beyond visual acuity such as the UFOV can be helpful in determining those at risk who may benefit from speed of processing training.[48] Older adults also need to be educated about these potential changes and their implications even in the presence of 'normal' visual acuity levels.

Peripheral Visual Fields, also referred to as side vision, are an important aspect of vision in maximizing safety and efficiency during daily activities. These skills include locating items when seated at a table such as utensils or serving dishes, finding items such as keys or eyeglasses and especially mobility related activities including walking in crowded areas, crossing busy intersections and driving.

Peripheral visual fields become slightly decreased in size with age.[52–55] Changes in visual fields are more evident when measured with moving targets (called kinetic perimetry). The loss is greater in the upper visual field (due to ptosis or droopy lids) and sensitivity to objects in the visual field decreases with increasing eccentricity (further from straight ahead gaze). A person with upper visual field loss associated with lid drooping (ptosis) may compensate by tilting their head back in order to reduce the effect of the ptosis. However, by doing so, they may cause limitations in their lower peripheral field regions, resulting in tripping and falling when encountering curbs and other mobility related obstructions. Slightly reduced fields as measured in a clinical setting can be a normal aging change or can indicate the presence of progressive eye disease (e.g. glaucoma). Any sudden changes are red flags for further investigation. For example, a sudden inability to see the right or left side of a person's surrounding environment, especially when accompanied by loss of motor function or sensation on the same side as the field loss, could be a sign of a stroke requiring immediate attention. It is important to check visual fields at regular intervals as even changes associated with normal aging may have implications for safety while walking, driving and other activities of daily living.[56]

Practical Pearls: An individual with lid ptosis that interferes with their upper field of vision should wear shatter resistant polycarbonate lenses to protect their eyes from injury by objects located higher in their field of view such as tree branches or upper cabinet corners. If a droopy lid covers a portion of the pupil, a surgical procedure called blepharoplasty may also be considered to elevate the lid and reduce or eliminate the upper visual field loss.

Stereoacuity also called stereopsis, binocular depth perception or 3D vision is a measure of how well the two eyes are working together. When we look at a scene or image binocularly, each eye is actually viewing from a slightly different perspective. These differences or disparities are then combined and processed in the seeing part of the brain (visual cortex) resulting in 3D vision. In order to have optimal levels of stereopsis, each eye must see clearly and the eyes need to be well aligned. While there are other monocular 'cues' to depth, such as size differences or linear perspective, stereopsis provides a view of images that may improve safety and quality of life. In the absence of disease, there is no decline in stereopsis up to at least 50 years of age.[15,16,57–59] There is, however, a marked decline in stereopsis after age 60 with an accelerating rate into the mid 80s.[40,60] While many individuals retain good stereopsis, studies such as the Salisbury Eye Evaluation Project (SEE Study) indicated that stereo blindness (inability to see depth at the largest stereoscopic disparity tested) is present in 10% of 65 to 69 year olds vs. 26% of 80 to 85 year olds.[60] This could have an impact on daily living activities and navigation.

Practical Pearls: Good stereoacuity helps individuals judge distance (e.g. parallel parking a car), depth of objects or gradings, detect drop-offs of curbs or edges of stairs, and complete fine motor tasks and other activities of daily living. An awareness of the normal decline in stereopsis in the absence of disease is important in order to increase safety, reduce risk of falls and allow more time for and avoid frustration with fine motor and other tasks.

Vernier Visual Acuity is referred to as a hyper or higher-level acuity involving neural in addition to ocular and retinal functions. It is a measure of the smallest difference in location between objects that can be discriminated (e.g. 'Are the shelves on a wall equally spaced, are stairs uneven or can we tell that two lines placed end to end are misaligned?'). While there is disagreement as to how vernier acuities change with age,[61,62,63] variations occur from individual to individual.

Practical Pearls: Although vernier acuities are not routinely evaluated in clinical settings, changes with age have been shown to occur. In the future, vernier acuities are an example of a vision test that may become useful as an assessment of neural functioning especially in individuals who complain of unexplained vision loss.

Color Vision allows us to see a 'rainbow' of colors across the spectrum of light that is visible to the human eye through the interactions of three types of cones (red sensitive, green sensitive and blue sensitive). The short wavelength sensitive or blue cones are the sparsest in normal human retinas. Up to approximately 5% of the population (8% of genetic males and .05% of genetic females) has an inherited color deficiency, meaning that one cone type is either missing or its sensitivity is diminished, with red-green deficiencies being the most prevalent. Normally, longer wavelength colors (e.g. reds) are more visible during the day, shifting to increased visibility of shorter wavelength colors (e.g. blues) at dusk and dawn. The ability to distinguish colors in the visual spectrum can begin to gradually decrease early after the third decade of life due to changes in the ocular media (e.g. lens of the eye), photoreceptors in the retina (cones) and post-receptoral (eye to brain) changes.[18,64] Increasing errors with age are seen on color vision testing, especially 'arrangement types of tests' that require putting colors in order.[65] Less sensitivity to blues and greens can occur due to normal aging changes and progressive yellowing of the lens that filters out blue light. These gradual changes are first seen after age 60 and are most noticeable after age 70. In addition, the smaller pupil that is often present in older adults directs the incoming light through the thickest part of the lens further exacerbating any changes.

While subtle changes in color vision may be part of the aging process, they can also indicate the onset or presence of an ocular or systemic disease process or be the result of drug toxicities. The ability to differentiate these types of acquired color vision deficiencies from normal age related changes are an important part of the clinical eye exam. Periodic baseline measures of color vision in each eye are important especially prior to beginning drug therapies known to affect color perception. For example, Sildenafil (brand name Viagra) can cause a blue-yellow color deficiency.

Practical Pearls: Changes in color vision can impact any daily task requiring color discrimination such as matching socks or other clothing or seeing discoloration or stains on clothing or utensils. More importantly with respect to safety, many older adults are taking multiple medications which may differ only slightly in shape or color. These are sometimes arranged in a 'pill box' on a weekly basis. As the print on packaging of medications is often below the visual resolution capability of many adults, the ability to accurately identify different pills, pill bottles or eye drop bottle caps can be affected by changes in color vision.[66] Increased lighting along with identifying, separating and relabeling bottles of possible similarly appearing medications can be helpful to avoid any mistakes.

Luminance Efficiency Functions are how well our retinal photoreceptors (rods and cones) respond to light levels. As we age, our overall ability to detect light across the visible spectrum (the portion of the electromagnetic spectrum that can be seen by the human eye) is reduced.[67,68] This is called our luminance efficiency function for rods and cones. We require more light to produce the same responses to many of the visual tests administered during an eye exam as well as to complete tasks for work or daily living.[69] As described earlier, changes in the surface of the cornea or the ocular media including the lens, and slower pupillary responses cause an increased scattering of light that enters the eye, resulting in an increased sensitivity to glare. The result is a delayed overall reaction time to light and a longer refractory time to recover from changes in lighting. This in turn can impact many of the visual skills discussed in this chapter.

Practical Pearls: As we age, we need more task-directed lighting. As an example, whereas a 20-year-old might need 100 watts of illumination to read a certain sized text, by the time an individual reaches 80 years of age they are likely to need four times or 400 watts to see the same sized text comfortably.[16] When glare is present, an increase in illumination is often needed for older individuals to detect objects and participate in activities of daily living. Task-directed lighting and the use of ambient window or room lighting that increases illumination and reduces sources of glare can be beneficial to meet the increased lighting needs of older adults.

Dark Adaptation is another measure of how our eyes respond to changes in light, specifically how rapidly an individual is able to adapt from an illuminated to a dim or dark environment (e.g. how quickly we can begin to see after entering a movie theater or darkened room). Clinically, this is determined by measuring how fast the photopigment in the rods and cones recover (regenerate) after being 'bleached' with a very bright light (which decreases their sensitivity and ability to be stimulated and respond to light). Dark adaptation represents a shift from cone mediated or photopic (daytime) vision to rod mediated or scotopic (nighttime) vision. As we age, there is a slowing in the rate of dark adaptation and the 'absolute level' of dark adaptation (how well we see in the dark after a period of adaptation) increasingly diminishes.[13,70] This occurs due to a reduction in the number of rods and their displacement throughout the retina along with miotic pupils, changes in the lens and/or changes in retinal and cortical processing. While changes in dark adaptation can be a normal sign of aging, they can also be related to medication use, an indication of vitamin A deficiency or a harbinger for new or progressing disease such as macular degeneration.[12]

Practical Pearls: The use of flashlights and motion detecting lights next to walkways and on stairs along with night lights in areas frequented at night such as bathrooms and kitchens can help minimize disorientation and improve confidence and safety for older adults who need to be outside after dark or wake up at night for a snack or personal needs. Driving at night should also be avoided if possible when there are concerns for reduced visibility in dimly lit environments.

Glare, which is sensitivity to stray light, is exacerbated by normal aging changes in the eye including increasingly miotic pupils, slower reacting pupils and early changes in the crystalline lens of the eye. Increased glare allows more stray light to enter the eye resulting in an increase in distractions and discomfort or disability glare.[17,71,72,73] Discomfort glare does not cause a loss in vision but is uncomfortable for the individual. In disability glare, there is an actual reduction in visual acuity and visual function due to the scattering of light or stray light reaching the retina. This

FIGURE 33.3   Photo of discomfort vs. disability glare and the effect of filters.

Examples of (A) Discomfort glare. (B) Disabling glare. (C) Improvement in glare with polarized lenses.

causes a veiling luminance (light) over the retina which reduces contrast, making it difficult to see important obstacles during indoor or outdoor activities, mobility and driving (Figure 33.3).

Sensitivity to glare increases with age[18] beginning for some individuals after age 40 and progressing dramatically after age 55.[71] This discomfort can occur from exposure to glare, both indoors and outdoors. Glare recovery is prolonged nearly twofold by age 66. Glare can also be experienced by individuals who have undergone refractive eye surgery.

Practical Pearls: The presence of glare contributes to an increased need to control illumination in daily activities such as reading, watching TV, using electronic devices or walking.[69] Glare from the sun in the early or late day along with oncoming headlights from cars or trucks causes an increased safety risk while driving. Removing additional sources of external glare created from eyeglasses, computer screens and windshields by keeping these items clean and smudge free can help reduce the effects of glare that are caused by light scattering in the eye itself. The use of sun visors, tints on windshields and eyeglasses to increase contrast and/or antireflective coatings on glasses can also help reduce the effects of glare. Avoid looking directly at the glaring light when possible. In addition, when possible, minimize driving in the direction of the rising or setting sun and avoid driving at night when oncoming glare producing headlights will be most noticeable to decrease visual discomfort and increase safety for older drivers.

Contrast Sensitivity is a measure of a person's ability to distinguish varied patterns of light and dark. It also includes the ability to notice the difference between an object and its background color or brightness such as the detection of edges and borders of curbs and stairs. Our ability to perceive contrast is best for medium sized objects, letters on an eye chart or medium spatial frequencies (number of pairs of black and white lines imaged at a given distance on the retina) compared to either large objects or letters and low spatial frequencies or to small objects or letters and high spatial frequencies. It is also better under normal or daytime lighting compared to dusk, nighttime or in poorly lit areas or hallways. Contrast sensitivity remains stable through middle age with changes beginning to occur after the fifth decade of life. Losses in contrast sensitivity with age are primarily at the medium and high spatial frequencies (medium and small objects or letters on an eye chart) due to reduced optical factors including reduced luminance, optical aberrations and increased light scatter. They also worsen with age under reduced lighting conditions due to loss of rods and ganglion cells and other neural factors.[74-76]

Practical Pearls: Many objects and surfaces in our environment are of low contrast. These can include low contrast reading material such as newsprint, facial features, light colored food on a white plate, objects in a room and walkways, curbs or stairs. Task lighting or colored filters may enhance contrast when reading low contrast text and when viewing other low contrast visual detail. Contrast is especially reduced in situations where glare is present (excessive lighting), in the presence of reduced illumination and at nighttime. As such, the normal aging changes in contrast can present safety risks especially in areas that are poorly lit or overly bright. Driving under conditions of poor contrast can increase risks for safety including the ability to detect objects on the road, street signs or persons in a crosswalk as examples. Recommendations to consider to address changes in contrast include avoiding or reducing driving speed when safe to do so and the use of high contrast lenses or tints when helpful. Marking edges, curbs or stairs with fluorescent tape or paint, especially in poorly lit areas can help improve safety (Figure 33.4).

Refractive Error occurs when the shape of the eye keeps light from focusing correctly on the retina (NIH website). Refractive errors include myopia (nearsightedness), hyperopia (farsightedness), astigmatism (abnormal focusing primarily due to the shape of the cornea that distorts vision) and anisometropia (unequal refractive error between the eyes). Refractive errors are addressed through the prescription of

FIGURE 33.4 Stairs showing low contrast (left) and enhanced contrast (right).

glasses, contact lenses and/or refractive surgery which eliminate blur by facilitating the optimal focus of images on the retina. Glasses can also protect the eyes from falls as well as the good eye if a fellow eye has poor vision.

Refractive errors, which are the most common conditions affecting vision at all ages, are present in many individuals throughout life and change from early infancy onward. Not all refractive errors impair vision or require corrective lenses. In adulthood, there are relative periods of stability where few changes occur (e.g. young adulthood in the 20s and 30s) with the exception of progression of nearsightedness (myopia). More significant changes occur during middle age (40s and 50s) when presbyopia occurs. Presbyopia is due to a progressive reduction in the focusing abilities of the lens of the eye and results in the need for reading glasses for near activities. After 55 years of age, small but often noticeable changes continue to occur in the absence of disease in some types of refractive error. These changes, which may be concerning to older adults who may worry more about loss of vision and blindness, are described below.

Individuals with either **myopia (nearsightedness) or hyperopia (farsightedness)** may show small changes in the power of their prescriptions. There can be a slight increase in myopia or a more moderate increase in hyperopia. In the latter case, the actual amount of hyperopia may not be increasing, rather the earlier focusing adjustments an individual was able to make to 'compensate' for uncorrected hyperopia are no longer possible due to the increasing inflexibility in the lens of the eye. This becomes most evident during the latter part of middle age (50s and early 60s) when any amount of hyperopia can no longer be compensated for through changes in the crystalline lens. In later years, there may be additional small changes in myopia and hyperopia due to normal age-related changes in the length and refracting power of various components of the eyes.[77,78]

After age 40, it is fairly common to see changes in **astigmatism** associated with subtle changes in the shape of the cornea. Astigmatism has the effect of blurring and/or distorting vision at both distance and near viewing and is a common type of refractive error seen at all ages. Both the type (a steepening of the horizontal meridian of the cornea resulting in a slight increase in 'against-the-rule' astigmatism) and amount of astigmatism can change in normal aging eyes.[17,18,79] The change in magnitude can be as much as one diopter (4 steps) of astigmatism between ages 40 and 80[80] but can vary from individual to individual. The number of people who have astigmatism also increases, making this type of refractive error very common in the aging population and thus important to evaluate for, and correct with, glasses or contact lenses.[79-82] There is also an increasing amount of unequal refractive error between the eyes (anisometropia) with advancing age, which can impact not only visual acuity and visual comfort but also binocular function and depth perception.[83]

Practical Pearls: Changes in refraction are common and normal, but can concern or worry older adults who may feel their vision is not what it used to be. It is important for them to know that these are changes that can be corrected through a careful refraction and new prescription eyeglasses when indicated. It is always important to provide a careful refraction and comfortable fitting frames along with discussions of bifocals, protective lenses and/or tint options to the older adult.

Eye movements include the ability to maintain steady fixation on an object, the ability to follow a moving object (versions) and the ability to move the eyes efficiently between two stationary objects (saccades). Fixational eye movements or the ability to maintain steady fixation are more difficult under scotopic (nighttime) conditions. As we age, there is little reported change in fixation under photopic (daytime) conditions, unless caused by disease. **Versions** or the ability to follow a moving object from point to point can change with age.[84] Changes in response time (latency or time to initiate the eye movement) and accuracy have been reported in the aging population, which are exacerbated with faster moving objects. In addition, the range of voluntary eye movements may be more limited with greater restrictions occurring on upward gaze.[85,86] When these are observed, pathology and neurological causes must always be ruled out. **Saccades** or the ability to move fixation from one stationary object to another show very small reductions in the peak velocity (speed) and latency in older adults. These changes, however, are minimal and not measurable clinically. When more significant changes in saccadic eye movements are observed, visual field deficits and neurological causes should be ruled out.

Practical Pearls: As older adults have increased difficulty with fixation under scotopic or dim illumination, the need for optimal task lighting, motion detection lighting and the use of flashlights or plug-in night lights can increase safety at night either in the home or outdoors. Older adults on fixed incomes may tend to turn off lights to save energy but should be discouraged from doing so if there are safety concerns. Younger family members should be informed of the need to 'keep' the lights on for older adults in the household. Items that are frequently accessed during everyday activities such as medication bottles, TV remotes or phones should be positioned at or close to eye level. In the absence of underlying neurological conditions which must always be ruled out, healthy older adults may benefit from vision training to improve eye movements.[17]

Vergence or the ability to **converge** (the inward repositioning of both eyes to view an object located at a closer distance) or **diverge** (look far away with both eyes) may show subtle changes with age. There is a decrease in our ability to effectively converge our eyes with age. These changes can typically occur between approximately 50 and 70 years of age and result in a slight tendency for some older individuals to show an outward eye drift (exophoria) or actual outward eye turn (exotropia) when reading or viewing objects at near.[87] Insufficient convergence could potentially result in eyestrain, double vision and discomfort with reading or close activities which can in turn impact an older person's ability to perform activities of daily living.

Practical Pearls: Convergence problems (e.g. convergence insufficiency) determined during an exam to be the result of normal aging changes can be treated with eye exercises (vision therapy/orthoptics)[17] and/or prism glasses.

Accommodation is the ability to focus images with our eyes when viewing shifts from distance to near in order to read and see clearly for close activities. Accommodation occurs due to changes in the shape of the lens, which is attached by a ring of fibrous strings (zonules) to the ciliary body, a muscle in the eye that contracts or relaxes when looking from a near object to one that is further away. Our accommodative abilities begin to show a noticeable and progressive decline between

40 and 50 years of age due to gradual sclerosis (hardening) of the lens and loss of flexibility resulting in 'presbyopia'. These changes are one of the first visual signs of normal aging eyes. A total loss of the ability to accommodate occurs by approximately 55 years of age[88] which means that this normal aging change is completed by the time a person is considered an older adult.[12] The onset of presbyopia occurs earlier in hyperopic individuals who are already focusing their eyes to compensate for the hyperopia.

Practical Pearls: Presbyopia is a decline in accommodation or focusing ability for near which is a normal part of aging that occurs in the middle, rather than later years. A small increase in the focusing power of reading lenses for older adults may be helpful, even in the presence of normal visual acuity, to improve contrast, comfort and access to a wider range of visual detail. Older individuals with low to moderate amounts of myopia (in the absence of astigmatism) are naturally near sighted and focused for close viewing activities without corrective lenses. They can therefore often simply remove their distance glasses for extended reading or other near tasks without the need for reading glasses or further magnification.

33.4 ADDITIONAL HELPFUL RECOMMENDATIONS

The following is a list of recommendations to consider when helping older adults compensate for normal age-related changes in vision:

- Assure that the individual is wearing the best optical correction (e.g. glasses, contact lenses);
- Consider the use of lightly tinted lenses indoors or sunglasses including polarized lenses outdoors;
- Consider photochromic lenses, which darken in sunlight, to help with adaptation outdoors. Certain photochromic lenses maintain a slight tint indoors and darken in the car while driving, both of which may be helpful to some individuals;
- Add hall lights in living areas with poor contrast;
- Use task directed lighting when possible while avoiding direct eye contact with the actual light bulb which can cause discomfort and after-images;
- Place transition lighting for outside to inside doors and walkways;
- Use outdoor lighting that minimizes veiling luminance; illuminating only the ground and walls;
- Add night lights and motion detecting lights to avoid disorientation in the dark or upon awakening during the night;
- Use window coverings to reduce glare from the sun;
- Use 'matte' finishes on countertops, for wall paint and on tables to reduce glare when indicated;
- When feasible, avoid high gloss floors to reduce reflective glare;
- Avoid driving at night or in the direction of the rising or setting sun if there are difficulties present due to glare or poor contrast;
- Mark the edges of stairs and walkways to improve contrast, especially the first and last stair;

- Use 'lime green' floor, wall and doorway markers, a color seen best both during the day and at night;
- Wear clothing colors and security belts that are easily seen when out for walks (e.g. bright red or orange during the day, light blues in the early evening and fluorescent lime green at any time during the day, evening or at night);
- Remove low hanging objects from living areas and walkways when possible including tree branches that are closer to the ground. Mark low hanging cabinets or doorway thresholds with fluorescent or brightly colored tape to avoid injury to the head, especially in the dark;
- Organize the home environment for easy location of items and enhanced contrast such as using a white plate on a dark place mat, cutting boards that contrast with the countertop, black TV remote on a white tray etc...;
- Remove obstacles that block indoor hallways and outdoor walkways;
- Remove throw rugs and other movable floor coverings which are tripping hazards especially those that blend in or have reduced contrast with the floor beneath;
- Allow more time to complete daily living activities such as food preparation, dressing, walking, parking the car or driving in general;
- Educate older adults regarding normal and abnormal visual changes and the need for regular eye care;
- Educate family members and friends about the potential visual changes we experience with age.

33.5 CONCLUSION

In summary, as we age, there are both normal structural (anatomical) and functional (physiological) changes that occur in the eye and throughout the visual pathway caused by optical, retinal and neural factors. These changes are typically gradual, and the older adult may or may not notice any differences in visual functioning. Increased age does not necessarily lead to worsening vision, and in fact many older adults maintain optimal vision. Lighting is a key factor for most of the visual skills discussed. Poor lighting or glare exacerbates any visual difficulties while glare-free lighting contributes to optimal visual function. Another important factor is that visual skills are also affected by increasing movement or motion which results in a degradation of visual acuity. Older individuals may need to slow down during some activities in order to better process visual information.

During routine eye examinations, there is increased variability of performance with age for most visual skills, and not every older adult with good eyesight will have 20/20 vision. This can present challenges to the practitioner in determining what are normal versus abnormal findings. For example, most primary eyecare doctors do not perform certain ancillary tests, such as contrast sensitivity or stereopsis that would detect decreased visual performance affecting activities of daily living in the normal aging eye.

Of note, all changes in vision due to aging can also be seen in the presence of disease, especially those that are more prevalent in the older adult population (e.g.

glaucoma, cataracts, age-related macular degeneration and diabetic retinopathy). Therefore, it is recommended to always consult an eye care provider to distinguish between normal aging changes versus those that are red flags for underlying ocular or systemic disease processes or medication toxicity. It is also important to have baseline visual skills testing at the onset of a disease process or at the commencement of a new medication. In addition to an ocular health examination, this can include measures of visual acuity, refractive error, visual fields, contrast, stereopsis, vergence, ocular motilities, color vision and other specialized testing.

In the absence of pathology, older adults will likely have some level of diminished visual function that may or may not affect daily living. The impact of these age-related changes may be exacerbated by reduced visual conditions including fog, twilight, glare, poor illumination, motion or movement and visual clutter or 'crowding' of objects in the person's field of vision. It is important to educate older adults about normal age-related changes, their functional impact and strategies that can help compensate for these changes in order to enhance safety and maintain quality of life.

REFERENCES

1. www.ihi.org/Engage/Initiatives/Age-Friendly-Health-Systems/Documents/ IHIAgeFriendlyHealthSystems_GuidetoUsing4MsCare.pdf.
2. Owsley C, McGwin G Jr. Association between visual attention and mobility in older adults. J Am Geriatr Soc. 2004 Nov;52(11):1901–6.
3. Wang YZ. Effects of aging on shape discrimination. Optom Vis Sci. 2001 Jun;78(6):447–54.
4. Künstler ECS, Penning MD, Napiórkowski N, Klingner CM, Witte OW, Müller HJ, Bublak P, Finke K. Dual task effects on visual attention capacity in normal aging. Front Psychol. 2018 Sep 3;9:1564.
5. Duncan MJ, Smith M, Clarke ND, Eyre EL, Wright SL. Dual task performance in older adults: Examining visual discrimination performance whilst treadmill walking at preferred and non-preferred speeds. Behav Brain Res. 2016 Apr 1;302:100–3.
6. Sepulveda JA, Anderson AJ, Wood JM, McKendrick AM. Differential aging effects in motion perception tasks for central and peripheral vision. J Vis. 2020 May 11;20(5):8.
7. Scialfa CT, Cordazzo S, Bubric K, Lyon J. Aging and visual crowding. J Gerontol B Psychol Sci Soc Sci. 2013 Jul;68(4):522–8.
8. Meng Q, Wang B, Cui D, Liu N, Huang Y, Chen L, Ma Y. Age-related changes in local and global visual perception. J Vis. 2019 Jan 2;19(1):10.
9. Wang YZ, Morale SE, Cousins R, Birch EE. Course of development of global hyperacuity over lifespan. Optom Vis Sci. 2009 Jun;86(6):695–700.
10. Weale RA. Aging and vision. Vision Res 1986;26:1507–12.
11. Weale RA. The Senescence of Human Vision. Oxford: Oxford University Press, 1992.
12. Owsley C, Ghate D, Kedar S. Vision and aging. Chapter 15 In: Rizzo M, Anderson S, Fritzsch B (eds.) The Wiley Handbook on the Aging Mind and Brain, First Edition. John Wiley & Sons LTD, Chichester, West Sussex 2018; 15: 296–314.
13. Owsley C. Vision and aging. Annual Rev Vis Sci. 2016 Oct 14;2:255–71.
14. Owsley C. Aging and vision. Vision Res. 2011 Jul 1;51(13):1610–22.
15. Garzia RP, Trick LR. Vision in the 90s: The aging eye. J Optom Vis Dev. 1992; 23:4–41.
16. Hagerstrom-Portnoy G, Schneck ME, Brabyn JA. Seeing into old age: Vision function beyond acuity. Optom Vis Sci. 1999;76:141–58.

17. Haegerstrom-Portnoy G, Morgan MW. Normal age-related vision changes. Chapter 2 In: Alfred A. Rosenbloom, Jr. (eds.) Rosenbloom & Morgan's Vision and Aging. 2007 Butterworth Heinemann Elsevier. St. Louis, Missouri. pp. 31–48.
18. Haegerstrom-Portnoy G. The Glenn A. Fry Award Lecture 2003: Vision in elders–summary of findings of the SKI study. Optom Vis Sci. 2005 Feb;82(2):87–93.
19. Wood JM. Aging, driving and vision. Clin Exp Optom. 2002 Jul;85(4):214–20.
20. Brabyn J, Schneck M, Haegerstrom-Portnoy G, Lott L. The Smith-Kettlewell Institute (SKI) longitudinal study of vision function and its impact among the elderly: an overview. Optom Vis Sci. 2001 May;78(5):264–9.
21. Erdinest N, London N, Lavy I, Morad Y, Levinger N. Vision through Healthy Aging Eyes. Vision (Basel). 2021 Sep 30;5(4):46.
22. Al-Mohtaseb, Z. et al. The relationship between dry eye disease and digital screen use. Clinical Ophthalmol (Auckland, N.Z.). 2021 Sep 10;15 3811–20.
23. Smith SC. Aging and vision. Insight. 2008 Jan–Mar;33(1):16–20; quiz 21-2.
24. American Optometric Association. Computer vision syndrome; 2020. Available from:www.aoa.org/healthyeyes/eye-and-vision-conditions/computer-vision-syndrome. Accessed November 13, 2020.
25. Millodot M. A review of research on the sensitivity of the cornea. Ophthalmic Physiol Opt. 1984;4(4):305–18.
26. Millodot M. The influence of age on the sensitivity of the cornea. Invest Ophthalmol Vis Sci. 1977 Mar;16(3):240–2.
27. Loewenfeld IE. Pupillary changes related to age. In: Thompson HS, Daroff R, Frisen L, Glaser JS, Sanders MD (eds.) Topics in Neuro-Ophthalmology. Baltimore, MD: Williams & Wilkins 1979. pp. 124–50.
28. Birren JE, Casperson RC, Botwineck J. Age changes in pupil size. J Gerontol. 5:267–71, 1960.
29. Weale RA. Presbyopia. Br J Ophthalmol. 46:660–8, 1962.
30. Weale R. Presbyopia toward the end of the 20th century. Surv Ophthalmol. 1989;34:15–30.
31. Sperduto RD, Seigel D. Senile lens and senile macular changes in a population-based sample. Am J Ophthalmol. 1980;90:86–91.
32. Kahn HA, Leibowitz HM, Ganley JP, et al. The Framingham Eye Study: I. Outline and major prevalence findings. Am J Epidemiol. 1977;106:17–32.
33. Tolentino FL, Schepens CL, Freeman HM. Vitreoretinal Disorders, Diagnosis and Management. Philadelphia, PA: W.B. Saunders 1976.
34. Curcio CA, Millican CL, Allen KA, Kalina RE. Aging of the human photoreceptor mosaic: Evidence for selective vulnerability of rods in central retina. Invest Ophthalmol Vis Sci. 1993 Nov;34(12):3278–96.
35. Patel NB, Lim M, Gajjar A, Evans KB, Harwerth RS. Age-associated changes in the retinal nerve fiber layer and optic nerve head. Invest Ophthalmol Vis Sci. 2014 Jul 22;55(8):5134–43.
36. Balazsi AG, Rootman J, Drance SM, Schulzer M, Douglas GR. The effect of age on the nerve fiber population of the human optic nerve. Am J Ophthalmol. 1984;97:760–6.
37. Curcio CA, Drucker DN. Retinal ganglion cells in Alzheimer's disease and aging. Ann Neurol. 1993 Mar;33(3):248–57.
38. Mikelberg FS, Drance SM, Schulzer M, Yidegiligne HM, Weis MM. The normal human optic nerve. Axon count and axon diameter distribution. Ophthalmology. 1989 Sep;96(9):1325–8.

39. Repka MX, Quigley HA. The effect of age on normal human optic nerve fiber number and diameter. Ophthalmology. 1989;96(1):26–32.

40. Rubin GS, West SK, Muñoz B, Bandeen-Roche K, Zeger S, Schein O, Fried LP. A comprehensive assessment of visual impairment in a population of older Americans. The SEE Study. Salisbury Eye Evaluation Project. Invest Ophthalmol Vis Sci. 1997 Mar;38(3):557–68.

41. Klein R, Klein BE, Linton KL, De Mets DL. The Beaver Dam Eye Study: visual acuity. Ophthalmology. 1991 Aug;98(8):1310–5.

42. Leibowitz HM, Krueger DE, Maunder LR, Milton RC, Kini MM, Kahn HA, Nickerson RJ, Pool J, Colton TL, Ganley JP, Loewenstein JI, Dawber TR. The Framingham Eye Study monograph: An ophthalmological and epidemiological study of cataract, glaucoma, diabetic retinopathy, macular degeneration, and visual acuity in a general population of 2631 adults, 1973–1975. Surv Ophthalmol. 1980 May–Jun;24(Suppl):335–610.

43. Long GM, Crambert RF. The nature and basis of age-related changes in dynamic visual acuity. Psychol Aging. 1990;5:138–43.

44. Burg A. Visual acuity as measured by dynamic and static tests: A comprehensive evaluation. J Appl Psychol. 50:460–6, 1966.

45. Reading VM. Visual resolution as measured by dynamic and static tests. Pflugers Archly fur die Gesante Physiologie 228:17–26, 1972.

46. Wood JM, Owsley C. Useful field of view test. Gerontology. 2014;60(4):315–8.

47. Owsley C. Vision and driving in the elderly. Optom Vis Sci. 1994 Dec;71(12):727–35.

48. Edwards JD, Ross LA, Wadley VG, Clay OJ, Crowe M, Roenker DL, Ball KK. The useful field of view test: Normative data for older adults. Arch Clin Neuropsychol. 2006 May;21(4):275–86.

49. Ball K, Owsley C. The useful field of view test: A new technique for evaluating age-related declines in visual function. J Am Optom Assoc. 1993 Jan;64(1):71–9.

50. Dukic Willstrand T, Broberg T, Selander H. Driving characteristics of older drivers and their relationship to the useful field of view test. Gerontology. 2017;63(2):180–8.

51. Toups R, Chirles TJ, Ehsani JP, Michael JP, Bernstein JPK, Calamia M, Parsons TD, Carr DB, Keller JN. Driving performance in older adults: current measures, findings, and implications for roadway safety. Innov Aging. 2022 Jan 7;6(1):1–13.

52. Johnson CA, Adams AJ, Lewis RA. Evidence for a neural basis of age-related visual field loss in normal observers. Invest Ophthalmol Vis Sci. 1989 Sep;30(9):2056–64.

53. Brenton RS, Phelps CD. The normal visual field on the Humphrey field analyzer. Ophthalmologica. 1986;193(1–2):56–74.

54. Drance SM, Berry V, Hughes A. Studies on the effects of age on the central and peripheral isopters of the visual field in normal subjects. Am J Ophthalmol. 1967 Jun;63(6):1667–72.

55. Rutkowski P, May CA. The peripheral and Central Humphrey visual field – morphological changes during aging. BMC Ophthalmol. 2017 Jul 17;17(1):127.

56. Freeman EE, Muñoz B, Rubin G, West SK. Visual field loss increases the risk of falls in older adults: The Salisbury eye evaluation. Invest Ophthalmol Vis Sci. 2007 Oct;48(10):4445–50.

57. Wright LA, Wormald RP. Stereopsis and ageing. Eye (Lond). 1992;6 (Pt 5):473–6.

58. Hofstetter HW, Bertsch JD. Does stereopsis change with age? Am J Optom Physiol Opt. 1976 Oct;53(10):664–7.

59. Schneck ME, Haegerstrom-Portnoy G, Lott LA, Brabyn JA. Ocular contributions to age-related loss in coarse stereopsis. Optom Vis Sci. 2000 Oct;77(10):531–6.

60. Lee SY, Koo NK. Change of stereoacuity with aging in normal eyes. Korean J Ophthalmol. 2005 Jun;19(2):136–9.

61. Lakshminarayanan V, Enoch JM. Vernier acuity and aging. Int Ophthalmol. 1995;19(2):109–15.
62. Li RW, Edwards MH, Brown B. Variation in vernier acuity with age. Vision Res. 2000;40(27):3775–81.
63. Garcia-Suarez L, Barrett BT, Pacey I. A comparison of the effects of ageing upon vernier and bisection acuity. Vision Res. 2004 May;44(10):1039–45.
64. Ichikawa K, Yokoyama S, Tanaka Y, Nakamura H, Smith RT, Tanabe S. The Change in color vision with normal aging evaluated on standard pseudoisochromatic plates Part-3. Curr Eye Res. 2021 Jul;46(7):1038–46.
65. Schneck ME, Haegerstrom-Portnoy G, Lott LA, Brabyn JA. Comparison of panel D-15 tests in a large older population. Optom Vis Sci. 2014 Mar;91(3):284–90.
66. Skomrock LK, Richardson VE. Simulating age-related changes in color vision to assess the ability of older adults to take medication. Consult Pharm. 2010 Mar;25(3):163–70.
67. Werner JS, Peterzell DH, Scheetz AJ. Light, vision, and aging. Optom Vis Sci. 1990 Mar;67(3):214–29.
68. Silvestre D, Arleo A, Allard R. Healthy aging impairs photon absorption efficiency of cones. Invest Ophthalmol Vis Sci. 2019 Feb 1;60(2):544–51.
69. Nylén P, Favero F, Glimne S, Teär Fahnehjelm K, Eklund J. Vision, light and aging: A literature overview on older-age workers. Work. 2014 Jan 1;47(3):399–412.
70. Jackson GR, Owsley C, McGwin G Jr. Aging and dark adaptation. Vision Res. 1999 Nov;39(23):3975–82.
71. Wolska A, Sawicki D. Evaluation of discomfort glare in the 50+ elderly: experimental study. Int J Occup Med Environ Health. 2014 Jun;27(3):444–59.
72. Collins M. The onset of prolonged glare recovery with age. Ophthalmic Physiol Opt. 1989 Oct;9(4):368–71.
73. Reading VM. Disability glare and age. Vision Res. 1968 Feb;8(2):207–14.
74. Owsley C, Sekuler R, Siemsen D. Contrast sensitivity throughout adulthood. Vision Res. 1983;23(7):689–99.
75. Higgins KE, Jaffe MJ, Caruso RC, deMonasterio FM. Spatial contrast sensitivity: Effects of age, test-retest, and psychophysical method. J Opt Soc Am A. 1988 Dec;5(12):2173–80.
76. Sekuler R, Owsley C, Hutman L. Assessing spatial vision of older people. Am J Optom Physiol Opt. 1982 Dec;59(12):961–8.
77. Haegerstrom-Portnoy G, Schneck ME, Brabyn JA, Lott LA. Development of refractive errors into old age. Optom Vis Sci. 2002 Oct;79(10):643–9.
78. Irving EL, Machan CM, Lam S, Hrynchak PK, Lillakas L. Refractive error magnitude and variability: Relation to age. J Optom. 2019 Jan–Mar;12(1):55–63.
79. Baldwin W, Mills D. A longitudinal study of corneal astigmatism and total astigmatism. Am J Opt Physiol Opt. 1981;58:206–11.
80. Hirsch MJ. Changes in astigmatism after the age of forty. Am J Optom Arch Am Acad Optom. 1959;36:395–405.
81. Anstice J. Astigmatism: Its components and their changes with age. Am J Optom Arch Am Acad Optom. 1971;48:1001–6.
82. Rozema JJ, Hershko S, Tassignon MJ; for EVICR.net, Project Gullstrand Study Group. The components of adult astigmatism and their age-related changes. Ophthalmic Physiol Opt. 2019 May;39(3):183–93.
83. Haegerstrom-Portnoy G, Schneck ME, Lott LA, Hewlett SE, Brabyn JA. Longitudinal increase in anisometropia in older adults. Optom Vis Sci. 2014 Jan;91(1):60–7.
84. Sharpe JA, Sylvester TO. Effect of aging on horizontal smooth pursuit. Invest Ophthalmol Vis Sci. 1978 May;17(5):465–8.

85. Leigh RJ. The impoverishment of ocular motility in the elderly; In: Sekuler R, Kline D, Dismukes K, eds. Aging and Human Visual Function; Symposium, Washington D.C. March 31–April 1. Alan R. Liss, Inc:. New York, 1982. pp. 173–80.

86. Lee WJ, Kim JH, Shin YU, Hwang S, Lim HW. Differences in eye movement range based on age and gaze direction. Eye (Lond). 2019 Jul;33(7):1145–51.

87. Sheedy JE, Saladin JJ. Exophoria at near in presbyopia. Am J Optom Physiol Opt. 1975 Jul;52(7):474–81.

88. Glasser A. In Adler's Physiology of the Eye (Kaufman PL, Adler FH, Levin LA, Alm A, eds) Philadelphia: Saunders/Elsevier, 2011.

Index

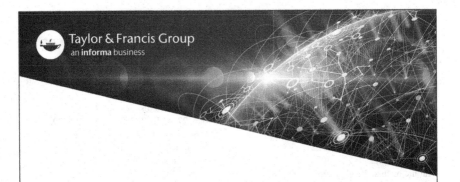

Printed in the United States
by Baker & Taylor Publisher Services